# 2nd Multiflow Summer School on Turbulence 2015



Journal of Physics: Conference Series Volume 708

Madrid, Spain
25 May – 26 June 2015

**Editor:**

**Javier Jimenez**

ISBN: 978-1-5108-2327-3
ISSN: 1742-6588

**Printed from e-media with permission by:**

Curran Associates, Inc.
57 Morehouse Lane
Red Hook, NY 12571

**Some format issues inherent in the e-media version may also appear in this print version.**



Printed by Curran Associates, Inc. (2016)

For permission requests, please contact the Institute of Physics at the address below.

Institute of Physics
Dirac House, Temple Back
Bristol BS1 6BE UK

Phone: 44 1 17 929 7481
Fax: 44 1 17 920 0979

techtracking@iop.org

**Additional copies of this publication are available from:**

Curran Associates, Inc.
57 Morehouse Lane
Red Hook, NY 12571 USA
Phone: 845-758-0400
Fax: 845-758-2633
Email: curran@proceedings.com
Web: www.proceedings.com

# Table of contents

## Volume 708

**2nd Multiflow Summer School on Turbulence**
**25 May to 26 June 2015, Madrid, Spain**

**Accepted papers received: 31 March 2016**
**Published online: 29 April 2016**

**Preface**

011001
OPEN ACCESS
Second Multiflow Summer School on Turbulence

Javier Jiménez

**Editorial**

011002
OPEN ACCESS
Editorial opinion: public dissemination of raw turbulence data

Juan A. Sillero and Javier Jiménez

**Peer review statement**

011003
OPEN ACCESS
Peer review statement

**Papers**

**THE ENERGY CASCADE AND REDUCED-ORDER MODELS**

012001
OPEN ACCESS
Homogeneous shear turbulence – bypass concept via interplay of linear transient growth and nonlinear transverse cascade

George Mamatsashvili, Siwei Dong, George Khujadze, George Chagelishvili, Javier Jiménez and Holger Foysi pg. 1

012002
OPEN ACCESS
A POD-based analysis of turbulence in the reduced nonlinear dynamics system

M-A Nikolaidis, B F Farrell, P J Ioannou, D F Gayme, A Lozano-Durán and J Jiménez pg. 19

012003
OPEN ACCESS
Unstable periodic orbits in plane Couette flow with the Smagorinsky model

Eiichi SASAKI, Genta KAWAHARA, Atsushi SEKIMOTO and Javier JIMÉNEZ pg. 33

012004
OPEN ACCESS
A new statistical tool to study the geometry of intense vorticity clusters in turbulence

Alberto Vela-Martin and Takashi Ishihara pg. 41

**WALL-BOUNDED FLOWS**

012005
OPEN ACCESS
Local topology via the invariants of the velocity gradient tensor within vortex clusters and intense Reynolds stress structures in turbulent channel flow

Abel-John Buchner, Adrián Lozano-Durán, Vassili Kitsios, Callum Atkinson and Julio Soria pg. 47

012006
OPEN ACCESS
Identifying coherent structures and vortex clusters in Taylor-Couette turbulence

Vamsi Spandan, Rodolfo Ostilla-Monico, Detlef Lohse and Roberto Verzicco pg. 61

012007
OPEN ACCESS
Space and time behaviour of the temperature second-order structure function in Rayleigh-Bénard convection

Riccardo Togni, Andrea Cimarelli, Adrián Lozano-Durán and Elisabetta De Angelis pg. 74

012008
OPEN ACCESS
Reynolds Stress Structures in the Hybrid RANS/LES of a Planar Channel

Jack Weatheritt, Richard Sandberg and Adrian Lozano-Durán pg. 84

012009
OPEN ACCESS
Geometry-induced fluctuations in the transitionally rough regime

Nabil Abderrahaman-Elena and Ricardo García-Mayoral pg. 99

012010
OPEN ACCESS
The minimal channel: a fast and direct method for characterising roughness

Michael MacDonald, Daniel Chung, Nicholas Hutchins, Leon Chan, Andrew Ooi and Ricardo García-Mayoral pg. 111

012011
OPEN ACCESS
Analysis and comparison between rough channel and pipe flows

David Sassun, Oscar Flores and Paolo Orlandi pg. 126

**COMPLEX FLOWS**

012012
OPEN ACCESS
Large-eddy simulations of adverse pressure gradient turbulent boundary layers

Alexandra Bobke, Ricardo Vinuesa, Ramis Örlü and Philipp Schlatter pg. 137

012013
OPEN ACCESS
Coherent structures in a zero-pressure-gradient and a strongly decelerated boundary layer

Mark P Simens, Ayse G Gungor and Yvan Maciel pg. 155

012014
OPEN ACCESS
On determining characteristic length scales in pressure gradient turbulent boundary layers

Ricardo Vinuesa, Ramis Örlü and Philipp Schlatter pg. 170

012015
OPEN ACCESS
Inception and evolution of coherent structures in under-expanded supersonic jets

P.C. Stegeman, J. M. Pérez, J. Soria and V. Theofilis pg. 179

Journal of Physics: Conference Series **708** (2016) 011001 doi:10.1088/1742-6596/708/1/011001

# Second Multiflow Summer School on Turbulence

**Editor: Javier Jiménez**

## PREFACE

Multiflow is a research program, funded by the European Research Council, whose goal is to improve our understanding of the multiscale dynamics of turbulence in fluids. Its second Summer School on Turbulence took place at the School of Aeronautics of the Technical University of Madrid from May 25 to June 26, 2015, with the goal of providing a meeting place for theoreticians, experimentalists and simulators, in which to develop and test new ideas on turbulence physics and structure. Around forty, mostly young, participants from twenty international groups met for five weeks of collaborative work, primarily using the computational data archived in the receiving institution but, in many cases, also contributing their own. Although the format included a few invited formal seminars and periodic plenary meetings, most of the work took place in small groups that, in many cases, changed their composition during the workshop. The papers in these proceedings reflect the results of the work of these groups which, in many cases, later continued in the form of new collaborations.

## Workshop organising committee:

Javier Jiménez
Vassilios Theofilis
School of Aeronautics, Technical University of Madrid
28040 Madrid, Spain

### Acknowledgements

The organizers are grateful to the European Research Council for their funding of this meeting, and to the School of Aeronautics of the Technical University of Madrid for their help in hosting it. The meeting would not have been possible without the dedication of those participants who acted as hosts to the visiting groups, facilitating everything from everyday problems to computer access and data usage: Guillem Borrell, José I. Cardesa, Siwei Dong, Miguel P. Encinar, Oscar Flores, Ricardo García-Mayoral, Ayse G. Gungor, Adrián Lozano-Durán, Yvan Maciel, Atsushi Sekimoto, Juan A. Sillero and Alberto Vela-Martín. Very especially, we are indebted to Aurora García Sarrión, without whose administrative help nothing would have worked.

### Picture credits

The cover figure displays the joint PDF of the mean inclination angle and the mean of the absolute value of the wall-normal velocity for transient structures found in a turbulent channel of $Re_\tau = 950$. Arrows represent the conditional evolution of the flow for a given inclination and intensity. Blue and red colours represent the local intensity of the streamwise and spanwise velocities respectively. The figure is due to Miguel P. Encinar. The participants picture was graciously provided by the press department of the School of Aeronautics–UPM.

Journal of Physics: Conference Series **708** (2016) 011001 doi:10.1088/1742-6596/708/1/011001

# Editorial opinion: public dissemination of raw turbulence data

**Juan A. Sillero and Javier Jiménez**

School of Aeronautics, Universidad Politécnica de Madrid, 28040 Madrid, Spain

E-mail: sillero@torroja.dmt.upm.es, jimenez@torroja.dmt.upm.es

**Abstract.** Many of the papers in this issue deal with processing of pre-existing large-scale turbulence data. We argue here that there is a certain urgency to the discussion of whether raw data should be made publicly available within the turbulence community, and of which are the best procedures, technology and rules for possible dissemination. Besides expressing the personal opinion that such sharing would be advantageous for the field, the urgency mostly arises from the danger that funding agencies and other institutions would otherwise set standards without proper community input. The experience of the Madrid School of Aeronautics with the dissemination of numerical simulation results is briefly reviewed, including the present technological solutions and usage statistics.

## 1. Introduction

Perhaps because of the natural tendency of engineering research to proprietary information, the traditional working model of fluid mechanics has been that data are generated and analysed within the same group. Data bases have always existed, and professional societies have specially contributed to generate and maintain them, but they usually contain processed results from which it is difficult to extract anything beyond what had been anticipated by the data compiler. The difference between raw data and what is eventually made public has worsened with the appearance of increasingly large simulation and experimental data sets. These are usually archived, but because they are bulky and difficult to read, they are often lost without further use. In some cases in which several interesting computations or experiments have been performed within the same institution, workshops and personal contacts have provided a mechanism for the use of unprocessed data by other groups, but the democratization of computer power, with data being generated at many places around the World, is making that model increasingly cumbersome and hard to justify.

Other communities have used a different working method for some time. High-energy physics, meteorology and molecular biology come to mind as examples in which there are specialists who generate data and others who exploit them, collaborating by means of more or less public data bases to which the former contribute and from where the latter draw their raw material. The process of collaboration is not trivial, and each community has developed its own protocols for data interchange, archiving, quarantine periods and attribution. The loss is a certain 'personal' way of doing science, which should probably not be completely abandoned, but the gain is a greater freedom in what can be observed, and the greater use to which the data can be put. An instructive case is astrophysics, where the division of labour dates at least to the analysis and publication by Kepler of Tycho Brahé's data in 1627. This eventually led to his planetary laws, and probably to Newton's dynamics, but it was not free of controversy; Kepler had to fight Tycho's heirs for ownership of the data, and eventually paid for the publication himself [1]. Astrophysical data are today mostly in the public domain under well-defined rules and formats.

Whether fluid mechanics should follow this model is open to discussion, and we believe that it should be discussed. Many groups make information public in their private websites, but there is no standard way of disseminating the availability of new data, no generally agreed way of storing or reading them and, in truth, little incentive to prepare the data for general use. An important problem is long-term curation and availability. In our experience, large data sets have a useful lifetime of around ten years, after which they are easier to recompute or remeasure than to retrieve from a database. During this time, they have to be maintained and, occasionally updated or reformated for purposes that were not anticipated when they were first stored. Most obvious are changes due to the evolution of the computational environment, and of the balance between central and user resources. An example could be the change in graphics requirements during the past few years due to the introduction of cheap graphic processors. As a forwards-looking example, the possible impact of cloud computing on future data use is unpredictable.

Even if limited, ten years is longer than the duration of most research contracts, and it is unclear how the cost of maintaining and preparing data is to be supported after the contract expires. Moreover, large simulation or experimental data have peculiarities that are in many cases only understood by their originator (typically a student or postdoc), and these constraints may be forgotten once the originator leaves the research group. For example, if the results of a spectral code are differentiated using low-order finite differences, the resulting errors are often unacceptable. Again, the useful lifetime of a data set typically exceeds the span of a thesis or a postdoctoral stay.

Broadly speaking, there are two models. The first one is a service centre offering data and on-site postprocessing facilities for a particular kind of problems [2]. The second is a more unstructured facility where data are stored and indexed with minimal on-site processing capabilities [3, 4, 5]. The former are very useful for the purpose for which they are developed but, in our personal experience, tend to be cumbersome when they are used for a very different purpose. They are also critically subject to the long-term funding issues mentioned above, especially if they are restricted to a particular problem. The unstructured services typically shift more of the required computer power to the side of the user. They also require continuing funding, but they are broader and easier to justify. The curation problem is also different in both cases. While the first one typically requires the continued attention of somebody who is familiar with the fluid mechanics of the problem, the second, if properly documented, only requires computational expertise. Thus, the first model tends to be permanently associated with the group that generated the data while, in principle, the second could be outsourced to a more generic institution.

There is a certain urgency to the discussion of whether and how our data should be shared. An increasing number of funding agencies and scientific journals are beginning to insist on open public access of research results (while others specifically forbid it, and the effect of intellectual property laws is uncertain). There are increasingly common warnings from the agencies that, if we do not decide on a viable dissemination policy and format, they will do it for us without regard to any issues that may differentiate our application from others. On the other hand, several supranational initiatives are starting to form for the long-term storage and standardised access of large scientific data [6, 7], and the time to start exerting our influence over them as a community is probably now.

## 2. The data base of the Universidad Politécnica de Madrid

To illustrate these concepts, we will briefly describe our experience with the public databases hosted by our research group, which mostly contain results of large-scale fluid-mechanical simulations. There are two types of data sets. The first ones are large collections ($\sim$ 100 TB) of raw flow snapshots. They may include as many as $10^5$ moderately large binary files ($\sim$ 1 GB), fewer ($\sim 10^3$) larger ones ($\sim$ 100 GB), or a mixture of both. The files can often only be used with the original numerical code that generated them, which is often unique to each set. These data are routinely mined by local users in a 'write once, read many' fashion, over a characteristic period of about ten years. Once checked, cleaned, and after initial publication from the data originator, most of them are made public without any particular access control [5]. The necessary codes are also made public under an open-source 'as is' license.

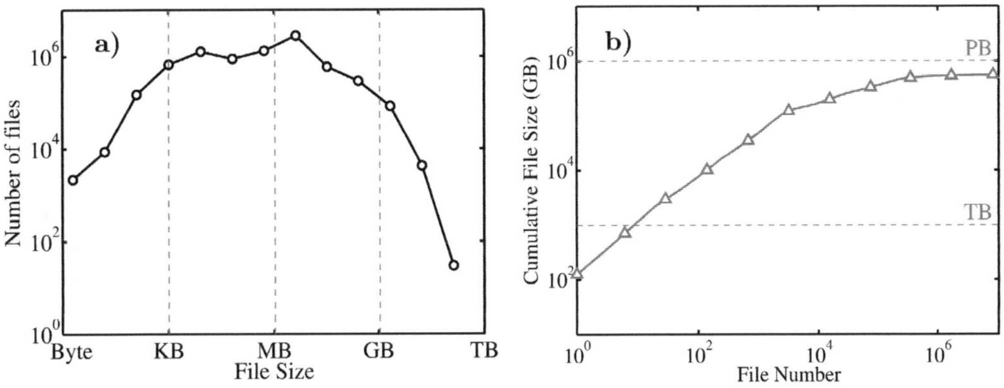


**Figure 1.** File size distribution in our current data centre storage. (a) Histogram of the file size distribution. (b) Sorted cumulative file size (GBytes) as a function of the total number of files.

The second type of files are obtained by postprocessing the original ones. They include flow fields expressed in primitive variables in physical space, rather than in the typically reduced set of variables and spectral representations of the original files. Another common type of postprocessed results is the temporal evolution of 'objects' defined in particular ways, some of them obtained on-the-fly during the simulations. The accumulated size of these data is typically several times larger than the original ones, and they are not intended for unrestricted public use. Although they are occasionally provided to outside researchers on the basis of personal collaborations, they are not carefully curated, and often not even well documented.

Some statistics of the file size are shown in figure 1. Figure 1(a) is a histogram of the size distribution of our current files, which we believe to be typical of those used in turbulence research. Figure 1(b) presents a compilation of the sorted cumulative file size versus the number of files. The figures show that there are $O(10^5)$ files larger than 1 GB, accounting for 65% of the total used storage ($\approx$ 550 TB). Traditional data bases are of limited use for this application. The size of a typical transaction is larger than those in most commercial services, such as financial ones, personal records, etc., whose characteristic entry size is a few KB, or even than those in scientific data bases, such as particle physics or genomics, where the characteristic entry size is a few MB.

Our postprocessing infrastructure is optimised for price/performance, but also, most importantly, for easy of maintenance. Besides larger computational clusters dedicated to simulations, the core of the system is a collection of Linux servers with relatively large memory ($\geq$ 128 GB). They are connected to a dedicated NFS (Network File System) server attached to a centralized NAS (Network Attached Storage) that requires a relatively powerful front-end with substantial processing capabilities. The hardware reserved for the open-access database is a high-density commodity JBOD disk enclosure with SAS interface, accepting sixty 3.5" drives in a 4U rack space. Because of the relatively large size of the files, they are mostly accessed sequentially rather than randomly, and our solution uses exclusively traditional mechanical 4 TB hard drives. Moreover, to test the long-term behaviour of a 'low-cost' storage system with decent performance, we have limited ourselves to SATA drives instead of the more expensive SAS ones.

This choice implies losing some of the benefits of the SAS interface, such as better command queuing and error checking algorithms, full duplex communications allowing multipath, higher mean time between failure (MTBF), or lower Bit Error Rate (BER). To compensate for this possible degradation of data integrity, we chose a ZFS file system [8] that guarantees integrity at disk-block level. This file system differs from others in being specifically designed to protect the data from silent data corruption [9] caused by multiple factors such as bit rot, current spikes, firmware bugs, etc. ZFS was conceived to control all the layers in a modern file system, from logical volumes down to block devices. For this reason, expensive hardware RAID cards are no longer needed to control the information that goes into each block device. ZFS offers its own software RAID solution using copy-on-write semantics (COW), avoiding problems such as the 'write hole' phenomenon [8] in traditional storage appliances. To protect the integrity of the data in the event of failure of one or more disks, we have configured ZFS in Raidz2 mode, allowing the loss of up to two disks for every group of ten. On top of that, ZFS allows the use of different lossless compression algorithms to perform on-the-fly compression of the data at block level. Depending on the installed capacity, the current price of such an storage system varies from 90 to 120 €/TB, which is within the funding range of individual research groups. It delivers I/O throughputs in excess of 2.5 GB/sec between the front-end node and the JBOD, and about 700 MB/sec to the NFS clients.

Although their large size makes the use of full data set by external researchers difficult, our experience is that downloads of individual files are reasonably common. Practical data transfer rates from our university to other academic institutions are about 1–2 TB per day, although it may be hard to sustain those velocities for several concurrent requests. After about a year of continued use, we have experienced no critical loss of data.

### *2.1. User interface*

Although data dissemination through personal contacts and visits remains important and will probably always remain so, there is an increased demand from external groups for a more impersonal interaction, which could not be satisfied until the system just described was implemented.

At present, the data open to unrestricted external use is about 150 TB, and provides the scientific community with an efficient way to download about 15 different DNS databases of incompressible channel and boundary layer flows at Reynolds numbers $Re_\tau \approx$180–4200. Information about specific databases can be found at http://torroja.dmt.upm.es/turbdata/index. The data base also provides utility tools to convert the data files between formats (e.g., from binary to HDF5), or to manipulate the original data.

Even if the requirements are simple in principle, their execution is challenging for several reasons. Firstly, there has to be a physical infrastructure able to host hundreds of terabytes, although not necessarily as fast as for a local service. This was already discussed in some detail in the previous section. Secondly, any centralized architecture in which one server interacts with many clients can strongly degrade the availability of the files to the local group, requires large bandwidth, and is expensive to run. Finally, it is important to retain the ability to resume partial file downloads when transferring large datasets, saving time and bandwidth resources.

Our first attempt was to adapt a peer-to-peer protocol such as BitTorrent to distribute large scientific files. This protocol benefits from several factors such as integrity, robustness, simplicity, and aggregate bandwidth capabilities, making it a reasonable choice to transfer large amounts of data from a local digital repository to the scientific community via the Internet network. We implemented a Torrent site for part of our databases without finding any unsurmountable technical difficulties. However, we encountered a number of practical ones that discouraged us from continuing its use: many institutions block P2P networks, the BitTorrent protocol listens from a large number of ports in the local machines (associated with the trackers), and it is not straightforward to automatise the indexation of hundreds of thousands of files. Most importantly, the essence of the P2P networks is to share a file with many peers simultaneously, and to complete a download in at most few hours. Neither has proved to be the case with our databases. The observed access pattern is most often a single user downloading a few Terabytes for a couple of days, and rarely there are several users simultaneously accessing the data. This motivated us to consider alternative solutions.

Our presently preferred solution has been to implement a simple but practical file repository using an Apache HTTP server. The files are served to the clients through a dedicated 1 Gbps Ethernet network without impacting on the availability of the files to the local group, which is served by a private high-speed QDR Infiniband network (40 Gbps). The HTTP protocol allows resumption of partial downloads in case of failure, although some local web servers are configured to avoid header range requests, preventing resumes. Our Apache web server is configured to accept such requests, and users should be able to continue unfinished downloads at any time, either using terminal commands like `wget`, or through a generic web browser.

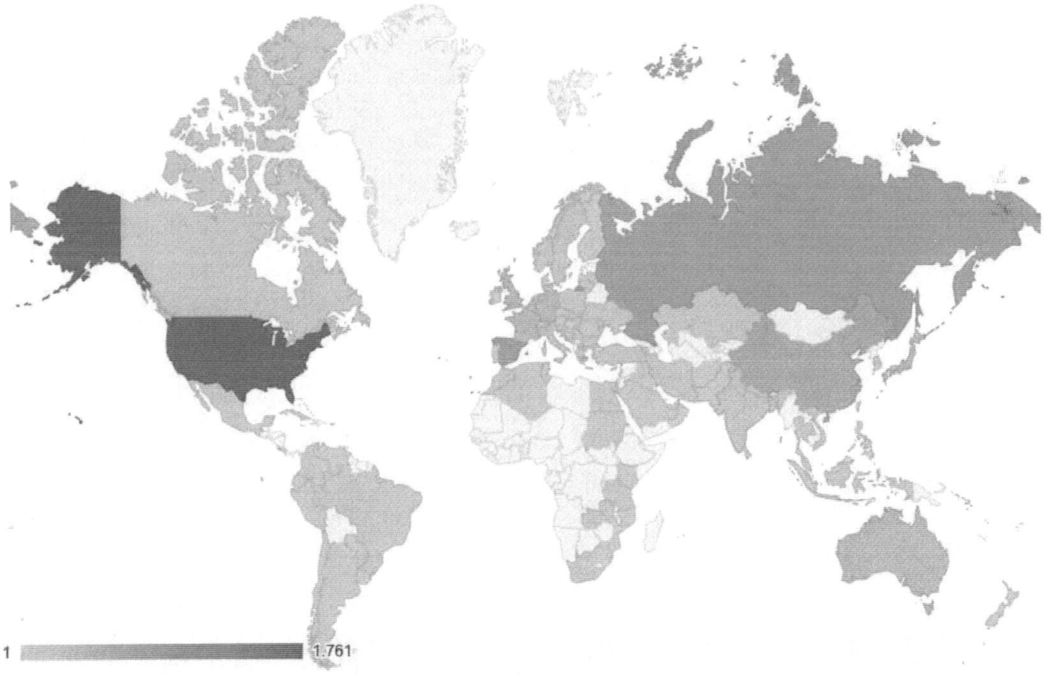


**Figure 2.** Statistics for the *post-processed* Boundary Layer databases based on the geographical location. Data provided by Google Analytics between October 2013 and November 2015. Source: http://torroja.dmt.upm.es/turbdata/blayers/

*2.2. Statistics of outside use of the data*

Besides the local data use by members of our group, data analytics collected in the open part of our website show the increasing interest of researches in raw and post-processed turbulence databases, which is also reflected in the increasing number of citations to the associated papers. Figure 2 shows a summary of the audience for the part of our website listing one- and two-point statistics for different turbulent boundary layers, based on geographical location. Since the open repository described above was brought online in February 2015 we have received about 600 visits per month: 48% correspond to direct searchs, 40% to referrals, and the remainder to organic searchs and social networks. Around 20% of the total hits are return visitors who check our databases often to either download new data or to check for content updates. For instance, 18% of the sessions come from a total of 251 universities from all around the world.

During the same period, the raw-data part of the web page averaged 100–150 sessions per month. About 37% of them correspond to higher education institutions, including more than 40 universities.

These statistics lead us to believe in the potential impact of a web-service that allows full access by the scientific community to raw databases, probably including applications beyond turbulence research.

**Acknowledgements**

This work was funded in part by the Multiflow programme of the European Research Council.

## References

[1] Hannan J 2009 *God's philosophers: how the medieval world laid the foundations of modern science* (Icon Books, London) ISBN 9781848310704

[2] Li Y, Perlman E, Wan M, Yang Y, Meneveau C, Burns R, Chen S, Szalay A and Eyink G 2008 *J. Turbul.* **9** N31

[3] `https://www.opensciencedatacloud.org/`

[4] `http://euhit.icm.edu.pl/services/turbase`

[5] `http://torroja.dmt.upm.es/turbdata/`

[6] `https://rd-alliance.org/`

[7] `http://eudat.eu/`

[8] ZFS `http://en.wikipedia.org/wiki/ZFS`

[9] Panzer-Steindel B 2007 Data integrity Tech. rep. CERN/IT

Journal of Physics: Conference Series **708** (2016) 011002 doi:10.1088/1742-6596/708/1/011002

# Peer review statement

All papers published in this volume of *Journal of Physics: Conference Series* have been peer reviewed through processes administered by the proceedings Editors. Reviews were conducted by expert referees to the professional and scientific standards expected of a proceedings journal published by IOP Publishing.

# Homogeneous shear turbulence – bypass concept via interplay of linear transient growth and nonlinear transverse cascade

**George Mamatsashvili[1,2,3], Siwei Dong[4], George Khujadze[3,5], George Chagelishvili[3,6], Javier Jiménez[4] and Holger Foysi[5]**

[1]Helmholtz-Zentrum Dresden-Rossendorf, P.O. Box 510119, D-01314 Dresden, Germany
[2]Department of Physics, Faculty of Exact and Natural Sciences, Tbilisi State University, Tbilisi 0179, Georgia
[3]Abastumani Astrophysical Observatory, Ilia State University, Tbilisi 0162, Georgia
[4]School of Aeronautics, Universidad Politécnica de Madrid, Madrid 28040, Spain
[5]Chair of Fluid Mechanics, Universität Siegen, Siegen 57068, Germany
[6]Institute of Geophysics, Tbilisi State University, Tbilisi 0128, Georgia

E-mail: g.mamatsashvili@hzdr.de

**Abstract.** We performed direct numerical simulations of homogeneous shear turbulence to study the mechanism of the self-sustenance of subcritical turbulence in spectrally stable (constant) shear flows. For this purpose, we analyzed the turbulence dynamics in Fourier/wavenumber/spectral space based on the simulation data for the domain aspect ratio $1:1:1$. Specifically, we examined the interplay of linear transient growth of Fourier harmonics and nonlinear processes. The transient growth of harmonics is strongly anisotropic in spectral space. This, in turn, leads to anisotropy of nonlinear processes in spectral space and, as a result, the main nonlinear process appears to be not a direct/inverse, but rather a transverse/angular redistribution of harmonics in Fourier space referred to as the nonlinear transverse cascade. It is demonstrated that the turbulence is sustained by the interplay of the linear transient, or nonmodal growth and the transverse cascade. This course of events reliably exemplifies the well-known bypass scenario of subcritical turbulence in spectrally stable shear flows. These processes mainly operate at large length scales, comparable to the box size. Consequently, the central, small wavenumber area of Fourier space (the size of which is determined below) is crucial in the self-sustenance and is labeled the vital area. Outside the vital area, the transient growth and the transverse cascade are of secondary importance - Fourier harmonics are transferred to dissipative scales by the nonlinear direct cascade. The number of harmonics actively participating in the self-sustaining process (i.e., the harmonics whose energies grow more than 10% of the maximum spectral energy at least once during evolution) is quite large - it is equal to 36 for the considered box aspect ratio - and obviously cannot be described by low-order models.

## 1. Introduction

In the 1990s, the nonnormal nature of shear flows and its consequences became well understood by the hydrodynamic community (see e.g., Refs. [1, 2, 3, 4, 5, 6, 7]). As a result, a new, so-called bypass concept emerged for explaining the onset and sustenance of turbulence in spectrally stable shear flows [8, 9, 10, 11, 12, 13, 14, 15, 16]. This concept is based on the linear transient, or nonmodal growth of vortical perturbations due to the nonnormality/shear, which is the only possibility for energy supply to turbulence in such flows [10]. In this case, the role of nonlinear processes becomes crucial - they underly self-sustenance of the turbulence. What is most important, a new type of nonlinear processes emerges which principally differs from the canonical ones, i.e., from direct and inverse cascades.

Actually, the shear-induced linear transient growth mainly depends on the orientation (and, to a lesser degree, on the value) of a perturbation wavevector (e.g., Ref. [17, 18]): the spatial Fourier harmonics of the perturbations having a certain orientation of the wavevector with respect to shear flow draw energy from the flow and amplify, whereas harmonics with other orientation of the wavevector give energy back to the flow and decay. In other words, the linear energy-exchange processes are strongly anisotropic in spectral **k**-space, which, in turn, leads to anisotropy of nonlinear processes in **k**-space. Specifically, as revealed in [17, 19, 20], in HD/MHD smooth shear flows the main nonlinear process is not a direct/inverse, but rather a transverse cascade, that is, angular redistribution of perturbation harmonics in **k**-space.

In the paper by Horton *et al.* [19], the nonlinear dynamics of coherent and stochastic vortical perturbations in 2D plane constant shear flow was studied numerically and the existence and importance of the nonlinear transverse cascade (NTC) for the self-sustaining dynamics of perturbations was demonstrated. Specifically, it has been shown that the interplay of NTC with linear nonnormality-induced phenomena becomes intricate: it can realize either positive or negative feedback. In the case of positive feedback, NTC repopulates those quadrants in **k**-space where the shear flow causes linear transient growth, i.e., repopulates harmonics with the potential of transient growth. In the paper by Mamatsashvili *et al.* [20], the dynamics of 2D perturbations in incompressible MHD flows with constant shear of velocity and uniform magnetic field parallel to the flow was considered. Investigating subcritical turbulence, the process of NTC – a keystone of the turbulence self-sustaining (non-decaying) dynamics in this simple open MHD flow system, was described in detail.

Studies of turbulence in constant mean shear flows, or homogeneous shear turbulence usually concentrate on dynamics and statistics in real space. Coherent vortical structures responsible for the self-sustenance of the turbulence were identified and described in [21, 22, 23, 24, 25], while the dynamics of these structures and hence the properties of turbulence at various box aspect ratios were extensively analyzed in [26, 27]. Compared to these studies, which mostly perform analysis in real/coordinate space (although the codes used are spectral), analysis of turbulence dynamics in Fourier space allows us to identify harmonics/modes that play a key role in the self-sustaining process of turbulence and define the range of aspect ratios for which the dynamically important modes are fully taken into account.

The main aim of the present paper is to extend the knowledge and results gained from the investigation of 2D flows to a more realistic 3D plane constant shear flow using direct numerical simulations (DNS). We consider the dynamics of homogeneous shear turbulence in Fourier/spectral space. An advantage of this idealized type of turbulence is that it includes the basic effects of shear, while avoiding the complication of more realistic cases (e.g., effect of boundaries, etc.). The investigation of this simple (constant shear) flow – (homogeneous) turbulence system allows to demonstrate clearly the scheme of realization of the bypass concept. The main idea of our investigation is the analysis of the flow dynamics in spectral space that allows us: (i) to follow directly the action of dynamical characteristics of linear and nonlinear processes (ii) to grasp NTC and (iii) to pin down the sequence of events underlying the self-sustenance of turbulence.

The paper is organized as follows. Sec. 2 is devoted to the physical model and derivation of dynamical equations in spectral space. The DNS of the turbulence dynamics and the analysis of the self-sustaining mechanism – the activity of linear and nonlinear processes in Fourier space are presented in Sec. 3. A summary and discussion are given in Sec. 4.

## 2. Physical model and equations

The motion of an incompressible fluid with constant viscosity, $\nu$, is governed by the Navier-Stokes equations

$$\frac{\partial \mathbf{U}}{\partial t} + (\mathbf{U} \cdot \nabla)\,\mathbf{U} = -\frac{\nabla P}{\rho} + \nu \nabla^2 \mathbf{U}, \tag{1}$$

$$\nabla \cdot \mathbf{U} = 0, \tag{2}$$

where $\rho$ is the density, $\mathbf{U}$ is the velocity and $P$ is the pressure.

Equations (1)-(2) have a stationary equilibrium solution – the flow in the $x$-direction with the linear shear of mean velocity in the vertical $y$-direction, $\mathbf{U}_0 = (Sy, 0, 0)$. Without loss of generality, the constant mean shear rate $S$ is chosen to be positive. Equilibrium density $\rho_0$ and pressure $P_0$ are spatially constant. Such an idealized configuration of a flow with a linear shear, despite its simplicity, allows us to grasp key effects of shear on the perturbation dynamics and ultimately on a resulting turbulent state in shear flows.

Consider 3D perturbations of the velocity and pressure, $\mathbf{u}$ and $p$, about the equilibrium. Representing the total fields as the sum of the equilibrium and perturbed values, $\mathbf{U} = \mathbf{U}_0 + \mathbf{u}, P = P_0 + p$, substituting these into Eqs. (1)-(2), we obtain the following system of equations governing the dynamics of perturbations with arbitrary amplitude

$$\left(\frac{\partial}{\partial t} + Sy\frac{\partial}{\partial x}\right) u_x = -Su_y - \frac{1}{\rho_0}\frac{\partial p}{\partial x} - \frac{\partial}{\partial x}u_x^2 - \frac{\partial}{\partial y}(u_x u_y) - \frac{\partial}{\partial z}(u_x u_z) + \nu\nabla^2 u_x, \tag{3}$$

$$\left(\frac{\partial}{\partial t} + Sy\frac{\partial}{\partial x}\right) u_y = -\frac{1}{\rho_0}\frac{\partial p}{\partial y} - \frac{\partial}{\partial x}(u_x u_y) - \frac{\partial}{\partial y}u_y^2 - \frac{\partial}{\partial z}(u_y u_z) + \nu\nabla^2 u_y, \tag{4}$$

$$\left(\frac{\partial}{\partial t} + Sy\frac{\partial}{\partial x}\right) u_z = -\frac{1}{\rho_0}\frac{\partial p}{\partial z} - \frac{\partial}{\partial x}(u_x u_z) - \frac{\partial}{\partial y}(u_y u_z) - \frac{\partial}{\partial z}u_z^2 + \nu\nabla^2 u_z, \tag{5}$$

$$\frac{\partial u_x}{\partial x} + \frac{\partial u_y}{\partial y} + \frac{\partial u_z}{\partial z} = 0, \tag{6}$$

From Eqs. (3)-(6) we derive a dynamical equation for the perturbation kinetic energy in order to investigate the energy balance in the self-sustained turbulent state. In the presence of the imposed shear-periodic boundary conditions (see Sec. III below), the evolution of the volume-averaged kinetic energy density, $E = \rho_0 \mathbf{u}^2/2$, is governed by the equation:

$$\frac{d}{dt}\langle E\rangle = -S\rho_0\,\langle u_x u_y\rangle - \rho_0\nu\langle(\nabla u_x)^2 + (\nabla u_y)^2 + (\nabla u_z)^2\rangle, \tag{7}$$

where the angle brackets, $\langle ... \rangle$ denote a volume average over an entire domain at a given time. The first term on the right hand side of Eq. (7) is the Reynolds stress and describes the exchange of energy between perturbations and the background flow. Note that this term originates from the linear term proportional to shear $(-Su_y)$ on the right hand side of Eq. (3). The second term is negative by definition and describes energy dissipation due to viscosity. It is evident that there is no contribution from the nonlinear terms in the total energy evolution Eq. (7). Thus, only Reynolds stress, if negative, can supply perturbations with energy, extracting it from the mean flow due to shear. In turbulence studied below, the Reynolds stress ensures energy injection into turbulent fluctuations, balancing viscous dissipation. The nonlinear terms, not directly tapping into the shear flow energy and therefore not changing the total perturbation energy, only redistribute energy gained by means of the stress among Fourier harmonics of perturbations with different wavenumbers (see below). Obviously, in the absence of shear $(S = 0)$, the contribution from the Reynolds stresses disappears in Eq. (7) and hence the total perturbation energy cannot grow, gradually decaying due to viscosity.

### 2.1. Spectral representation of the dynamical equations

We derive dynamical equations for the quadratic forms of velocity perturbations, ($u_x^2$, $u_y^2$, $u_z^2$), and kinetic energy density, $E$, in Fourier space to gain a deeper insight into the interplay of the flow shear and nonlinearity in the self-sustenance of perturbations (turbulence). For a spectral representation of the main equations, we decompose the perturbations into spatial Fourier harmonics

$$f(\mathbf{r},t) = \int \bar{f}(\mathbf{k},t) \exp\left(\mathrm{i}\mathbf{k}\cdot\mathbf{r}\right) d^3\mathbf{k} \tag{8}$$

where $f \equiv (\mathbf{u}, p)$ denotes the perturbations and $\bar{f} \equiv (\bar{\mathbf{u}}, \bar{p})$ is their corresponding Fourier transforms ($d^3\mathbf{k} \equiv dk_x dk_y dk_z$). Substituting decomposition (8) into Eqs. (3)-(6), we obtain the following equations governing the dynamics of perturbation harmonics in spectral space (the constant density is set to unity, $\rho_0 = 1$)

$$\left(\frac{\partial}{\partial t} - Sk_x\frac{\partial}{\partial k_y}\right)\bar{u}_x = -S\bar{u}_y - \mathrm{i}k_x\bar{p} + Q_x - \nu k^2\bar{u}_x, \tag{9}$$

$$\left(\frac{\partial}{\partial t} - Sk_x\frac{\partial}{\partial k_y}\right)\bar{u}_y = -\mathrm{i}k_y\bar{p} + Q_y - \nu k^2\bar{u}_y, \tag{10}$$

$$\left(\frac{\partial}{\partial t} - Sk_x\frac{\partial}{\partial k_y}\right)\bar{u}_z = -\mathrm{i}k_z\bar{p} + Q_z - \nu k^2\bar{u}_z, \tag{11}$$

$$k_x\bar{u}_x + k_y\bar{u}_y + k_z\bar{u}_z = 0, \tag{12}$$

where $k^2 = k_x^2 + k_y^2 + k_z^2$. These spectral equations contain the linear as well as the nonlinear, $\mathbf{Q}(\mathbf{k},t) = [Q_x(\mathbf{k},t), Q_y(\mathbf{k},t), Q_z(\mathbf{k},t)]$, terms that are the Fourier transforms of corresponding linear and nonlinear ones in the original Eqs. (3)-(6). The latter are given by

$$Q_i(\mathbf{k},t) = \sum_j k_j q_{ij}(\mathbf{k},t),$$

where

$$q_{ij}(\mathbf{k},t) = -\mathrm{i}\int \bar{u}_i(\mathbf{k}',t)\bar{u}_j(\mathbf{k}-\mathbf{k}',t)d^3\mathbf{k}', \qquad i,j = x,y,z$$

and describe nonlinear triad interactions among harmonics with different wavenumbers in Fourier $\mathbf{k}$-space. From Eqs. (9)-(12) one can eliminate the pressure

$$\bar{p} = 2\mathrm{i}S\frac{k_x}{k^2}\bar{u}_y - \frac{\mathrm{i}}{k^2}(k_xQ_x + k_yQ_y + k_zQ_z) = 2\mathrm{i}S\frac{k_x}{k^2}\bar{u}_y - \mathrm{i}\frac{\mathbf{k}\cdot\mathbf{Q}}{k^2},$$

and rewrite them in the following form:

$$\left(\frac{\partial}{\partial t} - Sk_x\frac{\partial}{\partial k_y}\right)\bar{u}_x = S\left(2\frac{k_x^2}{k^2} - 1\right)\bar{u}_y + Q_x - \frac{k_x}{k^2}(\mathbf{k}\cdot\mathbf{Q}) - \nu k^2\bar{u}_x, \tag{13}$$

$$\left(\frac{\partial}{\partial t} - Sk_x\frac{\partial}{\partial k_y}\right)\bar{u}_y = 2S\frac{k_xk_y}{k^2}\bar{u}_y + Q_y - \frac{k_y}{k^2}(\mathbf{k}\cdot\mathbf{Q}) - \nu k^2\bar{u}_y, \tag{14}$$

$$\left(\frac{\partial}{\partial t} - Sk_x\frac{\partial}{\partial k_y}\right)\bar{u}_z = 2S\frac{k_xk_z}{k^2}\bar{u}_y + Q_z - \frac{k_z}{k^2}(\mathbf{k}\cdot\mathbf{Q}) - \nu k^2\bar{u}_z, \tag{15}$$

or, for the amplitudes of the velocity perturbations

$$\left(\frac{\partial}{\partial t} - Sk_x\frac{\partial}{\partial k_y}\right)\frac{|\bar{u}_x|^2}{2} = S\left(1-2\frac{k_x^2}{k^2}\right)\mathcal{H}_k + \mathcal{N}_x - \nu k^2|\bar{u}_x|^2, \quad (16)$$

$$\left(\frac{\partial}{\partial t} - Sk_x\frac{\partial}{\partial k_y}\right)\frac{|\bar{u}_y|^2}{2} = 2S\frac{k_x k_y}{k^2}|\bar{u}_y|^2 + \mathcal{N}_y - \nu k^2|\bar{u}_y|^2, \quad (17)$$

$$\left(\frac{\partial}{\partial t} - Sk_x\frac{\partial}{\partial k_y}\right)\frac{|\bar{u}_z|^2}{2} = S\frac{k_x k_z}{k^2}(\bar{u}_z^*\bar{u}_y + \bar{u}_z\bar{u}_y^*) + \mathcal{N}_z - \nu k^2|\bar{u}_z|^2, \quad (18)$$

where $\mathcal{H}_k = -(\bar{u}_x\bar{u}_y^* + \bar{u}_x^*\bar{u}_y)/2$ is the spectrum of the Reynolds stress. Using Eqs. (16)-(18) together with the incompressibility condition (12), we obtain the evolution equation for the spectral energy density $\mathcal{E}_k = (|\bar{u}_x|^2 + |\bar{u}_y|^2 + |\bar{u}_z|^2)/2$,

$$\left(\frac{\partial}{\partial t} - Sk_x\frac{\partial}{\partial k_y}\right)\mathcal{E}_k = S\mathcal{H}_k + \mathcal{N}_k - \mathcal{D}_k, \quad (19)$$

where $\mathcal{D}_k = 2\nu k^2\mathcal{E}_k$ describes energy dissipation and $\mathcal{N}_x, \mathcal{N}_y, \mathcal{N}_z$ are the modified nonlinear transfer functions,

$$\mathcal{N}_i = \frac{1}{2}(\bar{u}_i^* Q_i + \bar{u}_i Q_i^*) - \frac{k_i}{2k^2}[\bar{u}_i^*(\mathbf{k}\cdot\mathbf{Q}) + \bar{u}_i(\mathbf{k}\cdot\mathbf{Q}^*)], \quad i = x, y, z$$

and their sum

$$\mathcal{N}_k = \mathcal{N}_x + \mathcal{N}_y + \mathcal{N}_z = \frac{1}{2}[\bar{\mathbf{u}}^*\cdot\mathbf{Q} + \bar{\mathbf{u}}\cdot\mathbf{Q}^*].$$

Equations (16)-(19) fully determine the nonlinear dynamics of the considered system in Fourier space and are the basis for subsequent analysis. According to them, four basic processes underly the dynamics of the perturbations and spectral energy:

(i) The second terms with $Sk_x\partial/\partial k_y$ in the brackets on the lhs of Eqs. (16)-(19), are the fluxes of the corresponding quantities parallel to the $k_y$-axis. These terms are of linear origin, coming from the convective derivative on the lhs of the main Eqs. (3)-(5) and therefore correspond to the advection by the mean flow. In other words, due to the background shear flow, the spectral quantities (Fourier transforms) "drift" in **k**-space: harmonics with $k_x > 0$ and $k_x < 0$ drift in opposite directions along the $k_y$-axis at a speed $S|k_x|$, whereas the ones with $k_x = 0$ are not advected by the flow. Since $\int d^3\mathbf{k}\partial(k_x\mathcal{E}_k)/\partial k_y = 0$, the drift only transports harmonics parallel to the $k_y$-axis without changing the total kinetic energy.

(ii) The first terms on the rhs of these equations are associated with shear, i.e., they originate from the linear term proportional to the shear rate on the rhs of Eq. (3) and describe influence of the mean flow on the dynamics of harmonics. Therefore, the first term on the rhs of spectral energy Eq. (19) describes exchange of energy between the mean flow and individual harmonics. This term in the spectral energy equation is related to the volume-averaged Reynolds stress entering Eq. (7) through

$$-\langle u_x u_y\rangle = \int \mathcal{H}_k(\mathbf{k}, t)d^3\mathbf{k},$$

and hence serves as a main source of energy for harmonics at the expense of which they can undergo amplification. This shear, or nonnormality induced process is in fact linear by nature and has a transient character due to the drift in **k**-space [28, 29, 30, 18]. In the case of turbulence studied in this paper, $\mathcal{H}_k$ describes injection of kinetic energy into turbulent fluctuations as a function of wavenumbers. The first terms on the rhs of Eqs. (16)-(18) describe the same process for the quadratic forms of perturbation velocity components. The harmonics, drifting parallel to the $k_y$-axis, go through dynamically important regions in spectral space at small and intermediate wavenumbers, where *energy-supplying* linear terms and *redistributing* nonlinear terms cooperate to sustain turbulence (see e.g., Fig. 8).

(iii) The second terms on the rhs of these equations, $\mathcal{N}_x$, $\mathcal{N}_y$ and $\mathcal{N}_z$, as mentioned above, describe, respectively, nonlinear transfers, or redistributions of the spectral amplitudes of the velocity components $|\bar{u}_x|$, $|\bar{u}_y|$ and $|\bar{u}_z|$ in **k**-space. Their net effect in the spectral energy budget over all wavenumbers is zero at all times:

$$\int [\mathcal{N}_x(\mathbf{k},t) + \mathcal{N}_y(\mathbf{k},t) + \mathcal{N}_z(\mathbf{k},t)] d^3\mathbf{k} = 0, \tag{20}$$

which is, in fact, a consequence of vanishing of the nonlinear terms in total energy equation in real space (Eq. 7). Although nonlinearity does not directly change the total (i.e., summed over all wavenumbers) spectral energy, it plays a central role in shear flow turbulence. $\mathcal{N}_k$ determines transfer, or cascade of the spectral energy in **k**-space. This energy is drawn from the mean flow by the linear stress term, leading to the development of its specific spectra. These transfer functions are one of the main focus of the present study. We aim to explore how they operate in the presence of shear by adopting the approach developed and used in [17, 19, 20], the main idea of which is performing a full 3D Fourier analysis of individual terms in the dynamical Eqs. (16)-(18), thus allowing for anisotropy of spectra and cascades. In particular, we will show that, like that in 2D shear flows considered in those studies, nonlinear transfers in the 3D case result in the redistribution of spectral energy among wavevector angles in **k**-space, which we refer to as NTC. We demonstrate that the interplay of NTC with linear phenomena (i.e., with the first rhs terms) becomes intricate: it realizes positive feedback, that is, repopulates those harmonics that have the potential of transient growth, thus contributing to the turbulence's self-sustenance.

(iv) The third terms on the rhs describe viscous dissipation of velocity. Comparing them with the energy-supplying linear terms, we can see that viscous dissipation becomes important at large wavenumbers $k \geq k_{\mathcal{D}} = \sqrt{S/\nu}$, where $k_{\mathcal{D}}$ denotes the effective wavenumber for which dissipation effects start to play a role.

## 3. Results

Now we turn to an analysis of nonlinear evolution of perturbations employing numerical methods. Main emphasis will be on the spectral aspect of the dynamics using the mathematical formalism outlined in the previous section. We solve full Navier-Stokes equations (3)-(6) in a rectangular box with size $0 \leq x \leq L_x, -L_y/2 \leq y \leq L_y/2, 0 \leq z \leq L_z$, in the streamwise ($x$), vertical/shearwise ($y$) and spanwise ($z$) directions, respectively, using $N_x \times N_y \times N_z$ number of grid points. Numerical simulations are performed with the code developed by Sekimoto et al. [27], which solves the Navier-Stokes equations in the velocity-vorticity representation [31]. The streamwise and spanwise directions are periodic so that the spectral method is applied. The vertical direction is shear-periodic, realized by adding "shift" factors in the compact-finite-difference matrices which guarantee spectral like resolution [32]. The method does not require the remeshing procedure leading to constant loss of enstrophy [33], which has to be avoided when studying vortical structures. To assess how well our simulations are resolved, we compare the Kolmogorov length scale $\eta = (\nu^3/\langle\varepsilon\rangle)^{1/4}$, where $\langle\varepsilon\rangle = \nu\langle(\nabla \times \mathbf{u})^2\rangle$ is the volume-averaged dissipation rate of the perturbation energy, to the grid cell sizes $\Delta x = L_x/N_x, \Delta y = L_y/N_y, \Delta z = L_z/N_z$ in the computation box.

Since in the $y$-direction the boundary conditions are shear-periodic, standard FFT technique cannot be applied for calculating Fourier transforms along this direction during post-processing. Due to shear flow, harmonics drift along the $k_y$-axis and, consequently, $k_y$ of an individual harmonic changes with time linearly, $k_y(t) = k_y(0) - Sk_xt$ (see e.g., Refs. [18, 28]) and remains consistent with the shear-periodic boundary conditions. Following [34, 35], we modified the standard FFT technique to take this effect into account as follows. As $k_y$ drifts, at each time $t$ in the corresponding computational domain in $(k_x, k_y)$-plane (at fixed $k_z$) there are only Fourier harmonics with wavenumbers $k_x = 2\pi n_x/L_x, k_y = 2\pi n_y/L_y - St'(2\pi n_x/L_x)$, where $t' = mod[t, L_x/(SL_y|n_x|)]$ and integers $n_x, n_y$ from the intervals $n_x \in [-N_x/2, N_x/2], n_y \in [-N_y/2, N_y/2]$. So, in this case Fourier transformation reduces to the decomposition into these wavenumbers. We calculate it by first transforming a physical quantity along $k_x$ and $k_z$ using the standard FFT, next by multiplying the result by $\exp(iSk_xt')$, that accounts for the time-dependence of $k_y$, and applying again FFT method along $k_y$.

The time is normalized by the dynamical, or shear time $S^{-1}$, while an arbitrary length scale can be used for normalizing distances and other physical variables, since the problem is in fact scale-free. The Reynolds number is defined in terms of $L_z$ as $Re_z = SL_z^2/\nu$. Together with the Reynolds number, the problem has two dimensionless free parameters – aspect ratios of the domain, $A_{xz} = L_x/L_z$, $A_{yz} = L_y/L_z$. In this paper, we consider the model with the cubic box $A_{xz} = A_{yz} = 1$, resolution $N_x \times N_y \times N_z = 128 \times 128 \times 128$ and the Reynolds number $Re_z = 4930$. In this case, we found that $\Delta x \approx 1.66\eta$, which implies that larger and intermediate scales, where, as we will show, the self-sustaining dynamics of turbulence is concentrated, are well resolved in this box. We normalize the wavenumbers $k_x, k_y, k_z$, respectively, by $\Delta k_x = 2\pi/L_x, \Delta k_y = 2\pi/L_y$ and $\Delta k_z = 2\pi/L_z$, i.e., $(k_x/\Delta k_x, k_y/\Delta k_y, k_z/\Delta k_z) \to (k_x, k_y, k_z)$. Since $\Delta k_x, \Delta k_y$ and $\Delta k_z$ are the grid cell sizes in Fourier space, the normalized streamwise and spanwise wavenumbers become integers $k_x, k_z = 0, \pm1, \pm2, ...$, while $k_y$, although changes with time due to drift, is integer at discrete moments $t_m = mA_{xz}/(|k_x|A_{yz})$, where $m$ is also a positive integer.

Starting from initial random perturbations, the flow settles into a self-sustained turbulent state. Our main goal is to investigate underlying self-sustaining process of the turbulence. We show that the turbulence is self-sustained by the interplay of linear and nonlinear processes and is characterized by quasi-periodic bursting events, which are related to the dominant mode. This scenario of a self-sustained state, based on a subtle cooperation between linear and nonlinear processes, is a keystone of the bypass concept of subcritical turbulence in spectrally stable shear flows [9, 11, 12, 13, 36, 14]. Finally, it should be stressed that we base our study on DNS results of turbulence and in this respect is general, not relying on simplifying assumptions usually made in low-order models of self-sustaining processes (e.g., [24, 25]).

It is well-known from the linear nonmodal theory of shear flows that optimal perturbations which are characterized by maximum transient growth during the dynamical – eddy turnover time, which can be assumed of the order of shear time, $t_d \simeq S^{-1}$ (or, $t_d \simeq 1$ in normalized units) are responsible for most of the energy extraction from the mean flow [5, 37]. In Fig. 1 we present the transient growth factor of energy for optimal harmonics during $t_d$, which is calculated from the linearized inviscid equations, i.e., using only the first linear terms on the rhs of Eqs. (13)-(15). These terms, and hence the dynamics of harmonics in the inviscid case, depends only on the ratio of wavenumbers, $k_y/k_x$ and $k_z/k_x$. In this figure, we show the growth factor as a function of these ratios. As one can see, the optimal growth factor is largest at $1 \leq k_z/k_x < 4$ and $0 < k_y/k_x < 1$. So, one can cover as many of these optimal perturbations as possible and thus better account for their major role in the energy exchange processes between turbulence and mean flow if the computational box aspect ratios satisfy the conditions $A_{xz} < A_{yz}$ and $A_{xz} \geq 1$. In agreement with the previous study of homogeneous shear turbulence by Pumir [26], in our simulations we also find that in addition to the non-symmetric ($k_x \neq 0$) optimal harmonics, a harmonic with a large scale spanwise variation, which

Journal of Physics: Conference Series **708** (2016) 012001 doi:10.1088/1742-6596/708/1/012001

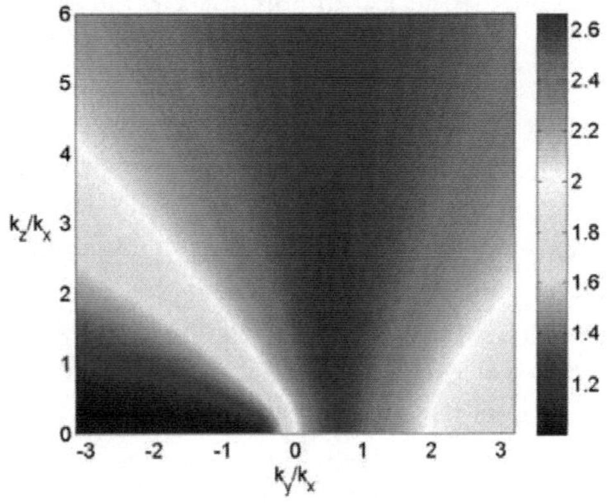


**Figure 1.** Transient growth factor of optimal harmonics' energy, $\mathcal{E}_k(t_d)/\mathcal{E}_k(0)$, as a function of wavenumbers. It is calculated during the dynamical time $t_d = S^{-1}$ with linearized equations of motion.

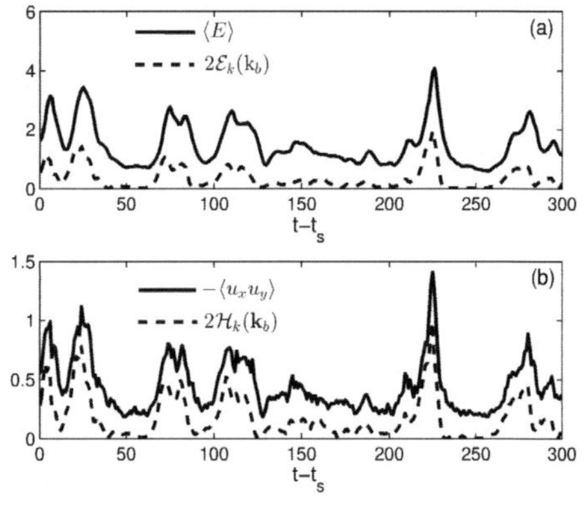


**Figure 2.** Evolution of (a) the volume-averaged total kinetic energy, $\langle E \rangle$, and the energy of the basic mode with $\mathbf{k}_b = (0, 0, \pm 1)$, $2\mathcal{E}_k(\mathbf{k}_b)$, as well as (b) the volume-averaged Reynolds stress, $-\langle u_x u_y \rangle$ and the stress corresponding to the basic mode, $2\mathcal{H}_k(\mathbf{k}_b)$, in the saturated state.

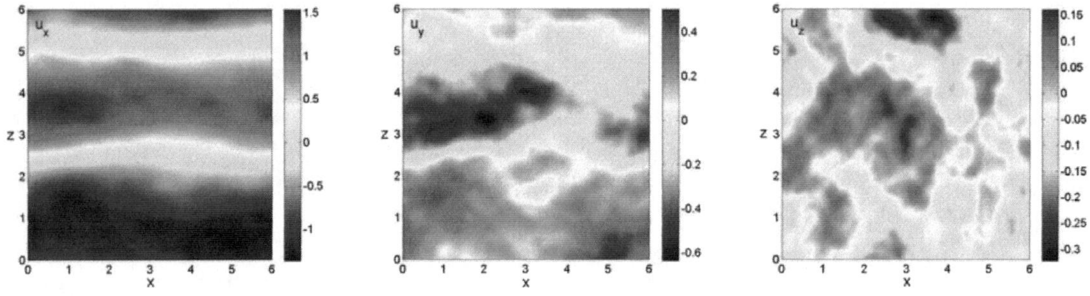


**Figure 3.** Distribution of the time-averaged velocity components in the fully developed turbulence. Slices in $(x, z)$-plane are at $y = 0$. $u_x$ component is larger than $u_y$ and $u_z$. In the plots of $u_x$ and $u_y$, a signature of the basic harmonic with large spanwise scale can be discerned.

is uniform in the streamwise and shearwise directions, $k_x = 0, k_y = 0, k_z = \pm 1$, plays a central in the dynamical processes. Due to its importance in turbulence dynamics, we call this harmonic as the basic mode. Other harmonics, whose energy grows more than 10% of the maximum value of the spectral energy at least once during the evolution, are considered to be actively participating in and forming the self-sustaining dynamics of turbulence. We show that quasi-periodic bursts of energy and Reynolds stress characteristic of homogeneous shear turbulence [26] are determined by a competition between the basic mode and other dynamically "active" harmonics. The number of these harmonics generally depends on the box aspect ratio and is quite large (in the considered here cubic box, it is 36, see Fig. 5, but can reach more than 200 for steamwise elongated boxes), indicating fairly complex dynamics, not amenable to reduced low-order description.

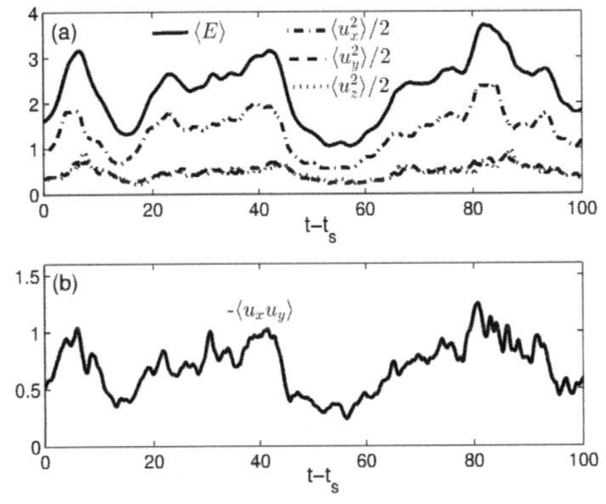


**Figure 4.** Evolution of (a) the volume-averaged perturbation kinetic energy (solid line) and quadratic forms of streamwise, $\langle u_x^2 \rangle/2$ (dot-dashed line), shearwise/vertical, $\langle u_y^2 \rangle/2$ (dashed line) and spanwise, $\langle u_z^2 \rangle/2$ (doted line) velocities as well as (b) the Reynolds stress.

Below we analyze in detail the specific energy spectra and self-sustaining dynamics of turbulence in wavenumber/spectral space by calculating the Fourier transforms of linear and nonlinear terms in governing equations using the simulation data.

### *3.1. Nonlinear evolution*

At the early stage of evolution, each harmonic contained in the initial perturbations grows transiently due to the flow nonnormality, leading to the growth of the total energy and Reynolds stress. Then after about $t_s = 30S^{-1}$, the amplitudes of the perturbations become sufficiently large in the nonlinear regime and the flow settles into a self-sustained turbulence. As was found in the paper by Pumir [26], in the case of cubic box, the basic harmonic/mode with wavenumber $\mathbf{k}_b = (0, 0, \pm 1)$, dominates over other ones, i.e., its spectral energy, $\mathcal{E}_k(\mathbf{k}_b)$, and stress, $\mathcal{H}_k(\mathbf{k}_b)$, are, respectively, maxima of $\mathcal{E}_k$ and $\mathcal{H}_k$ during an entire run. Figure 2 shows the time-evolution of the volume-averaged total energy and stress as well as the energy and stress of the basic mode in the saturated state, which appear to display similar patterns of quasi-periodic bursts in time. The volume-averaged Reynolds stress, $-\langle u_x u_y \rangle$, and the stress of the basic mode, $\mathcal{H}_k(\mathbf{k}_b)$ are positive at all times and non-decaying as required by the self-sustaining process according to Eq. (7). The bursts in the total energy and stress closely follow amplifications (peaks) of $\mathcal{E}_k(\mathbf{k}_b)$ and stress $\mathcal{H}_k(\mathbf{k}_b)$ in the basic mode. This indicates that these bursts are related to the dynamics of the basic mode, which thus appears to be mainly responsible for the energy pumping from the mean flow into turbulence. During the bursts, the contribution of $\mathcal{H}_k(\mathbf{k}_b)$ to the total Reynolds stress is as much as 60% - 70%, while other modes become important (although still remain smaller than the basic mode) in quiescent intervals between the bursts, when the total energy and stresses as well as the energy and stress of the basic mode are low. One can conclude, that most of the turbulent energy is primarily produced through the interaction of the basic mode with the mean shear flow.

The structure of the velocity components in the fully developed turbulence in physical space is presented in Fig. 3. The evolution of the volume-averaged perturbation kinetic energy and quadratic forms of streamwise, $\langle u_x^2 \rangle/2$, shearwise/vertical, $\langle u_y^2 \rangle/2$, and spanwise, $\langle u_z^2 \rangle/2$, velocities as well as the Reynolds stress are shown in Fig. 4. These plots show the dominance of the streamwise velocity over the shearwise and spanwise ones. In the $(x, z)$ slices of $u_y$ and especially of the $u_x$ velocity, we can clearly see the signatures of the dominant basic harmonic with small scale fluctuations superimposed on it due to other harmonics.

To understand the role of other modes in the turbulence dynamics and energy exchange process with the mean flow, it is important to define the main, energy-containing area in **k**-space, which we also refer to as *the vital area* of turbulence. For this purpose we identify two sets of harmonics whose energies grow from 5% to 10% and larger than 10% of the maximum

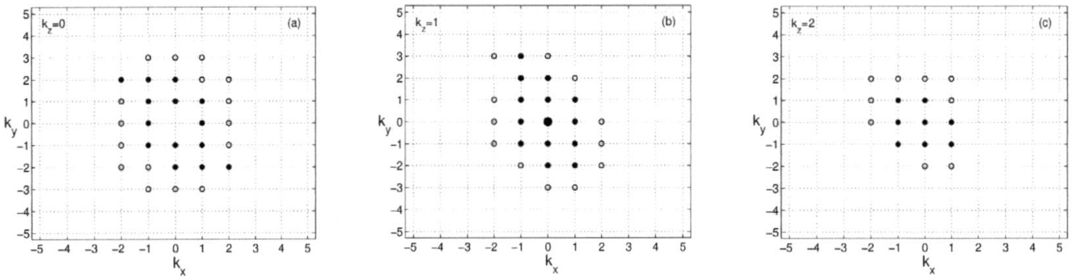

**Figure 5.** Distribution of the modes carrying dominant energy in **k**-space at (a) $k_z = 0$ (b) $k_z = 1$ and (c) $k_z = 2$. Black dots represent the harmonics whose energies grow more than 10% of the maximum spectral energy at least once during evolution. Bigger black dot corresponds to the basic mode having the maximum spectral energy at all times. Circles are the modes whose energies grow from 5% to 10% of the maximum spectral energy at least once during evolution.

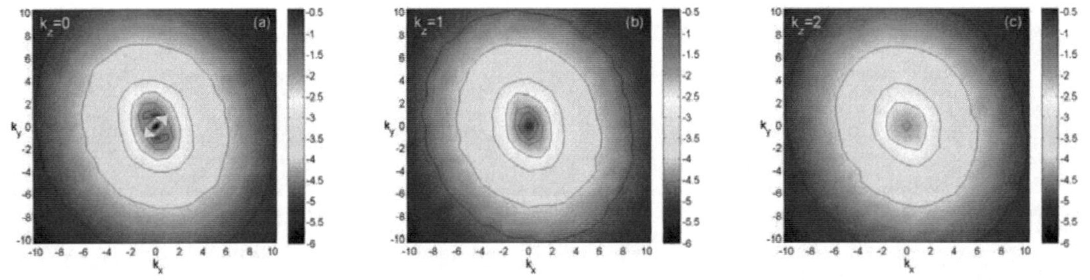

**Figure 6.** Logarithm of the time-averaged spectrum of the kinetic energy, $log_{10}(\mathcal{E}_k)$. The $(k_x, k_y)$-slices are at (a) $k_z = 0$, (b) $k_z = 1$ and (c) $k_z = 2$. The spectrum has an anisotropic character, having larger power in the $k_y/k_x < 0$ part of Fourier space at a given $k_x$. The spectral energy has maximum at $k_z = 1$ and decreases with increasing $k_z$.

spectral energy at each time at least once during the evolution. These dominant energy-carrying, or dynamically important harmonics are presented in Fig. 5. They have small wavenumbers $|k_x| \leq 2, |k_y| \leq 3, |k_z| \leq 3$ (at $|k_z| = 3$, not shown in this figure, mode energies always remain less than 10% of the maximum energy) which basically form the vital area of turbulence. Harmonics with larger wavenumbers $|k_x| > 2, |k_y| > 3, |k_z| > 3$ lie outside the vital area and always have energy and stress less than 5% of the maximum value, therefore, not playing a major role in the energy exchange process between the mean flow and turbulence. Essentially, only these large scale modes in the vital area take part in the self-sustaining dynamics of turbulence, but the basic mode/harmonic still remains dominant – the basic harmonic is most active among the active harmonics. We emphasize that the total number of large scale "active modes", which we define as the harmonics whose energies grow more than 10% of the maximum value at least once during the evolution (black dots in Fig. 5) is equal to 36, implying that the dynamics of homogeneous shear turbulence cannot by any means be reduced to low-order models of self-sustaining processes.

Figure 6 shows the time-averaged spectra of the kinetic energy in $(k_x, k_y)$-plane at different $k_z$. These time averages are done over an entire time of the saturated turbulence in the simulation. The spectra are anisotropic due to shear – the isolines are of elliptical shape inclined opposite to the $k_y$-axis. This fact indicates that on average modes with $k_y/k_x < 0$ have more energy than those with $k_y/k_x > 0$ at fixed $k_x$. The plots show that the time-averaged spectral kinetic

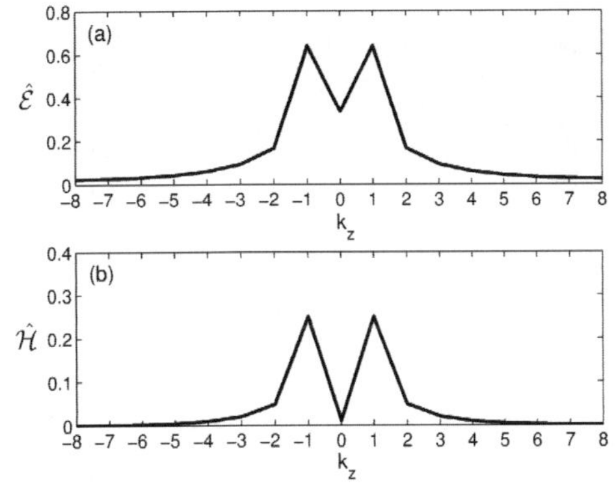


**Figure 7.** Integrated in $(k_x, k_y)$-plane time-averaged (a) kinetic energy spectrum, $\hat{\mathcal{E}}$, and (b) Reynolds stress, $\hat{\mathcal{H}}$ vs. $k_z$.

energy at fixed $k_z$ is larger for small $k_x$ and $k_y$ and decays at least by two order of magnitude at $\sqrt{k_x^2 + k_y^2} \approx 2$. So, the active modes are situated just within this range of small wavenumbers, as also seen in Fig. 5. As mentioned above, the maximum of the spectral energy is reached at $k_x = k_y = 0, k_z = \pm 1$ corresponding to the basic mode. The spectral energy is much reduced already for the next spanwise modes, $k_z = 2$, and further decreases with increasing $k_z$.

The integrated in $(k_x, k_y)$-plane the time-averaged spectral energy, $\hat{\mathcal{E}}(k_z) = \int \mathcal{E}_k dk_x dk_y$, and stress, $\hat{\mathcal{H}}(k_z) = \int \mathcal{H}_k dk_x dk_y$, as a function of $k_z$ are shown in Fig. 7. They both reach a maximum at $k_z = \pm 1$ and rapidly decrease with $k_z$.

The anisotropic nature of the kinetic energy spectrum, which clearly distinguishes it from typical isotropic turbulent spectra in the classical shearless case of Kolmogorov phenomenology, arises as a result of the specific interplay of linear and nonlinear processes in **k**-space. We show below that these processes are anisotropic over wavenumbers due to shear, resulting in a new phenomenon – the NTC of power in spectral space.

### *3.2. Energy injection, nonlinear transfers and their interplay in* **k***-space*

For the better understanding of the character of the anisotropic kinetic energy spectrum and nonlinear transfers, in Fig. 8 we present the distribution of the time-averaged energy injection, i.e., the Reynolds stress spectrum, $\mathcal{H}_k$, and the nonlinear transfer functions, $\mathcal{N}_x, \mathcal{N}_y$ and $\mathcal{N}_z$, in **k**-space. Since these quantities are symmetric with respect to a change $\mathbf{k} \to -\mathbf{k}$, without loss of generality, hereafter we concentrate on the upper part ($k_z > 0$) of **k**-space. From these figures it is clear that, like the spectral energies, all these quantities exhibit anisotropy over wavenumbers, that is, depend on wavevector angle, or orientation. $\mathcal{H}_k$ is primarily concentrated in the central part of spectral space, $|k_x| \leq 2, |k_y| \leq 3, |k_z| \leq 2$ and $k_y/k_x > 0$ – in the vital area, where it supplies the turbulence with energy via nonnormality-induced transient amplification of the dynamically active harmonics (see below). At $k_z = 0$, $\mathcal{H}_k$ is larger and mainly operates around $|k_y|, |k_x| \sim 1$ (Fig. 8a), it is positive at $k_x/k_y > 0$, supplying harmonics with energy there and negative at $k_x/k_y < 0$, returning energy to the flow. On the other hand, at $k_z \geq 1$, the distribution of $\mathcal{H}_k$ is changed – it is positive and appreciable around $k_x = k_y = 0$ (Figs. 8b and 8c). The Reynolds stress spectrum is highest, and therefore the energy extraction from the mean flow into turbulence occurs most intensively, at $k_z = 1$ with the maximum at $\mathbf{k_b}$ corresponding to the basic mode (Figs. 7b and 8b). Outside the vital area, at $|k_x| > 2, |k_y| > 3, |k_z| > 2$, the stress and hence the energy input rapidly decrease and vanish at large wavenumbers (Fig. 7b). The total energy input is negligible at $k_z = 0$, $\hat{\mathcal{H}}_k \approx 0$, as seen from Fig. 7(b), since $\mathcal{H}_k$ changes sign at this $k_z$ in $(k_x, k_y)$-plane (Fig. 8a). Thus, the main energy supply for turbulence comes from large-scale modes, with wavelengths comparable to the box size. These are the modes

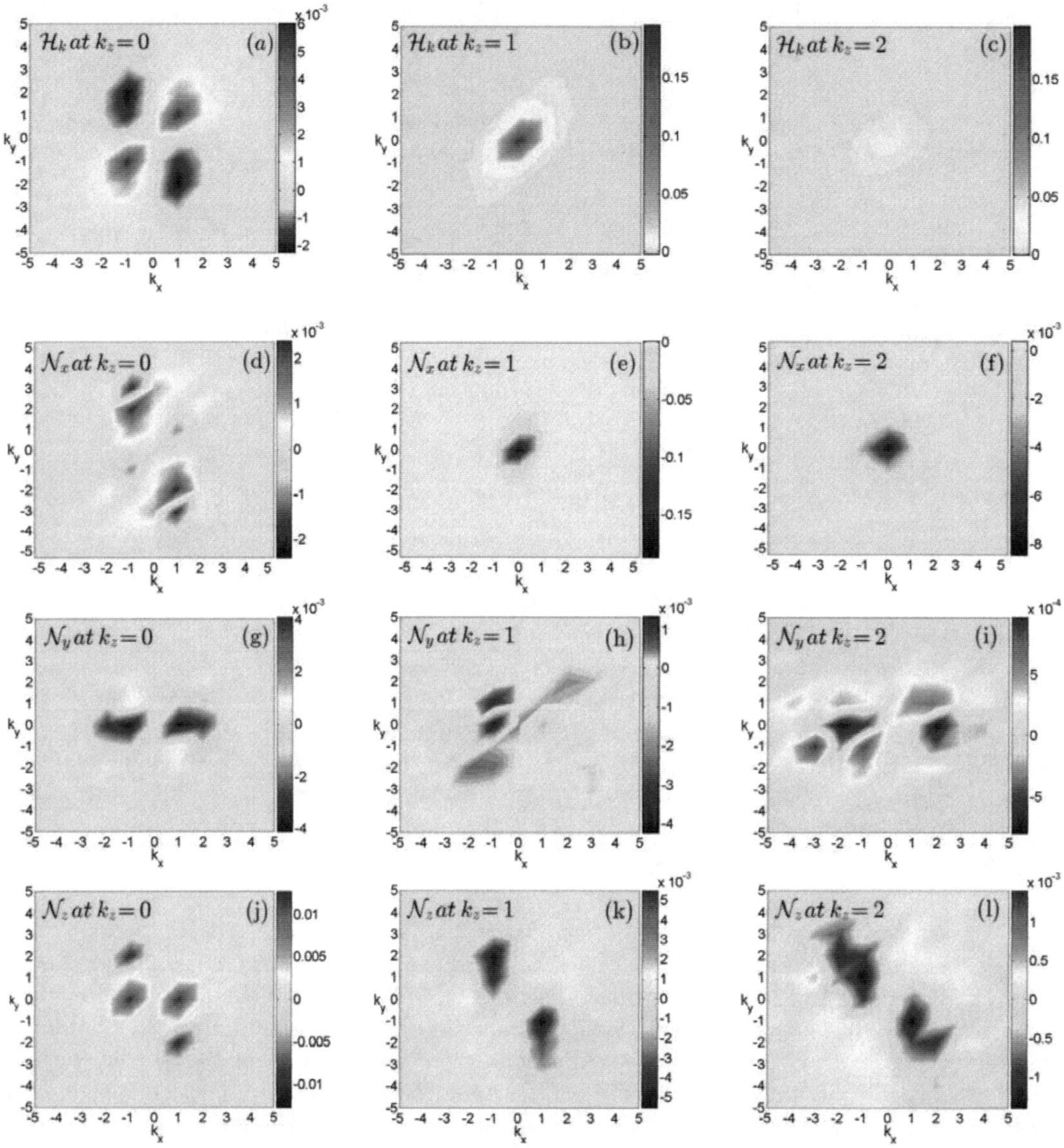

**Figure 8.** Maps of the time-averaged energy injection [(a), (b), (c)], $\mathcal{H}_k$, and nonlinear transfer, [(d), (e), (f)] $\mathcal{N}_x$, [(g), (h), (i)] $\mathcal{N}_y$, [(j), (k), (l)] $\mathcal{N}_z$ functions in **k**-space. Shown are $(k_x, k_y)$-slices of these quantities at $k_z = 0$ (left column), $k_z = 1$ (middle column) and $k_z = 2$ (right column). $\mathcal{H}_k$ is appreciable, and hence energy injection into turbulence mainly occurs, in the vital area at small wavenumbers, $|k_x| \leq 2, |k_y| \leq 3, |k_z| \leq 2$, on the $k_y/k_x > 0$ side. At $k_z = 1$, $\mathcal{H}_k$ reaches largest values and hence energy injection is most intensive at this spanwise wavenumber. $\mathcal{N}_x$, $\mathcal{N}_y$ and $\mathcal{N}_z$ terms transfer, respectively, the streamwise, shearwise and spanwise components of the spectral kinetic energy anisotropically, or transversely over wavevector angles in Fourier space, away from regions where they are negative $\mathcal{N}_x < 0$, $\mathcal{N}_y < 0$, $\mathcal{N}_z < 0$ (blue) to regions where they are positive $\mathcal{N}_x > 0$, $\mathcal{N}_y > 0$, $\mathcal{N}_z > 0$ (yellow and red).

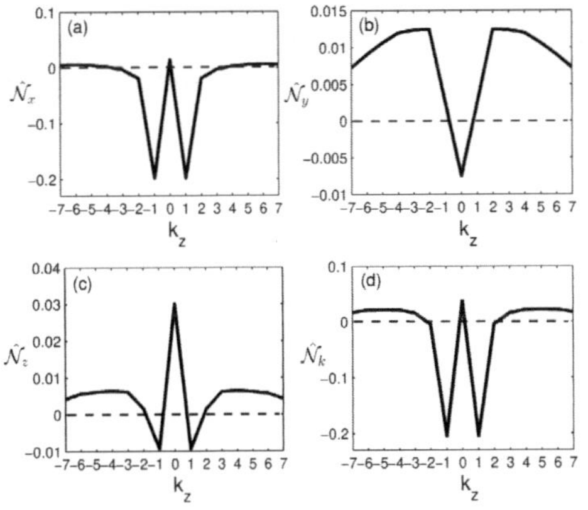

**Figure 9.** The time-averaged nonlinear transfer functions integrated in $(k_x, k_y)-$plane and represented as a function of $k_z$, (a) $\hat{\mathcal{N}}_x$, (b) $\hat{\mathcal{N}}_y$, (c) $\hat{\mathcal{N}}_z$ and their sum (d) $\hat{\mathcal{N}}_k$ describing nonlinear transfer of energy. These functions transfer corresponding quantities from $k_z$ at which these functions are negative to those $k_z$ at which they are positive.

which most effectively interact with the mean flow and gain energy, whereas smaller scale modes only receive energy from the former due to nonlinear transfers. Below we describe the character of nonlinear transfers in the presence of shear and their interplay with the energy-injecting stress term, leading to a self-sustained turbulent state. Note also that the energy injection into turbulence mediated by the stress spans a range wavenumbers and in this respect differs from classical cases of forced turbulence in flows without mean shear, where forcing is applied at a specific, or a narrow band of wavenumbers (see e.g., [38]).

The nonlinear terms $\mathcal{N}_x$, $\mathcal{N}_y$ and $\mathcal{N}_z$ describe, respectively, transfers of the energies in the streamwise, shearwise/vertical and spanwise velocity components. As noted above, they do not represent a source of the total energy for the turbulence, but only redistribute energy of harmonics, which is extracted from the mean flow, over wavenumbers and together with the energy-injecting stress term, determine the spectra. It is seen from Figs. 8(d)-(l) that different components of the nonlinear term operate differently in **k**-space and hence play different roles in the self-sustaining process. For this reason, we plot them separately instead of their sum, which describes nonlinear transfer of energy in Eq. (19). Since we are interested in the interplay of linear energy-injecting stress terms and nonlinear cascades, in this figure we show the nonlinear transfer terms also at $k_z = 0, 1, 2$ for which the stress is significant. First of all, these plots clearly demonstrate the anisotropic nature of the nonlinear processes in **k**-space. The transfer functions depend on wavevector angle and, consequently, redistribute the energies generally not only along wavevector, corresponding to direct/inverse cascades that are well-known in the classical turbulence theory, but also transverse to wavevector. This angular redistribution is the essence of NTC and occurs from the regions in **k**-space where the respective transfer functions are negative $\mathcal{N}_x,\mathcal{N}_y,\mathcal{N}_z < 0$ (blue areas in Figs. 8d-l) to regions where they are positive $\mathcal{N}_x,\mathcal{N}_y,\mathcal{N}_z > 0$ (yellow and red areas); at all other wavenumbers (light green areas) these functions are small. It is seen from Fig. 8(d)-(l) that the NTC primarily operates in the central vital area of the wavenumber space, $|k_x| \leq 2, |k_y| \leq 3, |k_z| \leq 2$, where, as described above, most of the energy supply process by $\mathcal{H}_k$ is concentrated. Outside the vital area at large wavenumbers, where the nonnormality-induced energy exchange between the mean flow and harmonics vanishes, the transfer functions become nearly independent of the wavevector angle and depend mostly on its magnitude, i.e., the cascade is no longer transverse but direct – the situation is similar to the classical homogeneous isotropic turbulence described by Kolmogorov phenomenology. Below we describe the interplay of the NTC with this energy-injecting linear phenomena that can realize either positive or negative feedback. In the case of positive feedback, the NTC repopulates harmonics having the potential of transient growth.

To characterize transfers also along $k_z$ due to the NTC, we integrate $\mathcal{N}_i$ over $(k_x, k_y)$-plane,

$\hat{\mathcal{N}}_i(k_z) = \int \mathcal{N}_i dk_x dk_y, (i = x, y, z, k)$, and represent them as functions of $k_z$ in Fig. 9. Figures 8(d)-(f) and 9(a) show that $\mathcal{N}_x$ is negative and large by absolute value in the vital area, where the stress, $\mathcal{H}_k$, is positive and appreciable, but it is small and positive at larger wavenumbers outside this area, $|k_x| > 2, |k_y| > 3, |k_z| > 3$, where the stress is vanishing. In other words, $\mathcal{N}_x$ provides a sink, or negative feedback for $\mathcal{H}_k$ by taking out energy in the streamwise velocity component gained from the mean flow and transferring it to large wavenumbers and part of it to $k_z = 0$, as evident from the dependence of $\hat{\mathcal{N}}_x$ on $k_z$. By contrast, $\mathcal{N}_y$ is positive in those regions of the vital area where $\mathcal{H}_k$ is also positive (Figs. 8g-8i and 9b). As a result, at these wavenumbers, it provides positive feedback by generating the energy in the shearwise velocity, $|\bar{u}_y|^2/2$, which, in turn, leads to the growth of the energy in the streamwise velocity, $|\bar{u}_x|^2/2$, due to the nonnormality/shear-induced linear growth mechanism (the first term of linear origin on the rhs of Eqs. 13 and 16, while $\mathcal{N}_x < 0$ at these wavenumbers and opposes this growth). This streamwise energy, which in fact appears to be largest among the other two components and hence giving the main contribution in the total energy, is then transferred to larger wavenumbers by $\mathcal{N}_x$. Note that this positive feedback by $\mathcal{N}_y$ – crucial to the self-sustenance of turbulence – is realized owing to the NTC, which ensures repopulation with the shearwise energy of those harmonics with the wavevector orientations ($k_y/k_x > 0$) having $\mathcal{H}_k > 0$ and hence the capability of nonmodal growth. Figures 8(j)-(l) and 9(c) show $\mathcal{N}_z$, which is responsible for the transfer of energy of the spanwise velocity. However, since the Reynolds stress, providing energy supply for the turbulence, is produced by the product (correlation) of $u_x$ and $u_y$, below we concentrate only on $\mathcal{N}_x$ and $\mathcal{N}_y$ functions governing nonlinear redistribution of energies in these velocity components. Figure 9(d) shows the nonlinear transfer function for the energy, $\hat{\mathcal{N}}_k = \hat{\mathcal{N}}_x + \hat{\mathcal{N}}_y + \hat{\mathcal{N}}_z$. It is dominated by $\hat{\mathcal{N}}_x$ and hence the nonlinear transfer of the total energy is similar to that of the streamwise energy, which in fact gives the dominant contribution to the former.

Thus, the transverse cascade of energy appears to be a generic feature of nonlinear dynamics of perturbations in spectrally stable shear flows. The conventional description of shear flow turbulence solely in terms of direct and inverse cascades, which leaves such nonlinear transverse cascade out of consideration, might be incomplete and misleading. It should be emphasized that revealing of the complete picture of these nonlinear cascade processes has become largely possible due to carrying out the analysis in Fourier space. Because of the shear-induced anisotropy of cascade directions, often used angle-integrated energy spectra and transfer functions alone, clearly, are not fully representative of the actual, more general, over wavevector angles, nonlinear redistribution of the spectral energies in **k**-space.

We have described above how the interplay of linear and nonlinear process happens in a time average sense in the fully developed saturated turbulence. To gain a better insight in its self-sustenance and appearance of the bursts in the total energy evolution seen in Fig. 2, below we analyze the dynamics of basic harmonic. Namely, we show that the alternating growth and decay intervals in its evolution defines the behavior of the bursts.

*3.2.1. The dynamics of the basic mode* We consider the dynamics of the dominant basic mode with $\mathbf{k}_b = (0, 0, \pm 1)$, which during the whole evolution corresponds to the maximum energy and stress in spectral space. For this mode, the dynamical Eqs. (16)-(18) with $S = 1$ are reduced to the following system:

$$\frac{\partial}{\partial t}\frac{|\bar{u}_x(\mathbf{k}_b, t)|^2}{2} = H_k(\mathbf{k}_b, t) + \mathcal{N}_x(\mathbf{k}_b, t) - \nu|\bar{u}_x(\mathbf{k}_b, t)|^2, \tag{21}$$

$$\frac{\partial}{\partial t}\frac{|\bar{u}_y(\mathbf{k}_b, t)|^2}{2} = \mathcal{N}_y(\mathbf{k}_b, t) - \nu|\bar{u}_y(\mathbf{k}_b, t)|^2, \tag{22}$$

$$\frac{\partial}{\partial t}\frac{|\bar{u}_z(\mathbf{k}_b, t)|^2}{2} = -\nu|\bar{u}_z(\mathbf{k}_b, t)|^2. \tag{23}$$

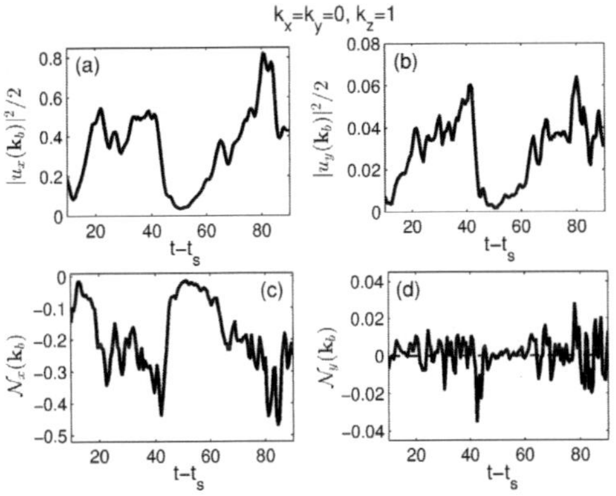


**Figure 10.** Evolution of (a) the streamwise, $|\bar{u}_x|^2/2$, and (b) shearwise, $|\bar{u}_y|^2/2$, spectral energies as well as the nonlinear transfer functions (c) $\mathcal{N}_x$ and (d) $\mathcal{N}_y$ for the basic mode with $\mathbf{k}_b = (0, 0, \pm 1)$. The streamwise energy is much larger than the shearwise one due to nonnormality-induced linear growth. $\mathcal{N}_x$ is negative during the whole evolution and acts as a sink for the streamwise energy, while $\mathcal{N}_y$ oscillates and, when positive, causes increase in $|\bar{u}_y|$; its time-averaged value over the evolution is positive $4.78 \cdot 10^{-4}$.

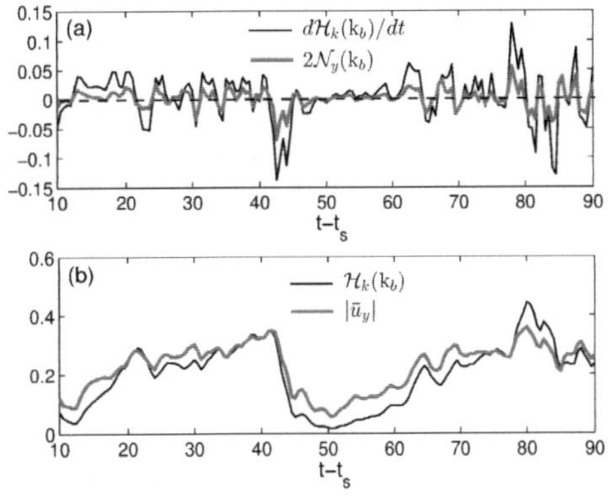


**Figure 11.** Evolution of (a) the stress production rate, $d\mathcal{H}_k(\mathbf{k}_b)/dt$, and nonlinear transfer term $\mathcal{N}_y(\mathbf{k}_b)$ as well as (b) the spectral Reynolds stress, $\mathcal{H}_k(\mathbf{k}_b)$, and shearwise velocity, $|\bar{u}_y(\mathbf{k}_b)|$ of the basic mode. The stress production rate closely follows $\mathcal{N}_y(\mathbf{k}_b)$ (we scaled it by a factor of 2 to make comparison easier) and, consequently, the amplifications (bursts) of both shearwise velocity and stress occur when $\mathcal{N}_y$ is positive and decrease when $\mathcal{N}_y$ is negative.

Note that $\mathcal{N}_z(\mathbf{k}_b, t) = 0$ and Eq. (23) can be solved directly to get exponential decay of spanwise velocity of the basic mode, $\bar{u}_z(\mathbf{k}_b, t) = \bar{u}_z(\mathbf{k}_b, 0)\exp(-2\nu t)$. So, after a sufficiently long period of time the spanwise velocity vanishes and the velocity field of the basic harmonic becomes two-dimensional. Figure 10 shows the time development of the squared velocity amplitudes, $|\bar{u}_x(\mathbf{k}_b)|^2/2$ and $|\bar{u}_y(\mathbf{k}_b)|^2/2$ and the nonlinear transfer functions $\mathcal{N}_x(\mathbf{k}_b)$ and $\mathcal{N}_y(\mathbf{k}_b)$ for the basic mode. Because it is of large scale and the Reynolds number is high, the dissipation terms in these equations appear to be much smaller than the transfer functions. The velocities exhibit a sequence of amplifications (bursts) and decays, with the streamwise velocity being always much larger than the shearwise one. $\mathcal{N}_x(\mathbf{k}_b)$ is always negative and therefore acts as a sink in the streamwise velocity evolution Eq. (21), while $\mathcal{N}_y(\mathbf{k}_b)$ oscillates irregularly with alternating sign. When $\mathcal{N}_y(\mathbf{k}_b)$ is positive, it amplifies the shearwise velocity, while when negative causes its decrease, but the time-average of $\mathcal{N}_y(\mathbf{k}_b)$ is positive, though small (see also Fig. 8h), indicating the growth trend of the shearwise velocity on average. The shearwise velocity, in turn, leads to the amplification of the streamwise velocity due to the shear-related linear first term on the rhs of dynamical Eq. (13). This is essentially the nonnormality induced linear nonmodal amplification process, resulting in the growth of streamwise velocity much larger than that of the shearwise one. Note that this is the only cause of amplification for $|\bar{u}_x(\mathbf{k}_b)|$, since the corresponding nonlinear term, $\mathcal{N}_x(\mathbf{k}_b)$, as mentioned above, is always negative and cannot amplify it. Then, $\bar{u}_y(\mathbf{k}_b)$ and the amplified $\bar{u}_x(\mathbf{k}_b)$ jointly give rise to positive stress, $\mathcal{H}_k(\mathbf{k}_b) > 0$ (Figs. 2b and 11b). As a result, the stress production rate, $d\mathcal{H}_k(\mathbf{k}_b)/dt$ (calculated by combining Eqs. 13 and 14), closely follows, or correlates with $\mathcal{N}_y(\mathbf{k}_b)$ (Fig. 11a, the correlation coefficient is $R(d\mathcal{H}_k/dt, \mathcal{N}_y) = 0.93$[1] and, consequently, the stress itself turns out to follow $|\bar{u}_y(\mathbf{k}_b)|$ (Fig. 11b, $R(\mathcal{H}_k, |\bar{u}_y|) = 0.97$), i.e., the peaks and dips in the time evolution of these quantities coincide. So, it can be said that transfer function $\mathcal{N}_y(\mathbf{k}_b)$ plays a central role in the dynamical behavior of the basic mode by providing a positive feedback to the nonmodal growth. When $\mathcal{N}_y(\mathbf{k}_b) > 0$, it causes the amplifications of $\bar{u}_y$, which via nonmodal growth mechanism leads to amplification of $\bar{u}_x$ and of the corresponding stress. This bursts alternate with the decaying intervals of these quantities corresponding to negative $\mathcal{N}_y(\mathbf{k}_b) < 0$. Because for the aspect ratio $(1,1)$, the basic mode is the dominant one among all the active modes, these bursts also manifest themselves in the total stress evolution, as evident in Fig. 2b.

The same cooperative action of linear transient growth and nonlinear cascade by $\mathcal{N}_y$ underlies the dynamics of other symmetric ($k_x = 0$) and non-symmetric ($k_x \neq 0$) active modes in the vital area, with wavenumbers in the vicinity of $\mathbf{k}_b$ (Fig. 5), however, its quantitative character depends on mode wavenumbers. From the time-averaged plots in Fig. 8, we see that on average the Reynolds stress of modes with $k_y/k_x > 0$ is larger than that of the modes with $k_y/k_x < 0$ and hence the former modes play a main role in supplying turbulence with energy. From these plots we also see that on average $\mathcal{N}_y > 0$ at $k_y/k_x > 0$ in the vital area, i.e., the NTC ensures continual repopulation of just these quadrants of Fourier space with new harmonics that contribute most to positive stress and energetically maintain the turbulent state.

## 4. Summary and discussion

The considered here background flow with constant mean shear is linearly stable (with asymptotically decaying linear perturbations) according to classical/modal stability theory. Therefore, the only energy source for turbulence's self-sustenance in such a flow comes from linear transient amplification of perturbations due to the nonnormality associated with the flow shear. To understand the mechanism of homogeneous shear turbulence's self-sustenance, we performed DNS and then analyzed the turbulence dynamics in Fourier/wavenumber space based on the DNS data. We Fourier transformed Navier-Stokes equations and derived evolution equations for the perturbed streamwise, shearwise and spanwise velocities in wavenumber space. In these spectral equations, using the simulation results, we calculated individual terms, which are classified into two types - terms of linear and nonlinear origin. The term of linear origin, specifically the Reynolds stress, is responsible for energy exchange between the turbulence and the mean flow through transient, or nonmodal amplification of perturbation harmonics due to the flow shear. The nonlinear terms, which do not directly draw the mean flow energy, act to transversely, over wavevector angles, redistribute this energy in Fourier space, continually repopulating those perturbation harmonics which can undergo transient growth. This new type of transversal redistribution of harmonics referred to as the nonlinear transverse cascade appears to be a generic feature of nonlinear dynamics of perturbations in spectrally stable shear flows. It was also demonstrated to take place in 2D HD and MHD shear flows [19, 20] and fundamentally differs from classical concepts of energy cascade processes (direct/inverse) in turbulence theory. So, the conventional description of shear flow turbulence solely in terms of direct and inverse cascades, which leaves such nonlinear transverse cascade out of consideration, might be incomplete and misleading. Thus, we have demonstrated that in the considered constant shear flow, turbulent state is sustained by the subtle interplay of linear and nonlinear processes - the first supplies energy for turbulence via shear-induced transient growth process (characterized by the Reynolds stresses) and the second plays an important role of providing a positive feedback that makes this transient growth process persist over long times and be self-sustaining against viscose dissipation.

[1] The correlation coefficient $R(a, b)$ between two time-dependent functions $a$ and $b$ is defined in a usual way as Pearson's product-moment correlation coefficient.

We found that harmonics that undergo significant transient amplification ("active modes"), and thus are responsible for the majority of energy supply for turbulence, have larger length scales, comparable to the box size. According to the presented analysis, the transverse cascade also operates in this range of length scales. So, this central, small wavenumber area of wavenumber space is the arena of activity of the energy supply linear and redistributing nonlinear processes and their interplay. Consequently, this area is the crucial in the self-sustenance of turbulence and we call it the vital area.

The number of these active harmonics (e.g., the harmonics whose energies grow more than 10% of the maximum spectral energy at least once during evolution) in the vital area generally depends on the box aspect ratio and is quite large (in the considered here box size, it is 36, Fig. 5), indicating fairly complex dynamics, which evidently cannot be described by low-order models. Thus, carrying out a full DNS, we demonstrated the realization of the transverse cascade and its role in the self-sustenance of homogeneous shear turbulence, without truncating the number of modes as usually done in low-order models of self-sustaining processes (e.g., Refs. [24, 25]).

We found that for the considered cubic box, the basic mode with the wavenumber $\mathbf{k}_b = (0, 0, \pm 2\pi/L_z)$ plays a special role. It yields the maximum Reynolds stress during the entire evolution and undergoes the largest transient amplification among the nearby active modes. The amplification of the basic mode occurs in bursts, alternating with decaying intervals. The basic mode contributes up to 70% of the total Reynolds stress during the bursts and thus largely determines the behavior of the turbulence's total energy - Fig. 2 shows that the time-history of the total energy and the stress closely follow those of the basic mode. However, at other aspect ratios $(A_{xz}, A_{yz}) = (3, 2), (1, 2)$ that we also examined, other nearby active modes grow comparably with the basic mode and hence their contribution to the total energy increases. Nevertheless, the largest bursts in the total energy and stress correspond to the growth (upsurge) of the basic mode for these aspect ratios too.

## Acknowledgments

GM, GK and GC are thankful for hospitality at the School of Aeronautics, Universidad Politécnica de Madrid, and for financial support within the Second Multiflow Summer Workshop, 25 May-26 June, 2015, funded in part by the Multiflow Program of the European Research Council. GM is supported by the Georg Forster Postdoctoral Research Fellowship from the Alexander von Humboldt Foundation. SD is supported by the China Scholarship Council.

## References

[1] Farrell B F 1988 *Phys. Fluids* **31** 2093–2102
[2] Butler K M and Farrell B F 1992 *Phys. Fluids* **4** 1637–1650
[3] Reddy S, Schmid P and Henningson D 1993 *SIAM J. Appl. Math.* **53** 15
[4] Trefethen L N, Trefethen A E, Reddy S C and Driscoll T A 1993 *Science* **261** 578–584
[5] Schmid P J and Henningson D S 2001 *Stability and Transition in Shear Flows* (Springer)
[6] Criminale W O, Jackson T L and Joslin R D 2003 *Theory and Computation of Hydrodynamic Stability* (Cambridge Univ. Press)
[7] Schmid P J 2007 *Annu. Rev. Fluid Mech.* **39** 129–162
[8] Farrell B F and Ioannou P J 1994 *Phys. Rev. Lett.* **72** 1188–1191
[9] Gebhardt T and Grossmann S 1994 *Phys. Rev. E* **50** 3705–3711
[10] Henningson D S and Reddy S C 1994 *Phys. Fluids* **6** 1396–1398
[11] Baggett J S, Driscoll T A and Trefethen L N 1995 *Phys. Fluids* **7** 833–838
[12] Grossmann S 2000 *Rev. Mod. Phys.* **72** 603–618
[13] Chapman S J 2002 *J. Fluid Mech.* **451** 35–97
[14] Eckhardt B, Schneider T M, Hof B and Westerweel J 2007 *Annu. Rev. Fluid Mech.* **39** 447–468
[15] Farrell B F and Ioannou P J 2012 *J. Fluid Mech.* **708** 149–196
[16] Brandt L 2014 *Eur. J. Mech. B-Fluids* **47** 80–96
[17] Chagelishvili G D, Chanishvili R G, Hristov T S and Lominadze J G 2002 *Sov. Phys. JETP* **94** 434–445
[18] Jiménez J 2013 *Phys. Fluids* **25** 110814

[19] Horton W, Kim J H, Chagelishvili G D, Bowman J C and Lominadze J G 2010 *Phys. Rev. E* **81** 066304
[20] Mamatsashvili G, Gogichaishvili G, Chagelishvili G and Horton W 2014 *Phys. Rev. E* **89** 043101–1
[21] Rogers M M and Moin P 1987 *J. Fluid Mech.* **176** 33–66
[22] Lee M J, Kim J and Moin P 1990 *J. Fluid Mech.* **216** 561–583
[23] Kida S and Tanaka M 1994 *J. Fluid Mech.* **274** 43–68
[24] Waleffe F 1995 *Phys. Fluids* **7** 3060–3066
[25] Waleffe F 1997 *Phys. Fluids* **9** 883–900
[26] Pumir A 1996 *Phys. Fluids* **8** 3112
[27] Sekimoto A, Dong S and Jiménez J 2016 *Phys. Fluids* **28**, 035101
[28] Chagelishvili G D, Zahn J P, Tevzadze A G and Lominadze J G 2003 *Astron. Astrophys.* **402** 401–407
[29] Tevzadze A G, Chagelishvili G D, Zahn J P, Chanishvili R G and Lominadze J G 2003 *Astron. Astrophys.* **407** 779–786
[30] Umurhan O and Regev O 2004 *Astron. Astrophys.* **427** 855–872
[31] Kim J, Moin P and Moser R 1987 *J. Fluid Mech.* **177** 133–166
[32] Lele S K 1992 *J. Comput. Phys.* **103** 16–42
[33] Rogallo R 1981 Numerical experiments in homogeneous turbulence *Tech. Memo* 81315, NASA
[34] Mamatsashvili G R and Rice W K M 2009 *Mon. Not. R. Astro. Soc.* **394** 2153–2163
[35] Heinemann T and Papaloizou J C B 2009 *Mon. Not. R. Astron. Soc.* **397** 64–74
[36] Rempfer D 2003 *Annu. Rev. Fluid Mech.* **35** 229–265
[37] Farrell B F and Ioannou P J 1996 *J. Atmos. Sci.* **53** 2025–2040
[38] Biskamp D 2003 *Magnetohydrodynamic turbulence* (Cambridge Univ. Press)

Journal of Physics: Conference Series **708** (2016) 012002 doi:10.1088/1742-6596/708/1/012002

# A POD-based analysis of turbulence in the reduced nonlinear dynamics system

**M-A Nikolaidis[1], B F Farrell[2], P J Ioannou[1], D F Gayme[3], A Lozano-Durán[4], J Jiménez[4]**

[1]University of Athens, Department of Physics, Panepistimiopolis, Zografos, 15784 Athens, Greece
[2]Department of Earth and Planetary Sciences, Harvard University, Cambridge, MA 02138, USA
[3]Department of Mechanical Engineering, Johns Hopkins University, Baltimore, MD 21218, USA
[4]School of Aeronautics, Universidad Politécnica de Madrid, 28040 Madrid, Spain

E-mail: pjioannou@phys.uoa.gr

**Abstract.** The structure of turbulence in a reduced model of turbulence (RNL) is analyzed by means of a Proper Orthogonal Decomposition (POD modes). POD analysis was carried out on two different components of the flow, the roll/streak and the perturbation structure. The POD structure in both RNL and direct numerical simulations (DNS) is similar and this correspondence suggests that the dynamics retained in RNL are the essential dynamical ingredients underlying the self-sustaining mechanism of the turbulent state.

## 1. Introduction

Proper Orthogonal Decomposition (POD) provides an orthogonal basis for the spatial structure, each component of which is ordered according to its contribution to the variance of the flow. This basis is optimally efficient for representing the variance in turbulent flows and allows a systematic reduction in complexity of the representation of the flow. The POD analysis can be performed separately on components of the flow in homogeneous directions including the streamwise-mean component of the flow (with streamwise wavenumber $k_x = 0$) and the streamwise-varying components of the flow with $k_x \neq 0$ [1–4]. In this work we exploit this freedom to efficiently identify the structures that dominate the spatio-temporal variability of the streamwise-mean flow as well as the structures that dominate the streamwise-varying $k_x \neq 0$ components. POD modes obtained from DNS are compared with those obtained from a closely related reduced nonlinear (RNL) model [5]. The purpose of this comparison is in part to elucidate the similarities and differences of the DNS with the RNL. The dynamics of RNL model turbulence involve interaction of the streamwise mean flow with few, even as few as one, streamwise flow harmonic. Despite this simplification, salient aspects of the turbulent dynamics are captured [6–8]. Given the similarity between the structures that dominate the variance in DNS and RNL turbulence, it is argued that RNL and DNS turbulence are sustained by a dynamically similar mechanism. RNL turbulence self-sustains through interaction between the time-dependent $k_x = 0$ roll/streak structure with the streamwise varying flow components through a parametric time-dependent instability mechanism. It is argued that the same type of dynamics are responsible for sustaining DNS turbulence.

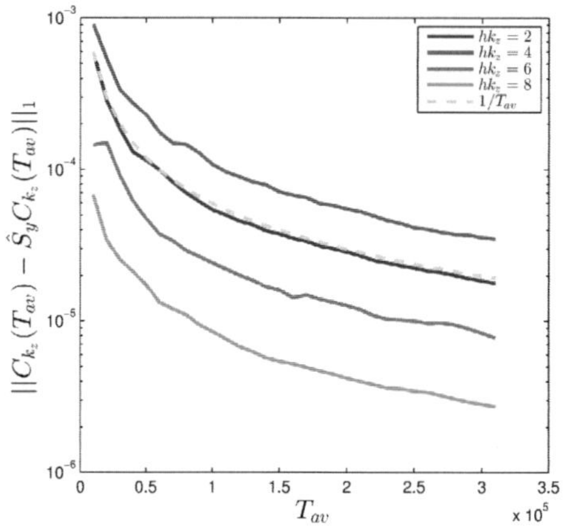

**Figure 1.** The 1-norm of the difference $C_{k_z} - \hat{S}_y C_{k_z}$ between the covariance matrix $C_{k_z}$ (6), and the covariance of the reflected flow about the $x-z$ plane (B.5), at the center of the flow ($y=1$) as a function of the averaging time, $T_{av}$, for $hk_z = 2, 4, 6, 8$ (top to bottom). This plot verifies that reflection symmetry about the centerline is a statistical symmetry of the flow and that this symmetry is approached at the rate $1/T_{av}$ consistent with the law of large numbers for quadratic statistics. Time is non-dimensionalized by $h/U$.

## 2. Procedure for obtaining the POD modes of the turbulent flow

Second-order statistics of flow structure can be obtained from the two-point same-time spatial covariance of the flow-field variables. Adopting the notation $\langle \cdot \rangle$ for the time average and $(\cdot)^T$ for transposition we form the covariance of the $k_x = 0$ components of the velocity,

$$C = \langle \mathcal{U}\mathcal{U}^T \rangle \ , \tag{1}$$

in which

$$\mathcal{U} = [U_s, V_s, W_s]^T \ , \tag{2}$$

is the column vector constructed of the deviations $(U_s(y,z,t), V_s(y,z,t), W_s(y,z,t))$ of the three components of the $k_x = 0$ Fourier mode of the streamwise ($x$), cross-stream ($y$) and spanwise ($z$) velocities $(U(y,z,t), V(y,z,t), W(y,z,t))$ from their spanwise averages, $([U](y,t), [V](y,t), [W](y,t))$.

In addition we form the covariance of the $k_x \neq 0$ components of the flow field,

$$c = \langle \mathcal{U}'\mathcal{U}'^T \rangle , \tag{3}$$

with

$$\mathcal{U}' = [u, v, w]^T , \tag{4}$$

the column of the three components of the $k_x \neq 0$ Fourier component of the flow velocity expanded at all points of the domain.

The POD basis is obtained by eigenanalysis of the two-point covariances, $C$ and $c$, of velocity components. The resulting orthogonal set of eigenvectors is then ordered descending in the magnitude of their eigenvalue to form the POD basis. The eigenvalue of each POD reveals its time-averaged contribution to the energy of the flow.

We have chosen for the study a plane Poiseuille flow in a doubly periodic channel in the streamwise, $x$, and spanwise, $z$, direction of size $L_x/h = 4\pi$, $L_z/h = \pi$, where $h$ is the channel's half-width in the wall-normal direction, $y$, in which $\mathrm{Re}_\tau = 100$ is retained in both the DNS and its reduced nonlinear model simulations (cf. Table 1). In order to obtain a converged POD basis a very long time series of a turbulent flow field is required. This convergence is facilitated by taking into account the statistical symmetries of the flow: homogeneity in the $x$ and $z$ direction, mirror symmetry in $y$ about the $x-z$ plane at the center of the channel, and mirror symmetry in $z$ about the $y-x$ plane at the center of the spanwise direction. The statistics of the turbulent flow have been verified to asymptote to the above statistical symmetries as the averaging time increases; convergence to the statistical symmetry is demonstrated for the case of the $y$ mirror-symmetry in Fig. 1. These statistical symmetries are not necessary consequences of the translation and mirror symmetry of the NS equations in the periodic channel; the solutions may undergo symmetry breaking and be inhomogeneous either in $x$ or $z$ or mirror asymmetric. An example of symmetry breaking of the statistical homogeneity in the meridional direction is apparent in the nearly time invariant zonal flows that are maintained in the turbulent atmosphere of Jupiter. This symmetry breaking in the pipe flow would amount to a turbulent state with temporally constant and spatially fixed streak structures that are aligned parallel to the streamwise axis (cf. Ref. [9]).

Statistical homogeneity in $z$ implies that the eigenvectors of $C$ are single harmonics of the form:

$$\Phi_{k_z} = \begin{pmatrix} A_{k_z}(y) \\ B_{k_z}(y) \\ \Gamma_{k_z}(y) \end{pmatrix} e^{ik_z z} , \tag{5}$$

where $A_{k_z}(y)$, $B_{k_z}(y)$ and $\Gamma_{k_z}(y)$ are $N_y$ dimensional column vectors, with $N_y$, the number of discretization points in $y$ [2]. At each time instant the $3N_y$ column vector of a $k_z \neq 0$ Fourier component, $\mathcal{U}_{k_z}(t)$, of the instantaneous flow field $\mathcal{U}$ is obtained and used to form $N_{k_z}$ average covariances:

$$C_{k_z} = \left\langle \mathcal{U}_{k_z}(t)\mathcal{U}^{\dagger}_{k_z}(t) \right\rangle , \tag{6}$$

where $N_{k_z}$ is the number of $k_z \neq 0$ Fourier components retained in the simulation and $\dagger$ is the Hermitian transpose. Eigenanalysis of these covariances determines $3N_y \times N_{k_z}$ eigenvectors and the POD orthogonal basis of the $k_x = 0$ flow field is this basis ordered decreasing in eigenvalue.

The POD modes for the $k_x \neq 0$ component of the flow are obtained similarly. The eigenvectors of $c$ are again single harmonics because of the statistical homogeneity in both $x$ and $z$ directions, of the form:

$$\phi_{k_x k_z} = \begin{pmatrix} \alpha_{k_x k_z}(y) \\ \beta_{k_x k_z}(y) \\ \gamma_{k_x k_z}(y) \end{pmatrix} e^{i(k_x x + k_z z)} , \tag{7}$$

where $\alpha_{k_x k_z}(y)$, $\beta_{k_x k_z}(y)$ and $\gamma_{k_x k_z}(y)$ are $N_y$ dimensional column vectors. To obtain the POD basis we determine at each instant the $3N_y$ column vector corresponding to the $k_x \neq 0$, $k_z \neq 0$ Fourier component of $\mathcal{U}'_{k_x k_z}(t)$, of the instantaneous flow field $\mathcal{U}'$ and then form $N_{k_x} \times N_{k_z}$ average covariances:

$$c_{k_x k_z} = \left\langle \mathcal{U}'_{k_x k_z}(t)\mathcal{U}'^{\dagger}_{k_x k_z}(t) \right\rangle , \tag{8}$$

where $N_{k_x}$ is the number of the $k_x \neq 0$ Fourier components retained in the simulation. Eigenanalysis of these covariances determines $3N_y \times N_{k_x} \times N_{k_z}$ eigenvectors and the POD orthogonal basis of the $k_x \neq 0$ flow field is this basis ordered decreasing in eigenvalue. A simpler calculation was carried out in this study, restricting the $k_x \neq 0$ components' POD analysis to a single plane, $y_n = 0.2066$. Note that because the physical flow field is real the Fourier components satisfy $\mathcal{U}_{-k_z} = \mathcal{U}^*_{k_z}$ where $*$ denotes the complex conjugate. Consequently $C_{k_z} = C^*_{-k_z}$ and covariances $C_{k_z}$ and $C^*_{-k_z}$ have the same eigenvalues, as the covariances are positive definite Hermitian matrices with real eigenvalues. Their corresponding eigenfunctions

**Table 1.** Simulation parameters. $[L_x, L_z]/h$ is the domain size in the streamwise, spanwise direction. $N_x$, $N_z$ are the number of Fourier components after dealiasing and $N_y$ is the number of Chebyshev components. $\mathrm{Re}_\tau$ is the Reynolds number of the simulation based on the friction velocity.

| Abbreviation | $[L_x, L_z]/h$ | $N_x \times N_z \times N_y$ | $Re_\tau$ | Re |
|---|---|---|---|---|
| NS100 | $[4\pi\ ,\ \pi]$ | $128 \times 63 \times 97$ | 100.59 | 1650 |
| RNL100 | $[4\pi\ ,\ \pi]$ | $3 \times 63 \times 97$ | 100.98 | 1950 |

are conjugate. This implies that the eigenvalues of the POD modes come in pairs corresponding to the $\sin(k_z z)$ and $\cos(k_z z)$ POD structure. When presenting the POD modes for each $k_z$ only one of this pair will be shown. Similarly for the POD modes of the $k_x \neq 0$ flow field: there is a four-fold degeneracy corresponding to the common eigenvalues of $c_{k_x k_z}$, $c_{-k_x k_z}$, $c_{k_x -k_z}$, $c_{-k_x -k_z}$ and the POD modes can be represented with $(x, z)$ structure either in the checkerboard form $\sin(k_x x)\sin(k_z z)$, $\cos(k_x x)\sin(k_z z)$, $\sin(k_x x)\cos(k_z z)$,$\cos(k_x x)\cos(k_z z)$, or in the oblique wave form: $\sin(k_x x + k_z z)$, $\cos(k_x x + k_z z)$, $\sin(k_x x - k_z z)$,$\cos(k_x x - k_z z)$, which will be chosen in our figures.

The complete set of POD modes form a basis orthogonal in the energy norm that spans the three components of the flow field. Although the POD basis is optimally efficient for representing perturbation energy, this basis is not optimal for representing the dynamics of the flow and care must be exercised in relating them to the dynamics of the turbulent state (for a discussion of the relation of the POD basis to the dynamics refer to Ref. [10]; construction of a basis for optimal representation of flow dynamics refer to Ref. [11]). In addition to non-optimality of the POD basis for representing dynamics is the potential for Galerkin projection of the dynamics using this basis to produce dynamical interaction inconsistent with NS which leads inter alia to support of unphysical instabilities [4]. What the POD basis does is to provide important information about flow structure that can be used to reconsrtuct the velocity fields with significantly lower degrees of freedom and therefore constrain the dynamics.

## 3. DNS and its RNL approximation

In this section we briefly describe the construction of the reduced nonlinear model (RNL) of Navier-Stokes plane channel turbulence. The incompressible Navier-Stokes equations in a constant mass-flux channel flow, with laminar flow in the streamwise $x$ direction, is decomposed into equations for the streamwise average flow, $\mathbf{U}$, and the perturbations, $\mathbf{u}'$, as follows:

$$\partial_t \mathbf{U} + \mathbf{U}\cdot\nabla\mathbf{U} - G(t)\hat{\mathbf{x}} + \nabla P - \nu\Delta\mathbf{U} = -\overline{\mathbf{u}'\cdot\nabla\mathbf{u}'}\ , \tag{9a}$$

$$\partial_t \mathbf{u}' + \mathbf{U}\cdot\nabla\mathbf{u}' + \mathbf{u}'\cdot\nabla\mathbf{U} + \nabla\mathbf{p}' - \nu\Delta\mathbf{u}' = -(\mathbf{u}'\cdot\nabla\mathbf{u}' - \overline{\mathbf{u}'\cdot\nabla\mathbf{u}'})\ . \tag{9b}$$

$$\nabla\cdot\mathbf{U} = 0\ , \quad \nabla\cdot\mathbf{u}' = 0\ . \tag{9c}$$

The pressure gradient $G(t)$ that drives the flow maintains constant mass flux. Quantities with an overline $\overline{\mathbf{u}'\cdot\nabla\mathbf{u}'}$ imply averaging over $x$. Capital letters indicate streamwise averaged quantities. No-slip, impermeable boundaries are placed at $y = 0$ and $y = 2h$.

The corresponding RNL equations are obtained by suppressing nonlinear interactions among streamwise non-constant flow components in the perturbation equations resulting in the right

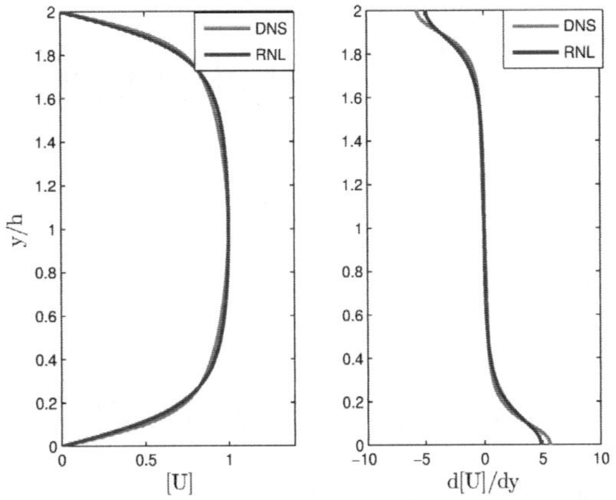


**Figure 2.** Left panel: The mean velocity profile of the DNS and RNL simulations. Right panel: The corresponding mean shear in the two simulations.

hand side of (9b) being set to 0. The RNL equations are:

$$\partial_t \mathbf{U} + \mathbf{U} \cdot \nabla \mathbf{U} - G(t)\hat{\mathbf{x}} + \nabla P - \nu \Delta \mathbf{U} = -\overline{\mathbf{u}' \cdot \nabla \mathbf{u}'} \ , \qquad (10a)$$

$$\partial_t \mathbf{u}' + \mathbf{U} \cdot \nabla \mathbf{u}' + \mathbf{u}' \cdot \nabla \mathbf{U} + \nabla \mathbf{p}' - \nu \Delta \mathbf{u}' = 0 \ . \qquad (10b)$$

$$\nabla \cdot \mathbf{U} = 0 \ , \quad \nabla \cdot \mathbf{u}' = 0 \ . \qquad (10c)$$

Under this restriction, the perturbation field interacts nonlinearly only with the mean, **U**, flow and not with itself. This quasi-linear restriction of the dynamics results in a turbulent state that is supported by a greatly restricted subset of streamwise Fourier components. The components that are retained by the dynamics identify the dynamically active components sustaining the turbulence. These components reveal the subset of streamwise harmonics that are are energetically active in the parametric instability that sustains the turbulent state [5–8].

Turbulence with realistic mean statistics is obtained with RNL dynamics even with the perturbation field further truncated to comprise a single wavenumber component. Simulations reported here retained only perturbations with streamwise wavelength $\lambda_x/h = 4\pi$, which correspond to the gravest streamwise mode in our computational channel of streamwise length $L_x/h = 4\pi$ and spanwise width $L_z/h = \pi$. These RNL simulations are compared with DNS simulations in this channel at $\mathrm{Re}_\tau = u_\tau h/\nu = 100$ with $u_\tau = \sqrt{\nu\, \mathrm{d}[U]/\mathrm{d}y|_{\mathrm{w}}}$ ($\mathrm{d}[U]/\mathrm{d}y|_{\mathrm{w}}$ is the shear at the wall) is the friction velocity. In order for the single component RNL simulation to obtain comparable $\mathrm{Re}_\tau$ with DNS the pressure gradient is increased appropriately. If the perturbation field in the RNL were not constrained to a single streamwise component but were allowed to evolve until its natural support in streamwise wavenumbers had been obtained, it would have retained three streamwise components, with wavelengths $\lambda_x/h = 4\pi, 2\pi, 4\pi/3$, and would sustain turbulence with the $\mathrm{Re}_\tau$ of the DNS without requiring adjustment of the pressure gradient. Numerical details can be found in Appendix A and a summary of the parameters of the simulations in Table 1. The time averaged streamwise mean flow $[U]$ and its associated shear in the DNS is shown in Fig. 2.

## 4. DNS and RNL POD modes

The POD basis for the $k_x = 0$ component of the flow in DNS and the RNL simulation will now be described. As explained earlier because of the statistical homogeneity of the flow in the $z$ direction, these POD modes come in $\sin(k_z z)$ and $\cos(k_z z)$ pairs. The first 8 POD modes of the

Journal of Physics: Conference Series **708** (2016) 012002 doi:10.1088/1742-6596/708/1/012002

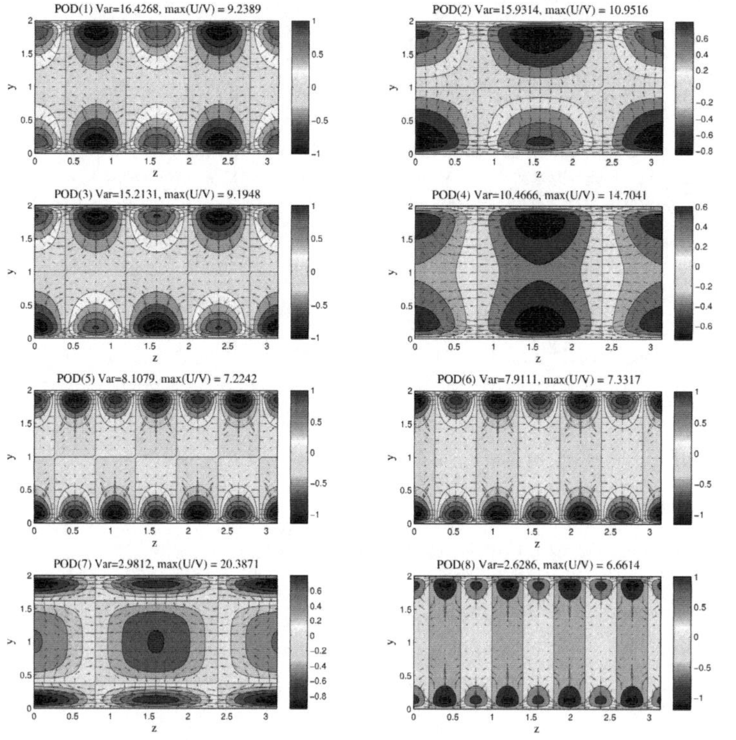


**Figure 3.** The first 8 POD modes of the turbulent channel flow. This basis has been obtained from data of a 270k advective ($ht/U$) time units simulation by exploiting the full symmetries of the flow as described in the text. There is a two-fold degeneracy such that each POD corresponds to two identical structures in $y$ with $\sin(k_z z)$ and $\cos(k_z z)$ structure in $z$. The contours show levels of the streamwise $U$ velocity and the arrows indicate the the vector velocity $(V, W)$. These structures and this ordering was verified to be converged by doubling and quadrupling the averaging time.

DNS simulation are shown in Fig. 3 and the corresponding POD modes for the RNL simulation in Fig. 4. Shown are both the streak velocity and the corresponding $(V, W)$ velocity field for each POD. Note that both the DNS and the RNL simulation produce rolls and streaks similarly collocated and with similar structure. The percentage variance explained by the first POD modes shown in Fig. 5 is similar in the DNS and RNL simulation.

Note that every POD component comes with its streamwise, wall-normal and spanwise velocity component. There is no a priori reason that these fields should be systematically correlated. However, a remarkable systematic correlation is seen between the wall-normal velocity $V$ of the roll and the streak velocity: positive $V$ is correlated with low speed streaks (defects in the streamwise flow) and vice versa. This correlation extends through to higher order POD modes. Each POD shows this remarkable correlation which indicates that the rolls and the streaks form a self-organized structure in which the lift-up mechanism is acting to maintain the streak. If instead the streaks were maintained directly by Reynolds stress forcing from the perturbation field no such systematic correlation between roll velocities and streak velocities would be expected. Analysis of the self-sustaining process (SSP) in RNL reveals that the streak is maintained by

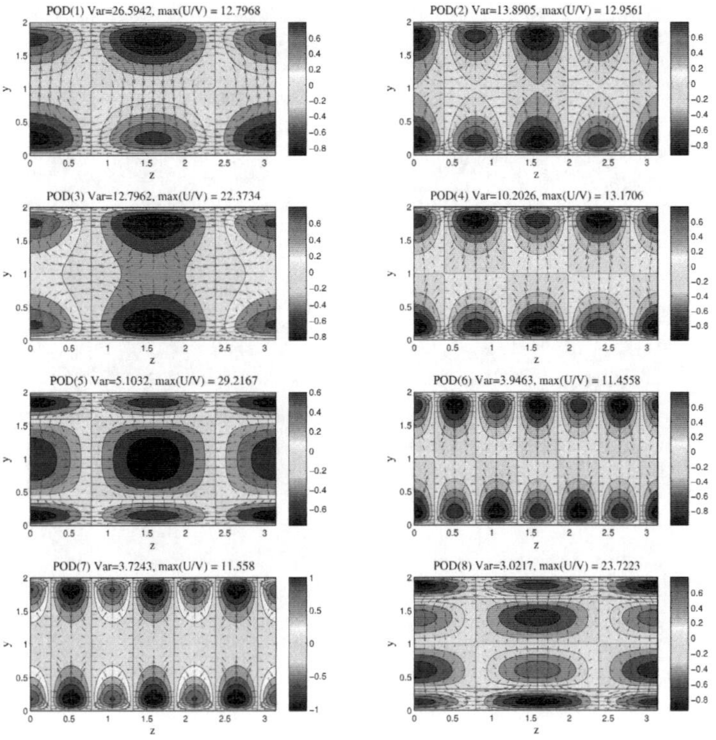


**Figure 4.** The first 8 POD modes of the turbulent channel flow under the corresponding RNL simulation as in Fig. 3.

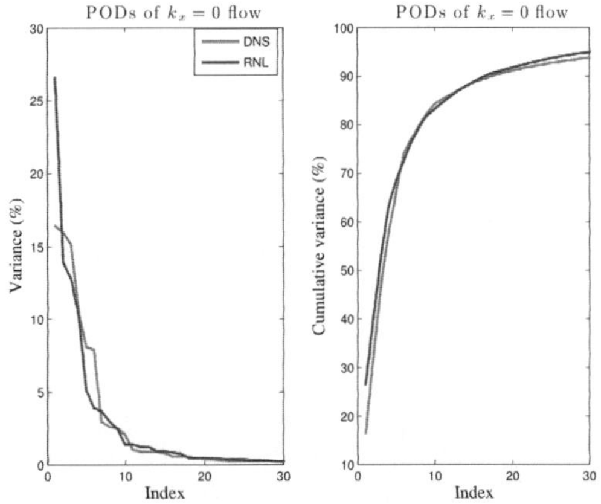


**Figure 5.** Left panel: Percentage energy (variance) of the $k_x = 0$ flow explained by each POD doublet in the DNS and RNL simulation. Right panel: The cumulative variance accounted for by the POD modes in the DNS and RNL simulation as a function of the number of POD modes included in the sum. The POD modes have been placed in descending order of the percentage variance that they account.

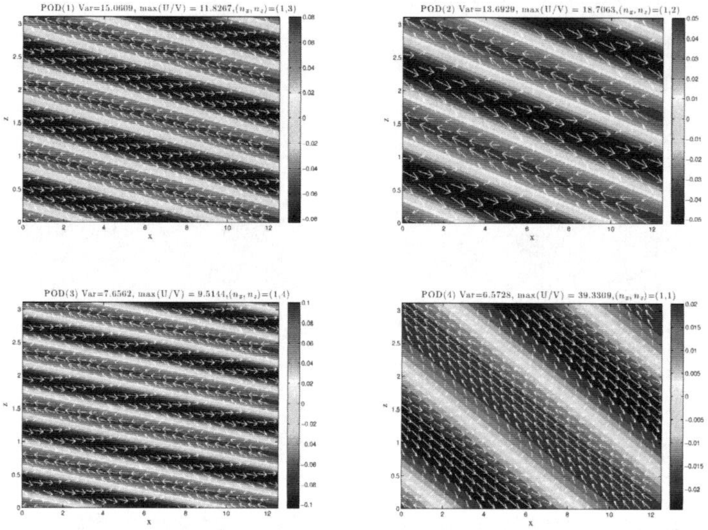

**Figure 6.** The first 4 POD modes of the DNS simulation at constant $y = 0.2066$. Contours are levels of the cross-flow velocity $v'$ on the plane, while the arrows indicate the vector velocity $(u', w')$ on the plane. Note that $u' > 0$ is associated with $v' < 0$ and vice versa an indication that the underlying dynamics are the lift-up mechanism as it operates in $k_x \neq 0$ perturbations.

the lift-up mechanism and that the eddies oppose the streak rather than maintain it. Note that the first 6 DNS POD modes of the roll/streak structure are reproduced in the RNL model, albeit with some reordering. The similarity of the RNL and DNS POD modes suggests that the same dynamics is operating and that the rolls and streaks self-organize so that the rolls form where the streaks are, in such a manner so as to reinforce the preexisting streak structure. This mechanism is evidently scale independent given that all the POD modes have this roll/streak velocity correlation. This scale independence suggests that the mechanism of maintaining the roll/streak structure is also scale independent.

The wall-normal $V$ velocities of the roll component of the POD modes are about 1/10 of the streak velocity which, assuming an average non-dimensional mean flow shear of magnitude 2 (cf. Fig. 2) is consistent with the emergence of the streak through the lift-up mechanism over an average 5 time units. Note that in the RNL the streak/roll velocity ratio increases, which is to be expected because RNL turbulence is less disrupted due to having far fewer perturbation structures and streaks can reach larger amplitude.

We now consider the $k_x \neq 0$ POD modes of the flow. These assume the form of harmonics in $x$ and $z$ given the statistical homogeneity in $x$ and $z$ and there is fourfold degeneracy in the POD spectrum as each of the four combinations of waves $(\pm k_x, \pm k_z)$ accounts for the same amount of energy. These POD modes were calculated at $y = 0.2066$. A plane wave representative of the top 4 POD quartets for the DNS and the RNL are shown in Fig. 6 and Fig. 7 and the variance accounted by the POD modes is shown in Fig. 8. As the RNL simulation was restricted to a single mode, $hk_x = 0.5$, the RNL and DNS POD sets are not as similar as they would be if the full support of RNL turbulence had been retained. Nevertheless, the first 4 POD quartets of both the RNL and the DNS are $hk_x = 0.5$ structures. For the RNL we can see in Table 3 that

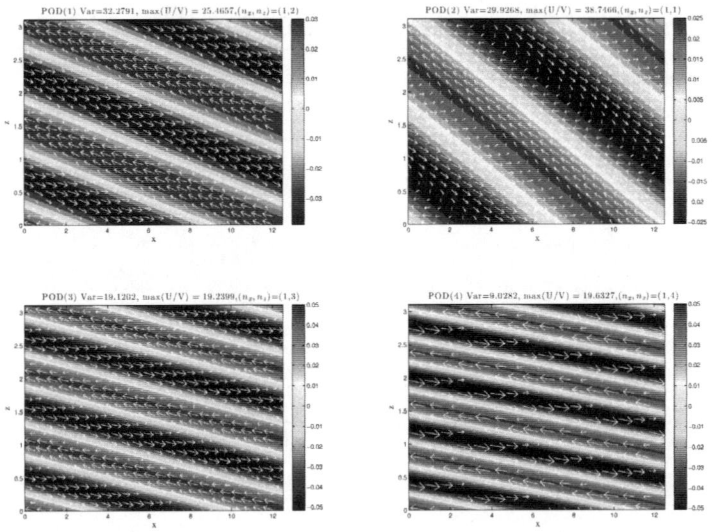

**Figure 7.** The first 4 plane POD modes of the RNL simulation as in Fig. 6.

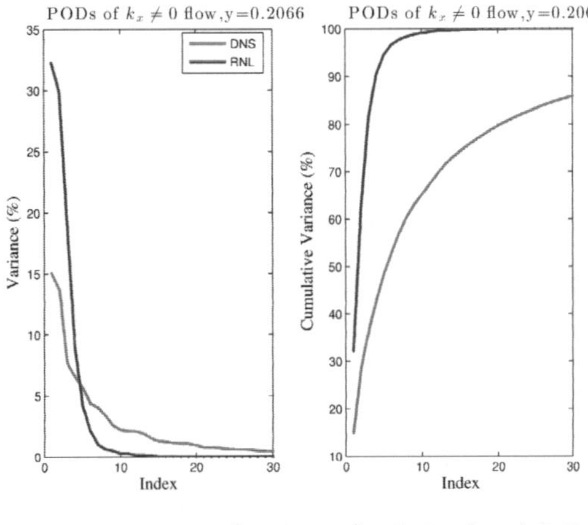


**Figure 8.** Left panel: Percentage energy (variance) of the $k_x \neq 0$ flow explained by each POD quartet in the DNS and RNL simulation. Right panel: The cumulative variance accounted for by the POD modes in the DNS and RNL simulation as a function of the number of POD modes included in the sum.

over 90% of the variance is explained by the first 4 modes, in fact the structure of the $hk_x = 0.5$ mode formed by using the first ten POD modes explains 99% of the variance. As expected, the DNS is more varied,with the $hk_x = 1$ mode making a comparable contribution (cf. Table 2).

As in the the case of the POD's at $k_x = 0$ the velocities $u'$ and $v'$ are correlated indicating that the lift-up mechanism is again at work, here however the lift-up mechanism is "supercharged" by the Orr-mechanism, which creates enhanced wall-normal velocity as the perturbations are sheared [12–15]. This supercharging mechanism associated with the eruptions in the flow does not operate at $k_x = 0$ and is distinct from the lift-up mechanism at $k_x = 0$. This distinction between the lift-up mechanisms operating at $k_x = 0$ and $k_x \neq 0$ is reflected in the ratio of the $u'/v'$ of the POD modes being approximately double the $U/V$ ratio in the $k_x = 0$ POD modes.

**Table 2.** POD modes of $y = 0.2066$ plane for DNS

| Index | $(\pm)n_x$ | $(\pm)n_z$ | Variance(%) | Cumulative Variance(%) |
|---|---|---|---|---|
| 1 | 1 | 3 | 15.06 | 15.06 |
| 2 | 1 | 2 | 13.69 | 28.75 |
| 3 | 1 | 4 | 7.66 | 36.41 |
| 4 | 1 | 1 | 6.57 | 42.98 |
| 5 | 2 | 3 | 5.67 | 48.65 |
| 6 | 2 | 2 | 4.40 | 53.05 |
| 7 | 2 | 4 | 4.04 | 57.09 |
| 8 | 1 | 5 | 3.37 | 60.46 |
| 9 | 3 | 3 | 2.60 | 63.06 |
| 10 | 2 | 1 | 2.21 | 65.27 |

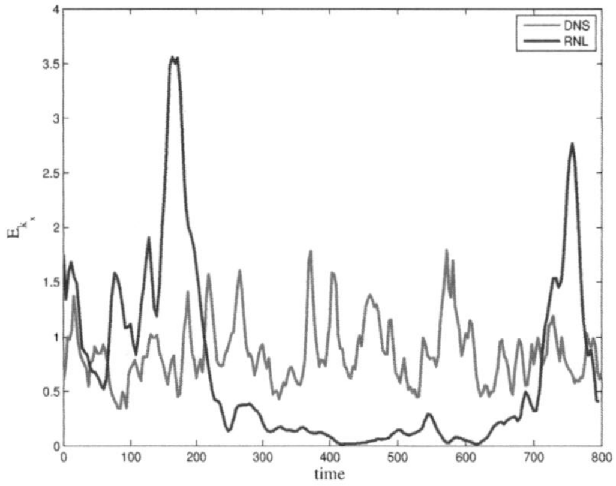


**Figure 9.** A segment of the energy time-series of $hk_x = 1$ on the plane $y = 0.2066$.

Also note that because the supercharge lift-up mechanism operating in the $k_x \neq 0$ perturbations scales inversely with the square of the total wavenumber of the perturbations the induced $u'$ is larger for the lower wavenumber POD modes (cf. Refs. [12, 15]).

Due to the statistical homogeneity, the $k_x \neq 0$ POD modes of individual planes give information only about the relations between the velocity fields in the assumed single wave harmonic form. In order to obtain an understanding of the perturbation structure it is necessary to project the instantaneous flow on the POD basis. The time evolution of the perturbation energy at the $y = 0.2066$ plane for a segment of the DNS and RNL simulation is shown in Fig. 9. Figure 10 shows 2 instantaneous snapshots of the perturbation structure projected on the top 10 POD modes of the $hk_x = 0.5$ basis, at the start of an energy rising event in the DNS and RNL simulation. Regions of $u'v' < 0$ with alternating $v'$ and $u'$ signs are collocated with the most prominent streak defects which in both cases are negative. On this plane the interaction of the streak with the perturbations is connected with the product $w'\partial_z U_s$, generating perturbation velocity in the process. The perturbation structure consists of oblique waves at the wings of the spanwise shear of the streak.

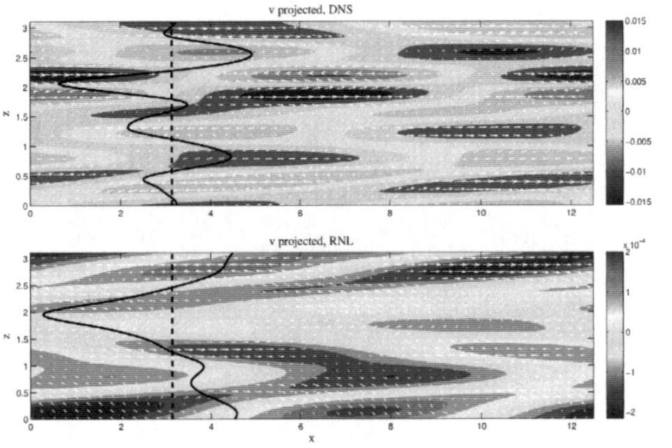


**Figure 10.** Instantaneous velocity fields of DNS and RNL projected on the first 10 POD quartets with $hk_x = 1$,showing the onset of an energy growth event. The contours show levels of the wall-normal velocity, the arrows the $(u, w)$ vectors of the perturbation field on the plane $y = 0.2066$. The black dashed line at $x = \pi$ is the 0 of the streak, $U_s$, at this plane and the black solid line is 10 times the value of the streak velocity $U_s$. DNS is taken from time= 356 and RNL from time= 704. The corresponding perturbation energy is shown in Fig. 9.

**Table 3.** POD modes of $y = 0.2066$ plane for RNL

| Index | $(\pm)\ n_x$ | $(\pm)\ n_z$ | Variance (%) | Cumulative Variance (%) |
|---|---|---|---|---|
| 1 | 1 | 2 | 32.28 | 32.28 |
| 2 | 1 | 1 | 29.93 | 62.21 |
| 3 | 1 | 3 | 19.12 | 81.33 |
| 4 | 1 | 4 | 9.03 | 90.35 |
| 5 | 1 | 5 | 4.20 | 94.55 |
| 6 | 1 | 6 | 2.14 | 96.70 |
| 7 | 1 | 7 | 1.03 | 97.73 |
| 8 | 1 | 1 | 0.63 | 98.34 |
| 9 | 1 | 8 | 0.51 | 98.87 |
| 10 | 1 | 2 | 0.28 | 99.15 |

## 5. Further truncation of the RNL dynamics

The possibility of reducing the RNL dynamics support even further by using projection on the POD's will now be examined. Projection on a basis comprised of all three components $(U_s, V, W)$ is constraining the dynamics in an undesirable way, since the streak $U_s \equiv U - [U]$ is the result of the $(V, W)$ operating on the mean profile $[U]$ and the $U_s$ itself. Therefore the streak dominated POD modes would not be able to correctly describe the roll dynamics since any roll is instantaneously associated with a much bigger streak [4]. For this reason we calculate the POD modes of only the streak portion of the flow and inquire whether we can obtain self-sustaining turbulence with a low order POD representation of the streak structure. It could be argued that a single pair of the POD basis cannot describe correctly the dynamics of the flow, since it would enforce symmetry between high and low speed streaks, and the mean flow would be consequently characterized by a nearly laminar unblunted profile.

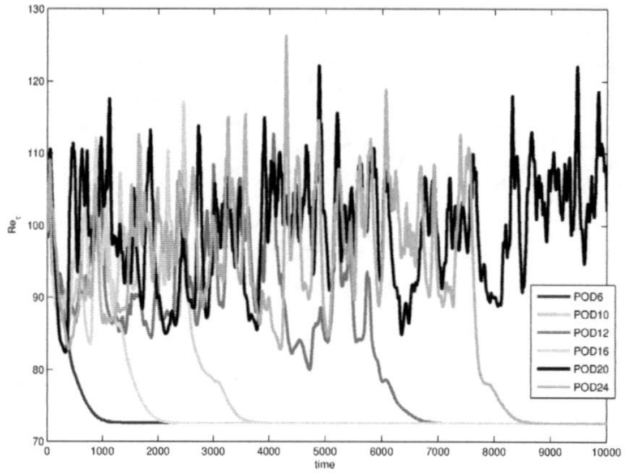


**Figure 11.** Friction Reynolds number $Re_\tau$ as a function of advective time for simulations truncating streak velocity $U_s$ to 6, 10, 12, 16, 20 and 24 POD pairs.

Figure 11 shows the time series of $\mathrm{Re}_\tau$ in simulations with a selection of POD pairs used for the truncation. The RNL system is otherwise unmodified in comparison to the one used for the calculation of the POD basis. When the lowest number of POD pairs is retained, the simulation laminarizes without exhibiting any dynamical behavior. An increase in the number of POD modes allows the system to retain its turbulent behavior. The lifetime of each simulation appears to be uncorrelated with the increase of the POD pair number. This suggests that the truncation likely does not cause relaminarization due to failure to represent the energy of the streak, which would be expected to produce laminarization far more quickly.

## 6. Discussion

Numerical data were obtained from a DNS of a turbulent channel and the corresponding RNL simulation. POD analysis was carried out on two different components of the flow, the roll/streak and the perturbation structure. Striking resemblance of the relations between the velocity fields was seen in the two simulations. This correspondence suggests a similar dynamics is at work in maintaining turbulence in these systems. Remarkably, the mechanism maintaining turbulence in the RNL system has been comprehensively characterized [5–8]. The close correspondence in structure and dynamics between RNL and NS dynamics invites the exploitation of the comprehensively characterized RNL turbulence to advance understanding of NS turbulence in channel flow. In particular, the lift-up mechanism operating at $k_x = 0$ and the associated "supercharge" mechanism operating at $k_x \neq 0$ have been clearly seen operating in both RNL and DNS including using the POD projections. As a further probe of the mechanism of RNL turbulence the POD modes were used to truncate the $k_x = 0$ velocity fields in a different RNL simulation, showing that the system sustains at a fairly low number of POD modes. This result provides an even more reduced turbulent dynamics in terms of the dimension of its support than the already highly reduced RNL turbulence which operates realistically supported by a single streamwise wavenumber.

The purpose of this comparison is in part to elucidate the similarities and differences of the DNS with the RNL. The dynamics of RNL model turbulence involve interaction of the streamwise mean flow with few, even as few as one, streamwise flow harmonic. Despite this simplification salient aspects of the turbulent dynamics are captured [5–8].

**Acknowledgments**
This work was funded in part by the Second Multiflow Program of the European Research Council. Brian Farrell was also supported by NSF AGS-1246929. We would like to thank Daniel Chung for his reviewing comments.

## Appendix A.

The data were obtained from a DNS of (10) and from the RNL that is directly associated with the DNS. Both the DNS and its directly associated RNL are integrated with no-slip boundary conditions in the wall-normal direction and periodic boundary conditions in the streamwise and spanwise directions. The dynamics were expressed in the form of evolution equations for the wall-normal vorticity and the Laplacian of the wall-normal velocity, with spatial discretization and Fourier dealiasing in the two wall-parallel directions and Chebychev polynomials in the wall-normal direction [16]. Time stepping was implemented using the third-order semi-implicit Runge-Kutta method.

## Appendix B. Calculating covariances for symmetrical states

Symmetries in reflections in $y$ about the centerline $x-z$ plane at $y=1$ and reflections in $z$ about the $y-x$ plane at $z=\pi/2$ add in the time series of the simulation 3 independent combinations of $A_{k_z}(y)$, $B_{k_z}(y)$ and $\Gamma_{k_z}(y)$ :

$$\hat{S}_y \Phi_{k_z} = \begin{pmatrix} A_{k_z}(-y) \\ -B_{k_z}(-y) \\ \Gamma_{k_z}(-y) \end{pmatrix} e^{ik_z z}, \hat{S}_z \Phi_{k_z} = \begin{pmatrix} A^*_{k_z}(y) \\ B^*_{k_z}(y) \\ -\Gamma^*_{k_z}(y) \end{pmatrix} e^{ik_z z},$$

$$\hat{S}_z \hat{S}_y \Phi_{k_z} = \begin{pmatrix} A^*_{k_z}(-y) \\ -B^*_{k_z}(-y) \\ -\Gamma^*_{k_z}(-y) \end{pmatrix} e^{ik_z z}. \tag{B.1}$$

To recalculate the additional covariances for the above arrangements of the velocity fields we can use the following separation of the covariance

$$C_{k_z} = \begin{pmatrix} C^{uu}_{k_z} & C^{uv}_{k_z} & C^{uw}_{k_z} \\ C^{vu}_{k_z} & C^{vv}_{k_z} & C^{vw}_{k_z} \\ C^{wu}_{k_z} & C^{wv}_{k_z} & C^{ww}_{k_z} \end{pmatrix}, \tag{B.2}$$

with $C^{u_i u_j}_{k_z} = (C^{u_j u_i}_{k_z})^\dagger$. In the following the $k_z$ subscript will be omitted and instead of $u_i u_j$ the superscript $ij$ will be used. So the covariance is written as:

$$C = \begin{pmatrix} C^{11} & C^{12} & C^{13} \\ C^{21} & C^{22} & C^{23} \\ C^{31} & C^{32} & C^{33} \end{pmatrix}. \tag{B.3}$$

Reflections of the velocities in $z$ result in the conjugation of each individual covariance, and placement of a minus sign in the $w$ components of the covariance, so that:

$$\hat{S}_z C = \begin{pmatrix} (C^{11})^* & (C^{12})^* & -(C^{13})^* \\ (C^{21})^* & (C^{22})^* & -(C^{23})^* \\ -(C^{31})^* & -(C^{32})^* & (C^{33})^* \end{pmatrix}. \tag{B.4}$$

Reflections in $y$ require to reverse the order of the row and column indexes in each individual covariance and if this operation is noted as $\hat{S}_y C^{ij} = C_R^{ij}$ we have:

$$\hat{S}_y C = \begin{pmatrix} C_R^{11} & -C_R^{12} & C_R^{13} \\ -C_R^{21} & C_R^{22} & -C_R^{23} \\ C_R^{31} & -C_R^{32} & C_R^{33} \end{pmatrix} . \tag{B.5}$$

Finally combined $z$ and $y$ reflections are obtained by the successive action of the individual operators on the initial covariance, and thus:

$$\hat{S}_z \hat{S}_y C = \begin{pmatrix} (C_R^{11})^* & -(C_R^{12})^* & -(C_R^{13})^* \\ -(C_R^{21})^* & (C_R^{22})^* & (C_R^{23})^* \\ -(C_R^{31})^* & (C_R^{32})^* & (C_R^{33})^* \end{pmatrix} . \tag{B.6}$$

## References

[1] Moin P and Moser R D 1989 Characteristic-eddy decomposition of turbulence in a channel *J. Fluid Mech.* **200** 471–509

[2] Berkooz G, Holmes P and Lumley J L 1993 The proper orthogonal decomposition in the analysis of turbulent flows *Ann. Rev. Fluid Mech.* **25** 539–575

[3] Sirovich L, Ball K S and Keefe L R 1990 Plane waves and structures in turbulent channel flow *Phys. Fluids A* **2** 2217–2226

[4] Moehlis J, Smith T R, Holmes P and Faisst H 2002 Models for turbulent plane Couette flow using the proper orthogonal decomposition *Phys. Fluids* **14** 2493–2507

[5] Farrell B F and Ioannou P J 2012 Dynamics of streamwise rolls and streaks in turbulent wall-bounded shear flow *J. Fluid Mech.* **708** 149–196

[6] Thomas V, Lieu B K, Jovanović M R, Farrell B F, Ioannou P J and Gayme D F 2014 Self-sustaining turbulence in a restricted nonlinear model of plane Couette flow *Phys. Fluids* **26** 105112

[7] Constantinou N C, Lozano-Durán A, Nikolaidis M A, Farrell B F, Ioannou P J and Jiménez J 2014 Turbulence in the highly restricted dynamics of a closure at second order: comparison with DNS *J. Phys. Conf. Ser.* **506** 012004

[8] Thomas V, Farrell B F, Ioannou P J and Gayme D F 2015 A minimal model of self-sustaining turbulence *Phys. Fluids* **27** 105104

[9] Avsarkisov V, Hoyas S, Oberlack M and Garcia-Galache J P 2014 Turbulent plane Couette flow at moderately high Reynolds number *J. Fluid Mech.* **751** R1

[10] Farrell B F and Ioannou P J 1993 Stochastic forcing of the linearized Navier–Stokes equations *Phys. Fluids A* **5** 2600–2609

[11] Farrell B F and Ioannou P J 2001 Accurate low-dimensional approximation of the linear dynamics of fluid flow *J. Atmos. Sci.* **58** 2771–2789

[12] Farrell B F and Ioannou P J 1993 Optimal excitation of three-dimensional perturbations in viscous constant shear flow *Phys. Fluids A* **5** 1390–1400

[13] Farrell B F and Ioannou P J 1993 Perturbation growth in shear flow exhibits universality *Phys. Fluids A* **5** 2298–2300

[14] Jiménez J, Del Álamo J and Flores O 2004 The large-scale dynamics of near-wall turbulence *J. Fluid Mech.* **505** 179–199

[15] Jiménez J 2013 How linear is wall-bounded turbulence? *Phys. Fluids* **25** 110814

[16] Kim J, Moin P and Moser R 1987 Turbulence statistics in fully developed channel flow at low Reynolds number *J. Fluid Mech.* **177** 133–166

# Unstable periodic orbits in plane Couette flow with the Smagorinsky model

**Eiichi SASAKI[1], Genta KAWAHARA[1], Atsushi SEKIMOTO[2] and Javier JIMÉNEZ[2]**

[1]Graduate School of Engineering Science, Osaka University, 1-3 Machikaneyama, Toyonaka, Osaka, Japan

[2]School of Aeronautics, U. Politécnica de Madrid, Madrid, Spain

E-mail: esasaki@me.es.osaka-u.ac.jp

**Abstract.** We aim at a description of the logarithmic velocity profile of wall turbulence in terms of unstable periodic orbits (UPOs) for plane Couette flow with a Smagorinsky-type eddy viscosity model. We study the bifurcation structure with respect to the Smagorinsky constant, arising from the gentle UPO reported by Kawahara and Kida [1] for the Navier-Stokes (NS) equation. We find that the obtained UPOs in the large eddy simulation (LES) system connect to those in the NS system, and that the gentle UPO in the LES system is an edge state branch whose stable manifold separates LES turbulence from an LES 'laminar' state. As the Reynolds number decreases this solution arises as the saddle solution of the saddle-node bifurcation. Meanwhile, the mean and root-mean-square velocity profiles of the node solution of the LES gentle UPO are in good agreement with those of LES turbulence.

## 1. Introduction

Near-wall turbulence can be found in daily life, which attracts many researchers' interests in the physical and engineering viewpoints, and its mean velocity has a well-known logarithmic profile that is considered the result of its multi-scale structure. Although there is a vast number of studies on near-wall turbulence, the dynamical origin of the logarithmic profile is not clear.

Plane Couette flow is one of the simplest shear flows, driven by parallel walls moving in opposite directions. Since the linear streamwise velocity profile of the laminar solution is linearly stable [2], there is no non-trivial solution bifurcating from the laminar one, and this problem is considered to be a typical example of subcritical transition to turbulence. Hamilton et al. [3] studied the minimal periodic domain where turbulence is sustained at low Reynolds numbers. In this domain, which we call the minimal flow unit [4] from now on, turbulence exhibits a quasi-periodic motion characterized by the formation and collapse of the streaks, i.e., the regeneration cycle. Kawahara and Kida [1] found two unstable periodic orbits (UPOs) in the minimal flow unit. Here we call them the vigorous and gentle UPOs. The vigorous UPO is embedded in turbulent state, its motion is characterized by the regeneration cycle, and the profiles of the mean and root-mean-square (RMS) velocities are similar to those of turbulence. On the other hand, the gentle UPO is an edge state that lies on the basin boundary between turbulent and laminar states. Other non-trivial solutions have been found, such as steady states [5, 6, 7], UPOs embedded in low-Reynolds-number turbulence [8] and localized solutions [9, 10], but all the found solutions correspond to transitional or turbulent flows at low Reynolds numbers rather than fully developed high-Reynolds-number turbulence.

Our aim is to find an explanation of developed wall turbulence in terms of dynamical properties of invariant sets. In order to decrease the number of degrees of freedom, we introduce the Smagorinsky-type eddy viscosity model. Large-eddy simulation (LES) can reproduce not only the profile of the mean flow, but its dynamics are similar to Navier-Stokes (NS) turbulence. In the case of NS system, the invariant solution found by Kawahara and Kida [1] describes the dynamics of low-Reynolds-number NS turbulence, i.e. the regeneration cycle. When we introduce the eddy-viscosity model, several questions arise: for example, whether or not there are invariant solutions representing LES turbulence, and whether we can obtain descriptions of developed turbulence using those solutions even the small-scale motion is modelled by an eddy viscosity. Having the similar questions, Rawat et al. [11] showed an LES steady state which represents the large-scale motion in plane Couette flow. Here we study whether the eddy viscosity model changes the phase-space structure of the NS system, and explore the bifurcation structure stemming from the gentle UPO with respect to the Smagorinsky constant.

In section 2 we explain the problem setting and show bifurcation structure arising from the gentle UPO in section 3. The gentle UPO in the NS system continuously connects to the LES system and the LES gentle UPO is the edge state at low Reynolds numbers. Those results suggest that the eddy viscosity model does not violate a phase-space structure of the NS system. As the Reynolds number decreases we find the saddle-node bifurcation, and the node solution of the LES gentle UPO is embedded in turbulent state. The profiles of the mean and RMS velocities of this solution are similar to those of turbulence, although its dynamics is simpler than the regeneration cycle observed in turbulent state. Section 4 is devoted to the conclusion and discussion.

## 2. Problem setting

We deal with the dimensionless LES equation with the Smagorinsky model [12, 13] given by

$$\frac{\partial u_i}{\partial t} + u_j \partial_j u_i = -\partial_i p + 2\partial_j \left\{ \left( \frac{1}{Re} + \nu_e \right) S_{ij} \right\}, \tag{1}$$

$$\partial_i u_i = 0, \tag{2}$$

where $t$ indicates time, $(u_1, u_2, u_3) = (u, v, w)$ is the velocity vector, and $p$ is the kinematic pressure. Here $\partial_i = \partial/\partial x_i$ stands for the spacial derivative with respect to $x_i$ and $S_{ij} = (\partial_i u_j + \partial_j u_i)/2$ is the strain rate tensor, where $(x_1, x_2, x_3) = (x, y, z)$ are the streamwise, wall-normal and spanwise coordinates, respectively. Quantities are normalized with the half-distance between walls, $h$, and the half-difference of the wall velocities, so that $Re = Uh/\nu$, where $\nu$ is the kinematic molecular viscosity. The dimensionless eddy viscosity $\nu_e$ is defined as

$$\nu_e = \{C_S \Delta(y) f_S(y)\}^2 |S|, \tag{3}$$

where $|S| = \sqrt{2S_{ij}S_{ij}}$. Here $C_S$ denotes the Smagorinsky constant, whose value is usually in the range of $0.1 \leq C_S \leq 0.15$, the filter width $\Delta(y) = \{\Delta_x \Delta_y(y) \Delta_z\}^{1/3}$ is defined from the grid spacings, and $f_S(y) = 1 - \exp(-y^+/A^+)$ is the van Driest damping function, with $A^+ = 25$. The $^+$ superscript denotes quantities rescaled with the friction velocity and the molecular viscosity. We impose periodic boundary conditions in the streamwise and spanwise directions, and the non-slip and impermeability conditions at the walls, $u|_{y=\pm 1} = \pm 1$ and $v|_{y=\pm 1} = w|_{y=\pm 1} = 0$. We set the streamwise and spanwise periods as the minimal flow unit, $L_x = 1.755\pi$, $L_z = 1.2\pi$ [3, 4].

We employ the Chebyshev-Fourier-Galerkin spectral method. For example, the wall-normal velocity is expanded as

$$v(x,y,z,t)=\sum_{k=-K}^{K}\sum_{l=-L}^{L}\sum_{m=0}^{M-4}\hat{v}_{k,l,m}(t)\exp(\mathrm{i}(\alpha kx+\beta lz))\Psi_m(y), \tag{4}$$

where $\hat{v}_{k,l,m}(t)$ is the expansion coefficient, $(\alpha,\beta)=(2\pi/L_x,2\pi/L_z)$ denote fundamental wavenumbers, and $\Psi_m(y)$ is the wall-normal Galerkin base defined by $\Psi_m(y)=T_m(y)-\frac{2m+4}{m+3}T_{m+2}(y)+\frac{m+1}{m+3}T_{m+4}(y)$, in terms of the $m$-th order Chebyshev polynomials, $T_m(y)$. Here $K,L$ and $M$ denote the truncation mode numbers, and we take the number of the grid points as $N_x\geq 3K+1, N_y=M+1, N_z\geq 3L+1$. In this paper, we set up the number of grid points and the truncation mode numbers as $(N_x,N_y,N_z)=(24,33,24)$ and $(K,L,M)=(7,32,7)$, respectively. The time integration is carried out using the second-order Crank–Nicholson and Adams–Bashforth methods. To obtain UPOs we employ the Newton-GMRES method. The periodic orbit with the period $T$ is the solution of the equation,

$$\boldsymbol{\phi}_T(\boldsymbol{x})-\boldsymbol{x}=\mathbf{0}, \tag{5}$$

where $\boldsymbol{x}\in\mathbb{R}^N(N=2(2K+1)(2L+1)(M-2)+2(M-1))$ is the $N$-dimensional real vector constructed by the real and imaginary parts of the expansion coefficients, and $\boldsymbol{\phi}_T(\boldsymbol{x})$ denotes the time-$T$ map given by the time integration up to $T$ from the initial condition $\boldsymbol{x}$. Substituting $\boldsymbol{x}=\boldsymbol{x}_0+\delta\boldsymbol{x}$ and $T=T_0+\delta T$ to (5) and linearizing the equations, we obtain

$$\left(\left.\frac{\partial\boldsymbol{\phi}}{\partial\boldsymbol{x}}\right|_{\boldsymbol{x}_0,T_0}-I\right)\delta\boldsymbol{x}+\boldsymbol{f}(\boldsymbol{x}_0)\delta T=-\boldsymbol{\phi}(\boldsymbol{x}_0)_{T_0}+\boldsymbol{x}_0, \tag{6}$$

where $\partial\boldsymbol{\phi}/\partial\boldsymbol{x}$ represents the monodromy matrix of the time-$T$ map, $I$ means the $N\times N$ identity matrix, and $\boldsymbol{f}(\boldsymbol{x})$ indicates the time derivative of $\boldsymbol{x}$. The linearized equation (6) has the $N$ equations in the $N+1$ unknowns. Then, we consider the additional condition [14],

$$(\boldsymbol{f}(\boldsymbol{x}_0),\delta\boldsymbol{x})=0, \tag{7}$$

where $(\boldsymbol{x},\boldsymbol{y})$ signifies the Euclidean inner product. To solve the equations (6) and (7), we carry out the GMRES computation [15], which is performed in the Krylov subspace without any inverse-matrix calculations. The stopping condition of the Newton iteration is

$$\frac{||\boldsymbol{\phi}_T(\boldsymbol{x})-\boldsymbol{x}||}{||\boldsymbol{x}||}<10^{-5}, \tag{8}$$

where $||\boldsymbol{x}||=(\boldsymbol{x},\boldsymbol{x})^{1/2}$ represents the Euclidean norm.

## 3. Bifurcation diagram of gentle UPO

Figure 1 shows the bifurcation structure arising from the gentle UPO reported by Kawahara and Kida [1] where the vertical axis indicates the maximal value of the cross flow energy defined by

$$E_{\mathrm{2D}}=\max_{t\in T}\int\frac{v^2+w^2}{2}\frac{\mathrm{d}V}{2L_xL_z}. \tag{9}$$

In the case of $C_S=0$, which represents the NS system, the gentle UPO is a saddle solution of the saddle-node bifurcation at $Re=236.1$ [16]. As $C_S$ increases the gentle UPO in the NS

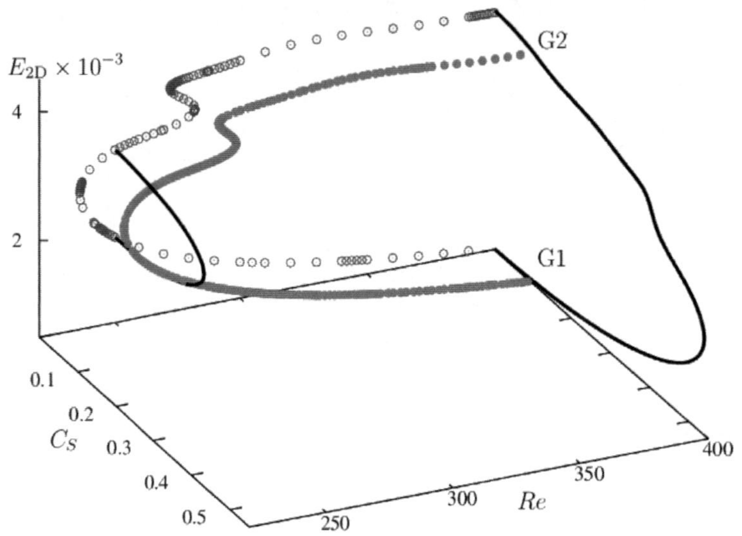


**Figure 1.** Bifurcation diagram of the gentle UPO. The vertical axis indicates the maximal value of the cross flow energy in the period. The blue and red points represent the cases of $C_S = 0$ (NS system) and 0.1, respectively. In the cases of $Re = 250$ and 400 the saddle-node bifurcation points are at $C_S = 0.2362$ and 0.5392, respectively.

system continuously connects with that in the LES system, and the branch finally reaches the node solution of the gentle UPO in the NS system through the saddle-node bifurcation at the finite $C_S$. In the case of $C_S = 0.1$, the gentle UPO appears from the saddle-node bifurcation at $Re = 238.4$. We have confirmed that the LES gentle UPO is also an edge state using the shooting method [17] (not shown). Hereafter we call the gentle UPO and the node solution G1 and G2, respectively.

In the case of $Re = 400$, the orbit projections on the total energy and the injection plane are shown in figure 2. Here the total energy $E_{3D}$ and the injection $I$ are given by

$$E_{3D} = \int \frac{u^2 + v^2 + w^2}{2} \frac{dV}{2L_x L_z}, \quad (10)$$

$$I = \frac{1}{2}\left( \left.\frac{d\langle u\rangle}{dy}\right|_{y=+1} + \left.\frac{d\langle u\rangle}{dy}\right|_{y=-1} \right), \quad (11)$$

where $\langle\ \rangle$ denotes the wall-parallel averaging. The turbulent orbits and UPOs in the LES system do not change a lot from the NS systems. Those results suggest that the eddy viscosity model does not violate the phase-space structure of the NS system. Similar results are reported by Rawat et al. [11] in the case of the steady states.

G2 is embedded in the turbulent state. The mean and RMS velocities of G2 are shown in figure 3, while figure 4 presents the snapshot of the streamwise velocity. G2 has the three symmetries,

$$\Sigma_1 : (u, v, w)(x, y, z, t) \to (u, v, -w)(x + L_x/2, y, -z, t), \quad (12)$$

$$\Sigma_2 : (u, v, w)(x, y, z, t) \to (-u, -v, w)(-x, -y, z + L_z/2, t), \quad (13)$$

$$\Sigma_3 : (u, v, w)(x, y, z, t) \to (u, v, w)(x + L_x/2, y, z, t + T/2). \quad (14)$$

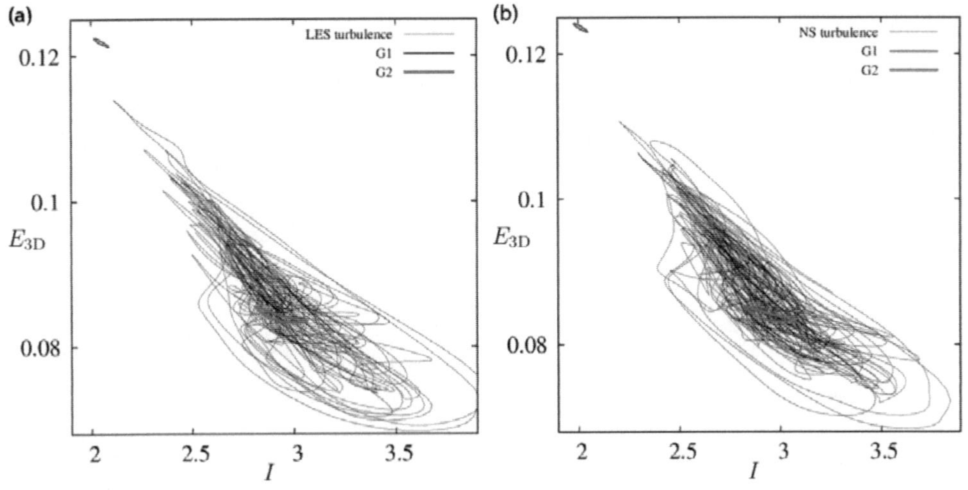


**Figure 2.** The projection on the total energy and the injection plane at $Re = 400$: (a) the LES ($C_S = 0.1$) and (b) the NS ($C_S = 0$) systems. The blue and red lines indicate G1 and G2, while the gray thin line means the turbulent orbit. In the case of the LES system, the periods of G1 and G2 are 84.13 and 63.23, respectively.

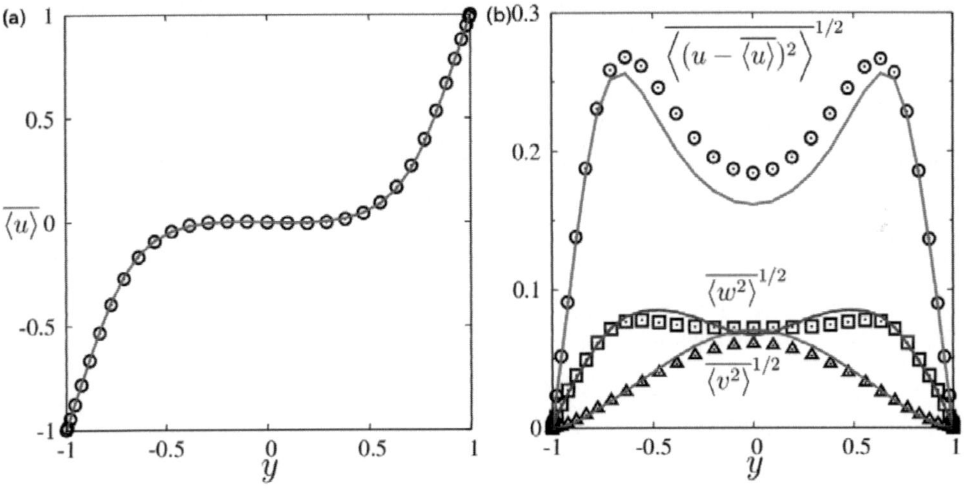


**Figure 3.** The mean velocity profiles of G2 at $C_S = 0.1$ and $Re = 400$: (a) the mean and (b) RMS velocities, where $\overline{\langle\ \rangle}$ signifies the time and the wall-parallel averaging. The symbols and lines denote the profiles of LES turbulence and G2, respectively.

The profiles of the mean and RMS velocities of G2 are similar to those of turbulence, but the dynamics is characterized only by a meandering of the streaks, which is simpler than the regeneration cycle, that is the formation and collapse of the low-speed streak found by Jiménez and Moin [4] and Hamilton et al. [3].

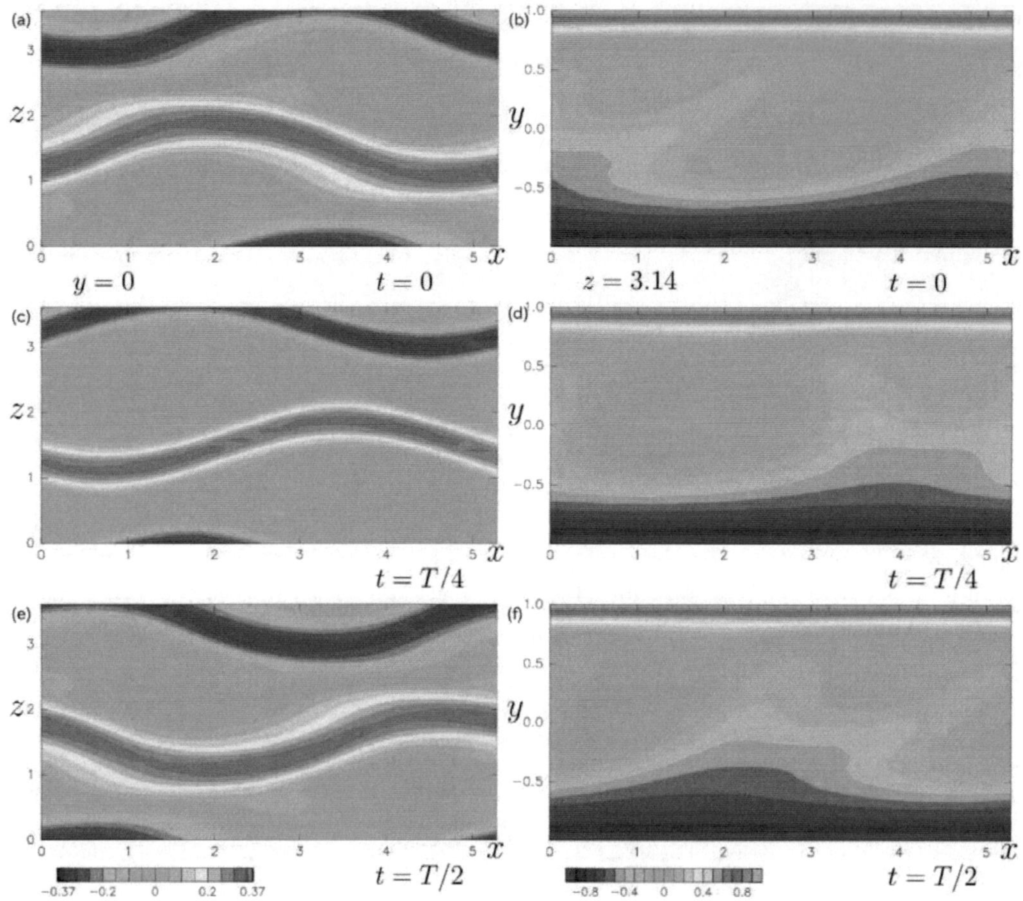


**Figure 4.** The time series of the streamwise velocity of G2 at $C_S = 0.1, Re = 400$. The left and right columns show the plane $y = 0$ and the plane $z = 3.14$, respectively. The top, middle and bottom rows show the snapshots at $t = 0, T/4$ and $T/2$, respectively.

## 4. Conclusion and discussion

We have studied the bifurcation structure arising from the gentle UPO found by Kawahara and Kida [1]. The gentle UPO is the edge state at $Re = 400$ in the NS system. We find that, as $C_S$ increases, the gentle UPO of the NS system connects with the LES branch which reaches the node solution of the NS system through a saddle-node bifurcation at a finite Smagorinsky constant. In addition, the saddle solution in the LES system is also an edge state. Those results suggest that the eddy viscosity model would not violate the phase-space structure of the NS systems. Moreover the node solution is embedded in the turbulent state, and the profiles of the mean and RMS velocities are similar to those of turbulence. On the other hand, the dynamics of the node solution is represented only by a meandering motion of the streaks, which is simpler than the regeneration cycle. It should be noted that Kawahara and Kida [1] found a different vigorous UPO that describes the statistical and dynamical properties of turbulence. Since the magnitude of the eddy viscosity is small at the low Reynolds number $Re = 400$ as shown in figure A2 in the appendix, the LES contribution is not significant. At higher Reynolds numbers we are now tracing the corresponding LES branches to the node solution and the vigorous solution, which could represent high-Reynolds-number turbulence.

It is also an interesting problem to characterize the nonlinear stability of the laminar state at high Reynolds number in terms of a periodic edge state and its stable/unstable manifolds. Peixinho and Mullin [18] examined experimentally the transition to turbulence in pipe flow, and found that the minimal amplitude of the disturbance leading to turbulence changes as $Re^{-m}$ with $m \approx 1.3 - 1.5$. Avila et al. [20] studied a transiently turbulent state, the so-called puff, and its characteristic lifetime. They found that the lifetime of this turbulent state is finite, which implies that perhaps the turbulent state at high Reynolds numbers might be a chaotic saddle in the dynamical systems theory. van Veen and Kawahara [19] showed the existence of the chaotic saddle using the homoclinic tangle arising from the gentle UPO at low Reynolds numbers.

**Acknowledgments**
The first author would like to thank Prof. Michio Yamada for fruitful discussions. This work was supported by JSPS KAKENHI Grant Number 50214672 and partially by the Multiflow program of the European Research Council.

## Appendix A. Statistical properties of LES turbulence

Here we briefly show the statistical properties of LES turbulence ($C_S = 0.1$) at high Reynolds numbers in the minimal flow unit, $(L_x, L_z) = (1.755\pi, 1.2\pi)$, to check the validity of the eddy viscosity model. Time averaging is carried out integrating up to $T_\mathrm{e} = 4000$, after discarding an initial transient up to $T_\mathrm{e} = 2000$.

Figure A1 shows the mean streamwise velocity. As we increase the Reynolds number with a fixed number of grid points, $(N_x, N_y, N_z) = (24, 33, 24)$ the mean streamwise velocity moves closer to the laminar profile. On the other hand, if we change the number of the grid points for each Reynolds number, the mean velocity profile reproduces the same Kármán constant of NS turbulence, $\kappa \approx 0.41$ shown by Pirozzoli et al. [21]. It should be noted that because of the small domain size of and low Reynolds number of the present LES, one should expect deviations from the logarithmic profile of the NS system in the overlap region. In fact, we have confirmed a good agreement between the profile of the mean streamwise velocity of the LES and NS systems in the case of $(L_x, L_z) = (2\pi, \pi)$ (not shown). As the Reynolds number and the number of the grid points increase, the filter width near the wall changes as $\Delta(1)^+ \simeq 1.88, 6.31, 11.0$ and $9.63$ for $Re = 400, 2000, 5000$ and $10000$, respectively. Figure A1 then suggests that to obtain an approximately constant Kármán constant the grid should be taken so as to satisfy $\Delta(1)^+ \leq 10$. Taking into account the numerical costs to obtain UPOs using the Newton method we anticipate that the practical limitation of the Reynolds number using present resources would be in the range of $Re \leq 5000$.

Figure A2 shows the ratio $\overline{\langle \nu_e \rangle}/Re^{-1}$ of the mean eddy viscosity to the kinematic molecular one. As the Reynolds number increases the ratio increases monotonically except for the case of $Re = 10000$ where we not only change the number of wall-normal grid points but also the wall-parallel one. Even at high Reynolds numbers the order of magnitude is moderate, less than unity. This implies that the eddy viscosity model might not significantly change the energy dissipation mechanism of the NS system. It should be noted that since the eddy viscosity is small at $Re = 400$, the LES contribution is unimportant at this low Reynolds number.

## References

[1] Kawahara G and Kida S 2001 *J. Fluid Mech.* **449** 291–300
[2] Romanov V A 1973 *Funct. Anal. Appl.* **7** 137–146
[3] Hamilton J M, Kim J and Waleffe F 1995 *J. Fluid Mech.* **287** 317–348
[4] Jiménez J and Moin P 1991 *J.Fluid Mech.* **225** 213–240
[5] Nagata M 1990 *J. Fluid Mech.* **217** 519–27
[6] Wang J, Gibson J and Waleffe F 2007 *Phys. Rev. Lett.* **98** 204501
[7] Itano T, Akinaga T, Generalis S C and Seki M S 2013 *Phys. Rev. Lett.* **111** 184502

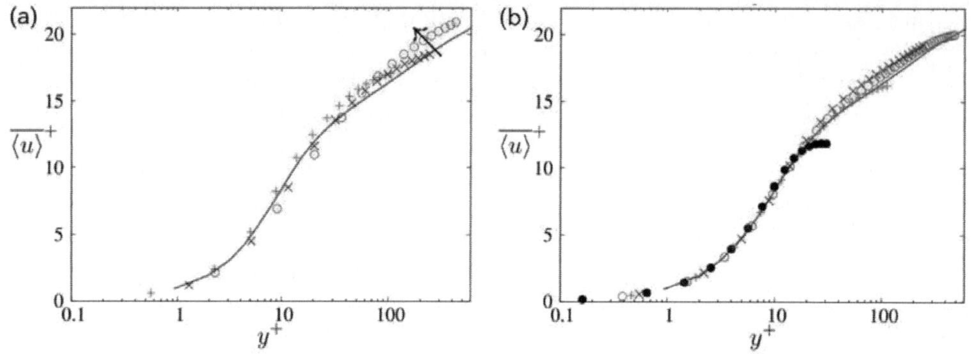


**Figure A1.** Mean streamwise velocity at several Reynolds numbers: ●, $Re = 400$; +, 2000; ×, 5000; ○, 10000. (a) The number of the grid points is fixed to $(N_x, N_y, N_z) = (24, 33, 24)$. (b) Grid increases with the Reynolds number. $(N_x, N_y, N_z) = (24, 33, 24), (24, 37, 24), (24, 49, 24)$ and $(48, 81, 48)$, where $h^+ = 33.7, 122, 258$ and 503, respectively. The gray line shows the result of a DNS of plane Couette flow in a periodic domain $(L_x, L_z) = (18\pi, 8\pi)$ at $h^+ = 550$ [22] We note that the profile at $Re = 400$ of (b) was shown in figure 3-(a) in outer scaling.

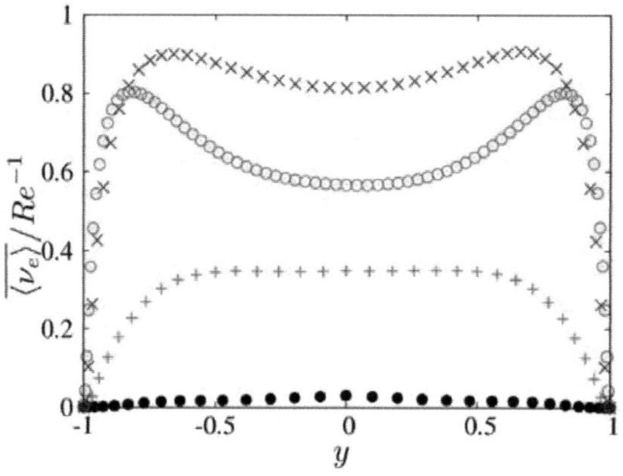


**Figure A2.** The ratio of the mean eddy viscosity to the kinematic molecular viscosity. The definition of symbols is the same as in figure A1.

[8] Viswanath D 2007 *J. Fluid Mech.* **580** 339–358
[9] Gibson J F and Brand E 2014 *J. Fluid Mech.* **745** 25–61
[10] Brand E and Gibson J F 2014 *J. Fluid Mech.* **750** R3
[11] Rawat S, Cossu C, Hwang Y and Rincon F 2015 *J. Fluid Mech.* **782** 515–540
[12] Deardorff J W 1970 *J. Fluid Mech.* **41** 453–480
[13] Moin P and Kim J 1982 *J. Fluid Mech.* **118** 341–377
[14] Mees A I 1981 *Dynamics Of Feedback Systems* (New York, Wiley)
[15] Saad Y 2003 *Iterative Methods for Sparse Linear Systems* 2nd ed. (SIAM)
[16] Shimizu M and Kawahara G 2014 *Euromech Coll.* **565**, Cargese, FR
[17] Toh S and Itano T 2003 *J. Fluid Mech.* **481** 67–76
[18] Peixinho J and Mullin T 2007 *J. Fluid Mech.* **582** 169–178
[19] van Veen L and Kawahara G 2011 *Phys. Rev. Lett.* **107** 114501
[20] Avila K, Moxey D, Lozar A, Avila A, Barkley D and Hof B 2011 *Science* **333** 192–196
[21] Pirozzoli S, Barnardini M and Orlandi P 2014 *J. Fluid Mech.* **758** 327–343
[22] Avsarkisov V, Hoyas S, Oberlack M and Garcia-Galache J P 2014 *J. Fluid Mech.* **751** R1

# A new statistical tool to study the geometry of intense vorticity clusters in turbulence

**Alberto Vela-Martin**[1], **Takashi Ishihara**[2]

[1]Universidad Politécnica de Madrid, 28040 Madrid, Spain

[2]JST CREST, Graduate School of Engineering, Nagoya University, Nagoya 464-8603, Japan

E-mail: alberto@torroja.dmt.upm.es

**Abstract.** Recent large-scale direct numerical simulations (DNS) of high-Reynolds number (high-$Re$) turbulence, suggest that strong micro-scale tube-like vortices form clusters in localized thin regions of space. However, to this date no thorough quantitative and statistical analysis of the geometry of such vortical clusters has been conducted. This study is intended to generate new statistical tools to study the shape and dynamics of these intense vorticity and strain regions. We first propose a new method for locating and analysing the geometrical properties of thresholded vortical clusters contained inside boxes of a given size. Second, we use this new tool to investigate the natural presence of intense shear layers and their relevance as geometrical features of high-$Re$ homogeneous turbulence. This new method is applied to the DNS of homogeneous incompressible turbulence with up to $4096^3$ grid points, showing that the geometry of high vorticity regions varies strongly depending on the threshold and on the size of the clusters. In particular for sizes in the inertial range of scales and high thresholds, approximately layer-like structures of vortices are extracted and visualized. Agreement of results with previous observations and known features of turbulence supports the validity of the proposed method to characterize the geometry of intense vorticity and strain regions in high-$Re$ turbulence.

## 1. Introduction

Vorticity and strain structures are known not to be randomly distributed in turbulence. Rather, they tend to be clustered in localized regions in space, where they are intense [1], and where the local dynamics are believed to be strongly influenced by strong velocity gradients. It has been suggested for some time [2] that these intense clusters tend to be located between large emptier regions in which velocity gradients are relatively weak, and large energy-containing eddies are dominant.

Recent theories [3] point at the possibility that these large empty regions play an important role in the dynamics of intense clusters, as they entail the formation of stable shear layers where intense vorticity is clustered and sheltered in thin layer-like regions of space. As the Taylor micro-scale Reynolds number ($Re_\lambda$) is increased, these layers become more relevant due to the inhibition of the Kelvin–Helmholtz mechanism responsible for their unstable behaviour [3]. When $Re_\lambda$ becomes large, this dynamical effect would imply the prevalence of layers as the relevant geometry for vortex clusters. Visual detection and analysis of individual layers in high-$Re_\lambda$ DNS, with typical thickness of the order of the Taylor micro-scale, stands as evidence to support this theory [5]. However, to the knowledge of the authors, no extensive statistical research has been conducted to properly identify and characterize these layers, and it is unclear how relevant they are in high-$Re_\lambda$ turbulence.

This work explores new statistical tools to study the geometrical distribution of intense vorticity and strain regions. We propose a new method for locating and analysing these clusters and their geometry through an extensive and quantitative statistical approach. Using this new tool, we expect to elucidate whether intense shear layers are naturally present in isotropic turbulence, and to what extent they represent a relevant geometric feature of high-Reynolds number turbulence.

## 2. Method

The method is intended as a reliable way to qualitatively and quantitatively describe the clustering geometry of intense vorticity regions. It is based on randomly probing the vorticity field to find the core of high density regions through an iterative method. The structure of the localized intense clusters is analysed using the inertia tensor of the set of thesholded points that form the cluster. Similar methods have been used to describe the structure of galaxy clusters [4].

The enstrophy field $|\omega|(\vec{x})$ is thresholded, and a field $P(\vec{x})$ is obtained such that,

$$P(\vec{x}) = 1 \text{ if } |\omega|(\vec{x}) \geq \alpha\omega',$$

$$P(\vec{x}) = 0 \text{ if } |\omega|(\vec{x}) < \alpha\omega',$$

where $\omega'$ is defined as the root-mean-square of the vorticity $\omega'^2 = \epsilon/\nu$, $\epsilon$ is the mean energy dissipation rate per unit mass and $\nu$ is the kinematic viscosity. The thresholded field is randomly covered with cubic boxes of size $\delta^3$. Each box, defining a domain $B$, is centred at $\vec{x}'$, and its centre of gravity $\vec{x}_{cg}$ is computed as

$$\vec{x}_{gc} = \frac{\sum_{\vec{x}\in B} \vec{x}P(\vec{x})}{\sum_{\vec{x}\in B} P(\vec{x})}, \tag{1}$$

where the sum extends to all points within the box. If $\vec{x}_{gc}$ is equal to the box centre $\vec{x}'$, the point is marked as the centre of a local cluster. Otherwise, the centre of the box is shifted to the centre of gravity of the thresholded field, $\vec{x}_{gc} \to \vec{x}'$, and the process is repeated to convergence.

After all the boxes converge, a set of local clusters is obtained. It is known that the thresholded enstrophy field contains large empty regions, and this algorithm occasionally detects some clusters that consist of a few isolated points within those empty regions. To avoid very low-density clusters, all boxes with a mean volume fraction $V_B = \sum_{\vec{x}\in B} P(\vec{x})/\delta^3$ lower than the average volume fraction of $P$ over the whole field are discarded. Moreover, since this method might identify some clusters more than once, repeated clusters are also discarded. The number of boxes used in our analysis is such that the total sum of their volume is eight times larger than the volume of the domain. Around 20% of the trials result in repeated clusters, suggesting that this number of boxes is large enough to properly sample the flow. The parameter $\delta$ is important, because it implicitly defines the maximum size of the structures contained in the box. It is through this parameter that we control the size of the clusters that we detect.

For each individual box $B$, the geometry of the cluster can be studied through the inertia tensor of the points within it,

$$I_{xx} = \frac{\sum_{\vec{x}\in B}\{(y-y')^2 + (z-z')^2\}P(\vec{x})}{\sum_{\vec{x}\in B} P(\vec{x})}$$

$$I_{yy} = \frac{\sum_{\vec{x}\in B}\{(z-z')^2 + (x-x')^2\}P(\vec{x})}{\sum_{\vec{x}\in B} P(\vec{x})}$$

$$I_{zz} = \frac{\sum_{\vec{x}\in B}\{(x-x')^2+(y-y')^2\}P(\vec{x})}{\sum_{\vec{x}\in B}P(\vec{x})}$$
$$I_{xy} = \frac{\sum_{\vec{x}\in B}(x-x')(y-y')P(\vec{x})}{\sum_{\vec{x}\in B}P(\vec{x})}$$
$$I_{yz} = \frac{\sum_{\vec{x}\in B}(y-y')(z-z')P(\vec{x})}{\sum_{\vec{x}\in B}P(\vec{x})}$$
$$I_{zx} = \frac{\sum_{\vec{x}\in B}(z-z')(x-x')P(\vec{x})}{\sum_{\vec{x}\in B}P(\vec{x})}$$

where primes denote the coordinates of the centre of each box, which is also the centre of gravity of the distribution of $P$ in the box. From the tensor, the three principal moments of inertial, $I_1 > I_2 > I_3 \geq 0$, can be obtained, as well as their respective inertia directions, $\vec{v}_1, \vec{v}_2, \vec{v}_3$.

These three quantities contain information on the shape and size of the cluster. The ratio $\rho = (I_1 + I_2 + I_3)^{1/2}/\delta$ describes the ratio between the radius of inertia of $P$ and the size of the box, and indicates how well we have achieved our aim of isolating clusters of a given size. The two aspect ratios $\epsilon_1 = I_2/I_1$ and $\epsilon_2 = I_3/I_1$, such that $1 \geq \epsilon_1 \geq \epsilon_2$, define the geometry of the structures. There are three ideal cases: $\epsilon_1 = \epsilon_2 = 1$ represents a perfect sphere; $\epsilon_1 = \epsilon_2 = 0.5$ represents a perfectly thin circular pancake or layer; and $\epsilon_1 = 1$ and $\epsilon_2 = 0$ represents an infinitely thin tube.

## 3. Dataset

The vorticity fields used in the analysis are obtained from integrating the incompressible Navier–Stokes equations in a triply periodic domain. The dataset contains flow fields from three simulations of forced homogeneous isotropic turbulence at different $Re_\lambda$. Further details about the simulations can be found in [6, 7]. For the lower-Reynolds number cases, enough time-decorrelated fields have been used to obtain a proper statistical representation of the flow. For the largest case, only one field is available but, given its high Reynolds number, the small scales are sufficiently represented to have converged statistics. Characteristic parameters of the simulations are presented in Table 1.

## 4. Results

When the algorithm is applied to the database mentioned above, it successfully detects a considerable number of unique dense clusters from which the geometric parameters $\rho$, $\epsilon_1$ and $\epsilon_2$ are extracted. Between $10^3$ and $10^7$ different acceptable clusters are identified, enough for properly converged statistics.

**Table 1.** Characteristic parameters of the DNS data. $N$ is the number of Fourier modes per dimension in each simulation, $k_{max}$ the largest resolved wavenumber, $\eta$ the Kolmogorov viscous scale, and $\lambda$ the Taylor micro-scale. ($N_{fields}$) is the number of independent fields used in the statistics. The details of the DNS are described in Ref. [6] for $N = 512, 1024$, and [7] for $N = 4096$.

| $N$ | $Re_\lambda$ | $k_{max}\eta$ | $L/\eta$ | $\lambda/\eta$ | $N_{fields}$ |
|---|---|---|---|---|---|
| 512 | 284 | 1.5 | 193 | 41.2 | 20 |
| 1024 | 384 | 1.5 | 386 | 43.9 | 10 |
| 4096 | 1131 | 1 | 2137 | 66.4 | 1 |

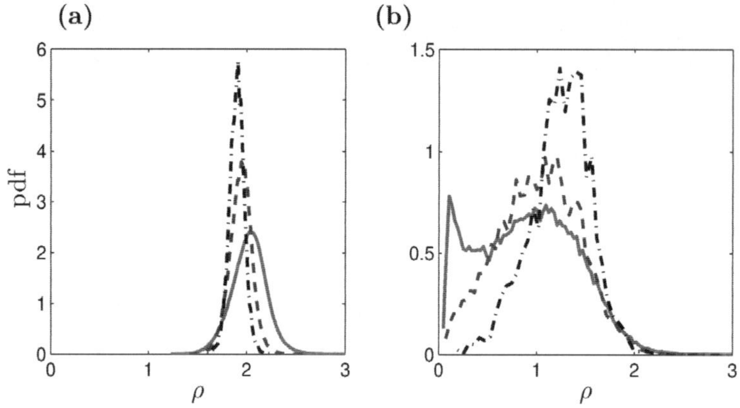


**Figure 1.** Probability density functions of the size ratio $\rho$. ——, $\delta = 19\eta$; - - -, $\delta = 70\eta$; -·-·-, $\delta = 136\eta$. (a) $\alpha = 1$. (b) $\alpha = 4$.

Figure 1 shows probability density functions (pdf) of the size ratio $\rho$. In general, they centre about unity, showing that the boxes identify clusters of roughly the intended size, but it is also clear that the higher thresholds in figure 1(b) contain a sub-population of clusters that are much smaller than the box size.

Figure 2 presents two-dimensional joint pdfs of the aspect ratios $(\epsilon_1, \epsilon_2)$. Only points inside the triangles drawn in each figure are possible. 'Spheres' are at the top-right corner, 'pancakes' are at the bottom centre, and 'tubes' are at the top left. The geometry of the clusters changes considerably with the threshold and with the clustering size. For low thresholds and/or large box sizes, the clusters are spherical, with $\epsilon_1 \approx \epsilon_2 \approx 1$. This reason is different in each case. If the size $\delta$ of the box is much larger than the size of a typical cluster, only the geometry of the far field of intense vorticity is captured, which is distributed roughly homogeneously with respect to the centre of the box. This can be appreciated in the tendency of pdfs to move towards the top-right corner as $\delta$ increases. On the other hand, the results for low thresholds represent the geometry of the weak background enstrophy, which is less intermittent and more evenly distributed in space than the intense structures.

The pdfs for small boxes and high $\alpha$ fall near $\epsilon_1 \sim 1$ and $\epsilon_2 \sim 0$, where thin elongated shapes are represented. This geometry is typical of single isolated strong vortical structures, which are known to be representative features of the intense vorticity field in turbulence [2]. That these two well-known features of the flow are identified by our algorithm gives us some assurance about the general soundness of the method.

In between the two limits, larger clusters with increasing $\alpha$ move from isotropic shapes towards intermediate aspect ratios. In this transition, the predominant cluster type moves towards a flatter geometry, which could be interpreted as being closer to layers. Figure 2 shows that this intermediate structures are more common for sizes in the inertial range of scales.

To reinforce the geometrical intuition of the shapes implied by the previous results, it is useful to relate the principal moments of inertia to those of an ellipsoid with semi-axes $(a > b > c)$. As before, pair-wise aspect ratios can be defined as $\tilde{b} = b/a$ and $\tilde{c} = c/a$, which are univocally related to $(\epsilon_1, \epsilon_2)$,

$$\tilde{b} = \left(\frac{1+\epsilon_2-\epsilon_1}{1+\epsilon_1-\epsilon_2}\right)^{1/2}, \quad \tilde{c} = \left(\frac{\epsilon_1+\epsilon_2-1}{1+\epsilon_1-\epsilon_2}\right)^{1/2}. \tag{2}$$

For the typical values found in large boxes and high $\alpha$ ($\epsilon_1 \approx 0.8$ and $\epsilon_2 \approx 0.5$), the aspect ratios of the equivalent ellipsoid are $\tilde{b} \approx 0.7$ and $\tilde{c} \approx 0.5$ which, although not strictly a thin layer, at least suggests a tendency of the vorticity clusters towards this type of geometry.

Journal of Physics: Conference Series **708** (2016) 012004 doi:10.1088/1742-6596/708/1/012004

A typical vorticity cluster with this 'layer-like' geometry is shown in figure 3. It is formed by tightly packed parallel vortices, and has $(\epsilon_1, \epsilon_2) = (0.79, 0.48)$ and $(\tilde{b}, \tilde{c}) = (0.73, 0.45)$. The three different views of this structure in figure 3 show that it is relatively wide in the direction normal to $\vec{v}_1$, and confined to a thinner layer in the other two directions. The joint pdfs in figure 2 show that this type of geometry is relatively common for intermediate values of $\delta$ and $\alpha$ at the three $Re_\lambda$ considered. Similar structures have been reported in [5].

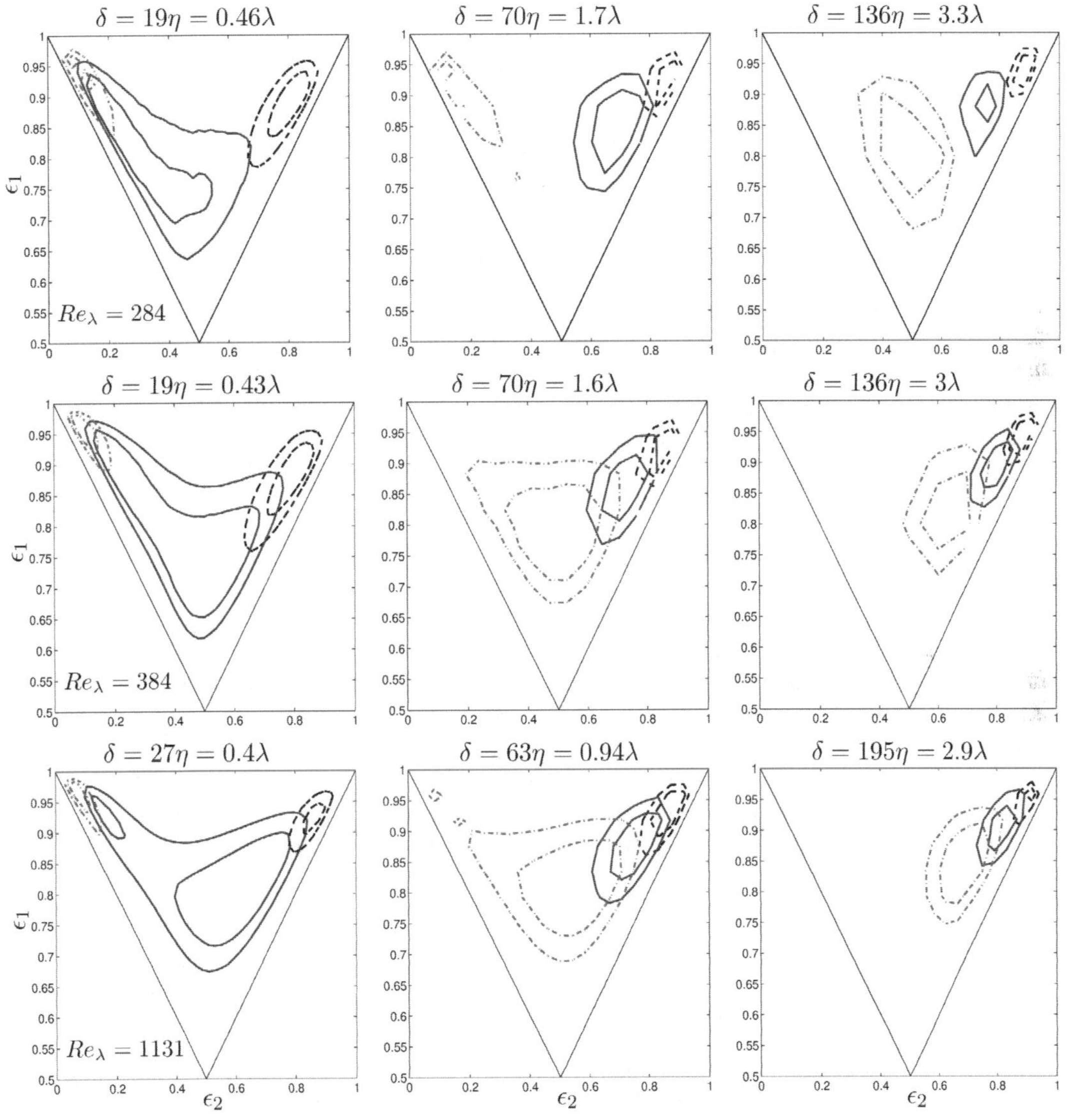


**Figure 2.** Isocountours at 0.6 and 0.3 of the maximum for the joint pdf of $\epsilon_1$ and $\epsilon_2$. The size of the clustering box is above each figure, in Taylor and Kolmogorov micro-scales. From top to bottom: $Re_\lambda = 284$, $Re_\lambda = 384$ and $Re_\lambda = 1131$. ---, $\alpha = 1$; ——, $\alpha = 2$; -·-·-, $\alpha = 4$

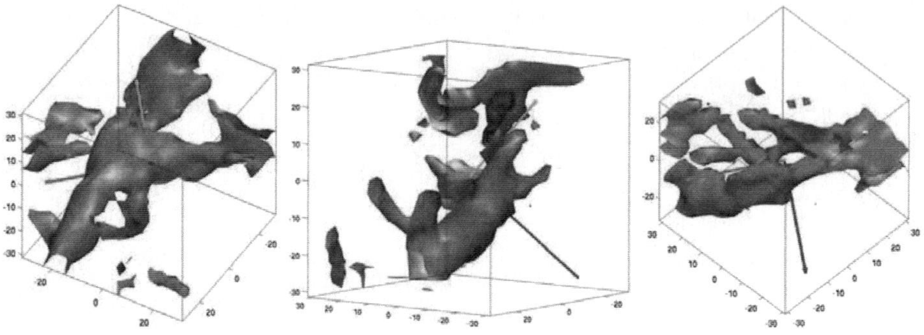

**Figure 3.** Typical intense vorticity structure at $Re_\lambda = 1131$, box size $\delta = 63\eta$ and threshold $\alpha = 4$. From left to right: view in the direction of the three main eigenvectors, blue $\vec{v}_1$, magenta $\vec{v}_2$ and green $\vec{v}_3$. For this structure: $\epsilon_1 = 0.79$ and $\epsilon_2 = 0.48$. Lengths in Kolmogorov micro-scale units.

## 5. Conclusions

We have presented a new method for the analysis of the geometry of intense vorticity. Dense clusters of size $\delta$ are located via an iterative method applied on thresholded enstrophy fields.

Clusters are analysed by means of the inertia tensor of the thresholded set of high-intensity points, and their geometry is studied considering the three principal inertia moments. An extensive application of this algorithm has been conducted for three different Reynolds numbers, using about $10^7$ clusters extracted from our database of isotropic turbulence.

The method is shown to be capable of detecting both locally strong tubular vortices for small clusters, and isotropically scattered vorticity for large ones, validating its ability to locate and characterize the geometry of known vorticity structures. For intermediate sizes within the inertial range of scales, a tendency of these clusters to form approximately layer-like objects is observed. Some of these objects have been extracted from the flow, showing what could be identified as potential shear layers. These results are in accordance with observations by [5], and suggest that thick layer-like vorticity structures might be naturally present in isotropic turbulence.

This analysis is related to the one in ref. [1], and produces similar results in that higher thresholds give rise to vorticity structures that move from sheets to tubes. The two analyses differ in that there was no explicit size control in [1], and that their clusters were geometrically connected, while the present ones are, in principle, general sets of points.

## Acknowledgments

This work is partially funded by the Multiflow program of the European Research Council.

## References

[1] Moisy F and Jiménez J 2004 *J. Fluid Mech.* **513**, 111–133.
[2] Jiménez J, Wray A A, Saffman P G and Rogallo R S 1993 *J. Fluid Mech.* **255** 65–90.
[3] Hunt J C R, Eames I, Westerweel J, Davidson P A, Voropayev S, Fernando J and Braza M 2010 *J. Hydro-environment Res.* **4**(2) 75–82.
[4] Plionis M, Barrow J D and Frenk, C S 1991 *Mon. Not. R. Astron. Soci.* **249** 662–677.
[5] Ishihara T, Kaneda Y and Hunt J C R 2013 *Flow Turbul. Combust.* **91** 895–929
[6] Cardesa J I, Vela–Martin A, Dong S and Jiménez J 2015 *Phys. Fluids* **27**, 111702.
[7] Ishihara T, Kaneda Y, Yokokawa M, Itakura K and Uno A 2007 *J. Fluid Mech.* **592** 335–366.

# Local topology via the invariants of the velocity gradient tensor within vortex clusters and intense Reynolds stress structures in turbulent channel flow

**Abel-John Buchner[1], Adrián Lozano-Durán[2], Vassili Kitsios[1], Callum Atkinson[1], Julio Soria[1,3]**

[1]Laboratory for Turbulence Research in Aerospace and Combustion (LTRAC), Department of Mechanical and Aerospace Engineering, Monash University, Clayton, Australia
[2]School of Aeronautics, Universidad Politécnica de Madrid, Madrid, Spain
[3]Department of Aeronautical Engineering, King Abdulaziz University, Jeddah, Kingdom of Saudi Arabia

E-mail: abel-john.buchner@monash.edu

**Abstract.**
Previous works have shown that momentum transfer in the wall-normal direction within turbulent wall-bounded flows occurs primarily within coherent structures defined by regions of intense Reynolds stress [1]. Such structures may be classified into wall-attached and wall-detached structures with the latter being typically weak, small-scale, and isotropically oriented, while the former are larger and carry most of the Reynolds stresses. The mean velocity fluctuation within each structure may also be used to separate structures by their dynamic properties. This study aims to extract information regarding the scales, kinematics and dynamics of these structures within the topological framework of the invariants of the velocity gradient tensor (VGT). The local topological characteristics of these intense Reynolds stress structures are compared to the topological characteristics of vortex clusters defined by the discriminant of the velocity gradient tensor. The alignment of vorticity with the principal strain directions within these structures is also determined, and the implications of these findings are discussed.

## 1. Introduction

Turbulent wall-bounded flows contain features at a range of scales and with various topological properties and dynamic behaviours. In studying the motion of fluid within a turbulent wall-bounded flow, it is common to separate the flow into distinct regions of spatially and/or temporally coherent velocity or vorticity. These are often labelled "coherent structures" [2, 3]. This has the advantage of defining a tangible entity whose topology and evolution may be quantitatively studied. There is however significant ambiguity and confusion about how to formally define these coherent structures, and the physical relevance of each proposed method. Commonly, coherent vortical structures may be defined by the second invariant of the velocity gradient tensor (VGT), $Q_A$ [4], or by contiguous positive regions of its discriminant, $D$ [5, 6]. Lagrangian methods such as the Finite Time Lyapunov Exponents (FTLE) have also been used [7, 8] and, more recently, Lozano-Durán *et al.* [1] studied the dynamics of structures defined based on regions of intense Reynolds stress. Cucitore *et al.* [9] and Chakraborty *et al.* [10] present comprehensive descriptions and analyses of popular coherent structure identification criteria.

This paper focusses on the study of the intense Reynolds stress structures of Lozano–Durán *et al.* [1] in a turbulent channel flow, and their comparison with structures defined based on the VGT discriminant, as in del Álamo *et al.* [6]. The term "vortex cluster" is used here to identify structures defined based on the VGT discriminant.

The overall topology and dynamics of intense Reynolds stress structures and vortex clusters has been characterised previously [1, 6], but questions remain regarding the topology and alignment of the flow within these structures on the local level. It is unknown what the most likely topological states are within these structures, whether this differs from the the whole flow, and how structures delineated based on different criteria differ from one another locally.

To address these unknowns, intense Reynolds stress structures and vortex clusters are compared here in the topological framework of the invariants of the VGT [5], which may be used to elicit information regarding the local topology within a velocity field. A comparison is also made regarding the likely local alignment of the strain and vorticity fields within each type of structure. Chong *et al.* [5], Soria *et al.* [11], Perry and Chong [12], Blackburn *et al.* [13], and Ooi *et al.* [14] present comprehensive discussion on the physical interpretation of the VGT invariants.

## 2. Intense Reynolds stress structures

Lozano–Durán *et al.* [1] investigated the three-dimensional structure of momentum transfer in turbulent channels by studying the structure of the intense Reynolds stresses that carry most of the wall-normal flux of momentum. Particular emphasis was placed on the logarithmic and outer layers. The flow in this study was separated into coherent structures based on contiguous regions of Reynolds stress values above a threshold, $H$,

$$|\tau(\mathbf{x})|/\left(u'(y)\,v'(y)\right) > H \tag{1}$$

where the Reynolds stress, $\tau$, is normalised by $u'$ and $v'$, the root–mean–square (RMS) of each velocity component at each wall–normal location, $y$. An apostrophe is used throughout this paper to denote the RMS of a quantity. A percolation analysis was used to set a normalised Reynolds stress threshold of $H_1 = 1.75$. Individual structures are defined by spatially connecting neighbouring points satisfying this criterion.

Sub–classification of these coherent structures is possible, using the quadrant analysis of Wallace *et al.* [15], Willmarth and Lu [16], and Lu and Willmarth [17]. This analysis categorizes the intense Reynolds stress structures as sweeps, ejections, and inward and outward interactions based on the sign of the fluctuating streamwise and spanwise velocity components, $u$ and $v$, averaged within each structure as

$$u_m = \frac{\int_\Omega u \cdot dV}{\int_\Omega dV} \tag{2}$$

and

$$v_m = \frac{\int_\Omega v \cdot dV}{\int_\Omega dV}, \tag{3}$$

where integration is performed over the points, $\Omega$, belonging to each individual structure's volume. Lozano-Durán *et al.* [1] label these as $Q_1$, $Q_2$, $Q_3$, and $Q_4$ structures. Note that $Q_i$ refers to quadrant, in contrast to the $Q_A$ used before to denote the second invariant of the VGT. This terminology, illustrated in figure 1, is used here to separate intense Reynolds stress structures into families which travel towards the wall, or away from the wall, and those which move faster or slower than the local mean velocity in the streamwise direction.

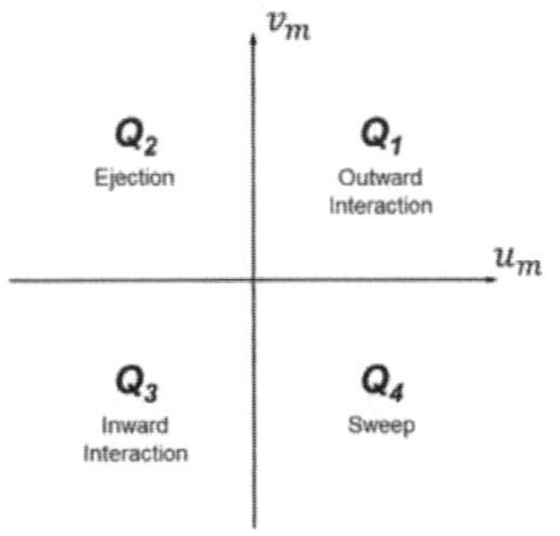


**Figure 1.** Quadrant based classification of structures. $u_m$ and $v_m$, as defined in equations 2 and 3, are the average fluctuating streamwise and wall–normal velocity within individual structures.

It has also been previously shown that a natural separation exists in the classification of structures based on the wall-normal location of the structures' minimum extent. Those structures whose minimum bound lies within twenty wall units of the wall may be considered "wall–attached" structures, while all others may be considered "wall–detached". This separation is based on a probability density function (PDF) of structure heights and occurs for both vortex clusters [6] and intense Reynolds stress structures [1] at the same wall-normal location, $y^+ = 20$. Each of these families of structures display different properties; wall–detached structures appear isotropically oriented, with their size depending on the distance to the wall indirectly through the Kolmogorov length–scale $\eta$, although this dependence is very weak, varying as $\eta \sim y^{1/4}$. Wall–attached structures on the other hand tend to be much taller, in some cases extending from near the wall, well beyond the centreplane of the channel. Lozano-Durán *et al.* [1] showed that the predominant structure within the logarithmic layer of a turbulent channel flow is a pair of wall–attached intense Reynolds stress structures (one sweep and one ejection), with an associated vortex cluster embedded in the ejection. These tall–attached sweeps ($Q_4$'s) and ejections ($Q_2$'s) account for a disproportionately large volume fraction of the flow, when compared to the inward and outward interactions ($Q_1$ and $Q_3$ structures), and are the focus of the topological analysis presented here.

## 3. Topological analysis via the velocity gradient tensor

The invariants $P_A$, $Q_A$, and $R_A$ of the VGT may be computed from

$$P_A = -A_{ii} \tag{4}$$

$$Q_A = -\frac{1}{2}A_{ij}A_{ji} \tag{5}$$

$$R_A = -\frac{1}{3}A_{ij}A_{jk}A_{ki} \tag{6}$$

where $A_{ij}$ represents the VGT [5]. The $P_S$, $Q_S$, $R_S$, $P_W$, $Q_W$, and $R_W$ invariants are similarly defined from the rate of strain $S_{ij}$ and rate of rotation $W_{ij}$ tensors, respectively. The rate of strain and rate of rotation tensors are the symmetric, and skew–symmetric, parts of the velocity gradient tensor, such that

$$A_{ij} = S_{ij} + W_{ij}. \tag{7}$$

The relative dominance of each of the invariants indicate the local topology of the flow [5], as classified in figure 2. The first invariant, $P_A$, is identically zero due to incompressibility. The second invariant, $Q_A$, relates to the rotational nature of the flow, with values greater than zero indicating dominance of enstrophy over strain, while negative values imply the opposite.

The topological effect of the third invariant, $R_A$, depends strongly on the value of $Q_A$. For $Q_A > 0$ and large, $R_A < 0$ represents vortex stretching and $R_A > 0$ vortex compression. In the limit of large negative values of $Q_A$ however, $R_A$ is dominated by strain self–amplification, $R_A \approx R_S$, and positive and negative values of $R_A$ are associated with sheet-like and tube-like structures respectively. The tent–like curve in figure 2(a) is the curve along which the discriminant,

$$D_A = \frac{27}{4} R_A^2 + Q_A^3 \tag{8}$$

of the velocity gradient tensor is equal to zero. This curve separates focal from non–focal local topologies, as illustrated on the figure.

In addition to the $Q_A - R_A$ plane, we examine in this paper the probabilistic topological distribution of the flow in the $Q_S - Q_W$ and $Q_W - \Sigma$ planes, where $\Sigma$ is the normalised vortex stretching rate given by

$$\Sigma = \frac{R_S - R_A}{Q_W}. \tag{9}$$

The second invariant of the rate of strain tensor, $Q_S$, is exclusively negative, and when its magnitude is large compared to $Q_W$, the flow topology is dominated by irrotational dissipative topology. On the other hand, when $Q_W$ is large and $Q_S$ is close to zero the flow takes on highly enstrophic topology with little dissipation, such as in the solid body rotation at the centre of vortex–tubes. When $Q_S$ and $Q_W$ are in balance, $-Q_S = Q_W$, the value of $Q_A$ is low and vortex sheets dominate.

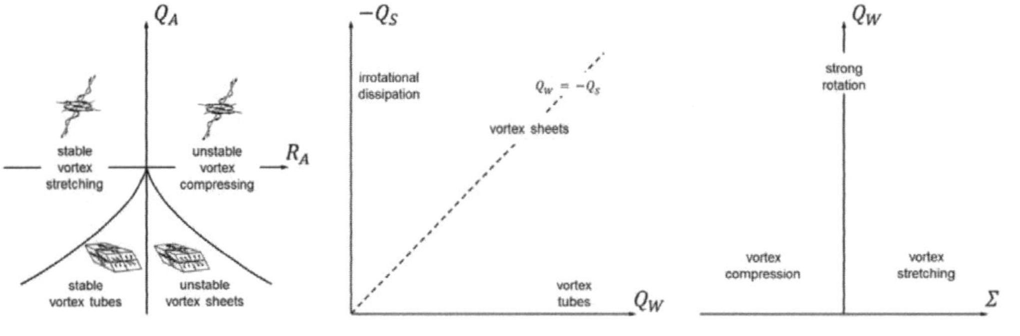


(a) (b) (c)

**Figure 2.** Local flow topology classification based on invariants of the velocity gradient tensor, in (a) the $Q_A - R_A$ invariant plane, (b) the $Q_S - Q_W$ invariant plane, and (c) the $Q_W - \Sigma$ invariant plane. The tent–like curve in (a) is the zero–contour of the VGT discriminant, $D_A$. Cartoons are taken from Soria *et al.* [11].

## 4. Numerical simulation and statistics

The direct numerical simulation (DNS) of a turbulent channel flow, performed by del Álamo *et al.* [18] provides us with a database for the calculation of intense Reynolds stress structures and vortex clusters and the study of the local topology therein. The Navier–Stokes equations are solved in the manner of Kim, Moin, and Moser [19] for the wall–normal vorticity, $\omega_y$, and the wall-normal velocity Laplacian, $\nabla^2 v$. Dealiased Fourier expansions are used in the streamwise and spanwise directions, and spatial discretisation is implemented by Chebyshev polynomials in the wall-normal coordinate. Integration in time is via the third order semi-implict Runge-Kutta method of Moser, Kim, and Mansour [20].

The friction Reynolds number of the simulation is $h^+ = 934$ and the simulation domain extends to $L_x \times L_z = 8\pi \times 3\pi$ channel half heights, $h$, in the streamwise, $x$, and spanwise, $z$, directions. $N_x = 3072$, $N_z = 2304$, and $N_y = 385$ grid points are used in the streamwise, spanwise, and wall-normal directions respectively. The viscous–scaled grid resolution, $\Delta x^+ = 7.6$, after dealiasing in the streamwise direction is equal to the wall–normal spacing, $\Delta y_c^+$, at the channel centre–plane. In the spanwise direction, it is $\Delta z^+ = 3.8$. The statistics presented here are accumulated from six statistically independent velocity fields containing a total of $8.8 \times 10^5$ vortex clusters and $4.0 \times 10^5$ intense Reynolds stress structures, of which 28% are $Q_2$'s and 32% are $Q_4$'s. The basic numerical parameters of the simulation are given in table 1.

**Table 1.** Relevant simulation parameters. In the table, $h$ is the channel half–height, $L_x$ and $L_z$ are the streamwise and spanwise lengths of the simulation domain. $\Delta x$, and $\Delta z$ are the dealiased grid resolutions in the streamwise and spanwise directions respectively. At the channel centreplane, the wall–normal resolution is $\Delta y_c$. In number of grid points in the streamwise and spanwise directions are represented by $N_x$ and $N_z$, while $N_y$ is the number of Chebyshev collocation points in the wall–normal direction. The + exponent indicates viscous scaling.

| $h^+$ | $L_x/h$ | $L_z/h$ | $(\Delta x^+, \Delta z^+, \Delta y_c^+)$ | $(N_x, N_z, N_y)$ |
|---|---|---|---|---|
| 934 | $8\pi$ | $3\pi$ | $(7.6, 3.8, 7.6)$ | $(3072, 2304, 385)$ |

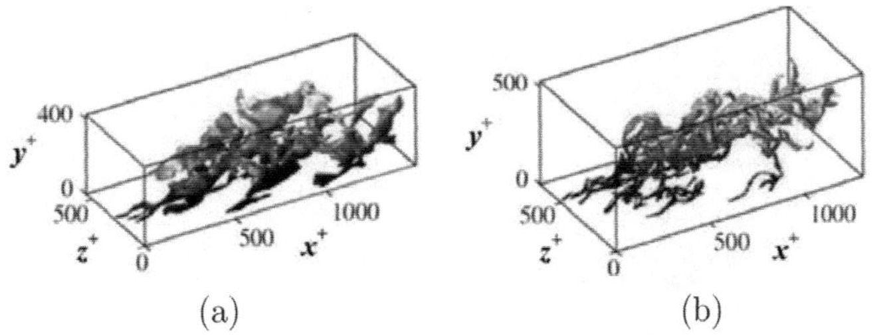


**Figure 3.** Examples structures, coloured by wall–normal location. Taken from Lozano–Durán *et al.* [1] (a) Attached $Q_2$ (ejection) structure, and (b) attached vortex cluster. Axes are labelled in wall units.

Illustrative examples taken from Lozano–Durán *et al.* [1] of a wall–attached $Q_2$ intense Reynolds stress structure, and a wall attached vortex cluster taken from the simulation are shown in figure 3. The surfaces shown are the threshold isosurfaces of the normalised Reynolds stress and the VGT discriminant, respectively. The axis scales are labelled in wall units, and colour represents wall–normal location. Qualitatively, it may be observed that the structures display a complex, fractal geometry, containing detail at a range of scales. As stated by Lozano–Durán *et al.* [1], the immediate impression is that vortex clusters are composed of "sponges of strings" while intense Reynolds stress structures are composed of "sponges of flakes".

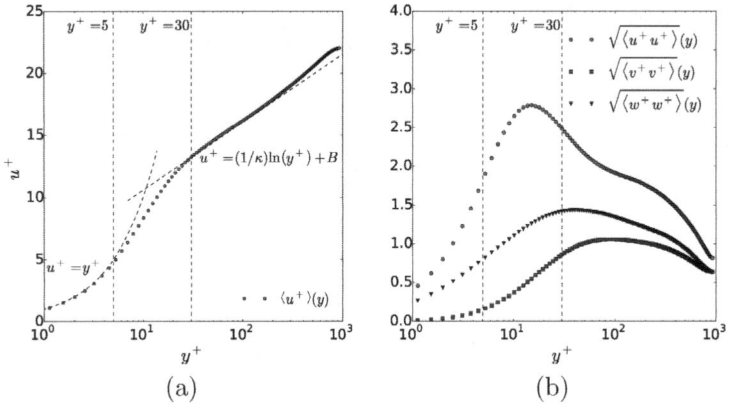


**Figure 4.** (a) Mean streamwise velocity profile, and (b) RMS of velocity components as a function of wall–normal location. Inner scaling applied.

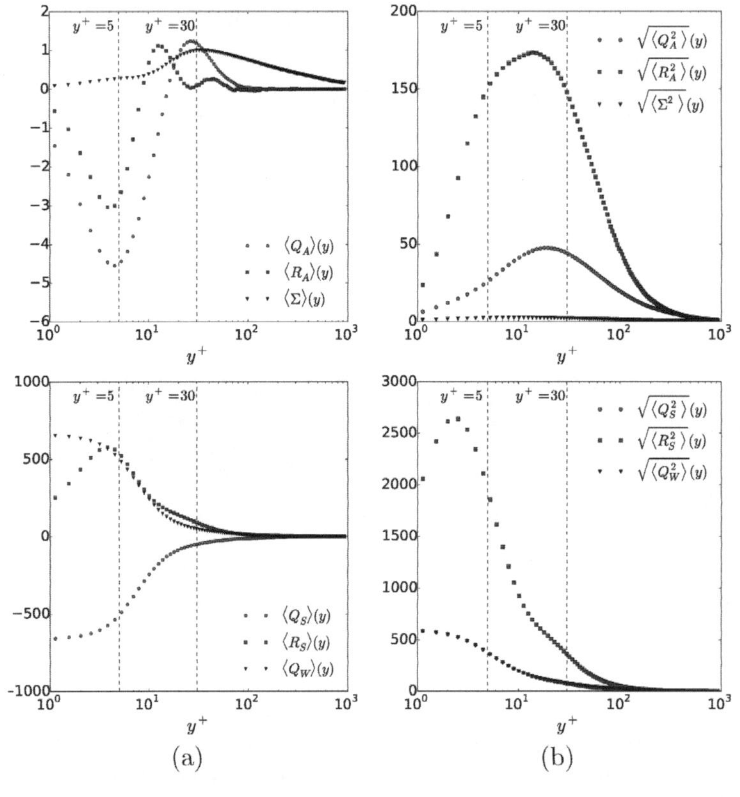


**Figure 5.** (a) Mean, and (b) RMS of VGT invariants and vortex stretching rate as a function of wall–normal location.

Mean and RMS profiles of each velocity component are shown in figure 4 as a function of wall–normal location, $y$. Averaging is performed in each wall–parallel $x-z$ plane, and fluctuating velocity components are defined as the departure from the mean in each $x-z$ plane also. The mean streamwise velocity profile displays a logarithmic behaviour in the approximate range $30^+ < y < 0.2h$, taking $\kappa = 0.42$ and with an additive constant of $B = 5.1$. This logarithmic relation and also the linear law of the wall in the viscous sublayer are represented by dashed curves on the figure. The vertical dashed lines, repeated in each pair of axes, delineate the viscous, buffer, and logarithmic layers at $y^+ = 5$ and $y^+ = 30$. The RMS profile of the streamwise component peaks at $\sqrt{\langle u^+u^+\rangle} = 2.8$ at $y^+ \approx 10.5$, while the spanwise component fluctuations are stronger than those in the wall–normal component.

The statistics of each tensor invariant are also plotted, in figure 5. Near the wall both $Q_A$ and $R_A$ are on average strongly negative, implying predominance of stable flow topologies as described in figure 2(a). Above some wall–normal location within the buffer layer both become positive. The mean vortex stretching rate, $\langle\Sigma\rangle$, is positive at all wall-normal locations, peaking in magnitude at the lower bound of the logarithmic layer. The RMS of both $Q_A$ and $R_A$ peaks within the buffer layer, with the $R_A$ fluctuations being stronger than the fluctuations in $Q_A$. The RMS of $\Sigma$ is much smaller than the RMS of both $Q_A$ and $R_A$, despite its mean magnitude being of the same order. The curves of $Q_S$ and $Q_W$ appear to nearly mirror one another about zero, since by definition $Q_A = Q_S + Q_W$ and the mean magnitude of $Q_A$ is much smaller than that of $Q_S$ and $Q_W$. This observation also accounts for the close match between the RMS curves of $Q_S$ and $Q_W$. The magnitude of both the mean and fluctuating values of $Q_S$ and $Q_W$ is large near the wall and monotonically decreases with $y$, while the mean and RMS values of $R_S$ peak within the viscous layer before also decreasing with distance from the wall.

## 5. Joint probability density functions

Joint probability density functions (JPDFs) in the $Q_A - R_A$ invariant plane are given in figure 6. The joint PDFs are separately calculated and plotted for each type of structure, and for the whole channel flow domain. The joint PDFs are normalised such that their integral equals 1, and contour lines are logarithmically spaced in increments of a decade, with the outermost contour shown indicating a normalised JPDF level of $10^{-3}$.

The $Q_A - R_A$ joint PDF for the whole channel exhibits the classic "teardrop" shape documented in Soria *et al.* [11] for a plane mixing layer, and also by Blackburn *et al.* [13] for turbulent flow in a channel, Ooi *et al.* [14] for isotropic turbulence, and Atkinson *et al.* [21] in a turbulent boundary layer. The teardrop shaped joint PDF in all of these cases has a strong preferential alignment towards positive $R_A$ values along the zero-discriminant line. Compared with the whole channel, the conditional joint PDF for vortex clusters displays a greater likelihood for values of the invariants $Q_A$ and $R_A$ in the upper left quadrant, indicating strong vortex stretching. This is especially the case for wall-detached vortex clusters. By definition the topology within vortex clusters does not include negative discriminant values, and this is reflected in the joint PDFs. Additionally, it appears that small positive values of discriminant are more likely to occur within wall–attached vortex clusters than in the wall–detached variety, where the joint PDF is shifted upwards away from the zero discriminant curve. This suggests stronger gradients associated with the local topology in the detached case.

The joint PDFs of $Q_A - R_A$ conditional on topology within high Reynolds stress structures are a less radical departure from the joint PDFs calculated from whole channel domain. Each displays a teardrop shape and it is clear that, unlike vortex clusters, the high Reynolds stress structures include topology with both positive and negative discriminant. The joint PDFs for the $Q_2$ structures display proportionally larger $Q_A$ and $R_A$ discriminants than the channel as a whole, but not to the same extent as the vortex clusters. The asymmetric teardrop tail in the lower right quadrant is especially pronounced and this is especially the case for wall-attached $Q_2$

Journal of Physics: Conference Series **708** (2016) 012005 doi:10.1088/1742-6596/708/1/012005

structures. The joint PDFs for $Q_4$ structures exhibit significantly less asymmetry and a greater tendency towards small magnitudes of the $Q_A$ and $R_A$ invariants. In general, wall–attached $Q_4$ structures tend to contain smaller invariant values than their wall-detached equivalents. It is well-known than $Q_2$'s are associated with stronger $uv$ values than $Q_4$'s [1], but the present results indicate that this is also the case for gradients. The small tail or bump along the zero discriminant curve in the lower left quadrant is also more pronounced in the $Q_4$ distributions than in the other joint PDFs presented here.

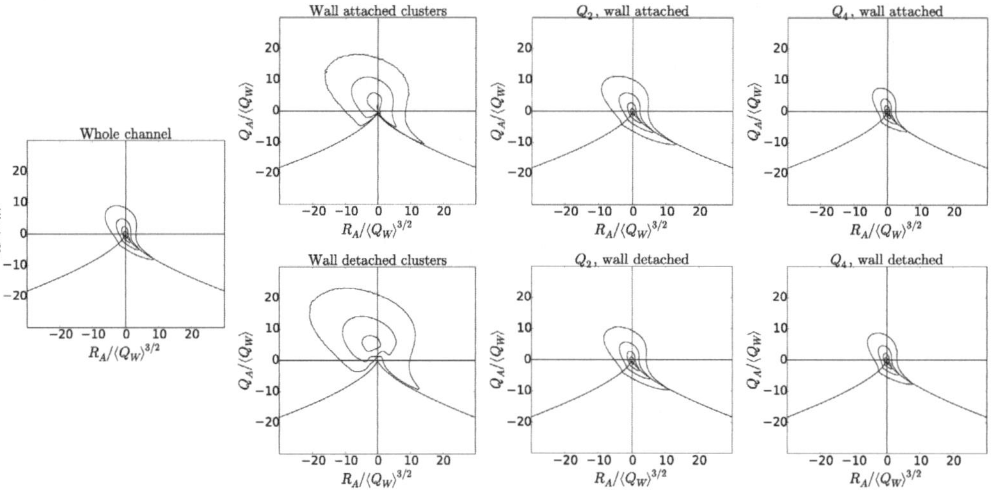


**Figure 6.** Joint PDFs of $Q_A$ vs $R_A$ for each structure type. Decade–spaced contour intervals, with outermost contour at $10^{-3}$.

Figure 7 illustrates the probabilistic topological behaviour, conditional on structure type in the $Q_S - Q_W$ plane. Physically, $Q_W$ correlates with enstrophy density and high values indicate strongly rotational topology, while values of $-Q_S \gg Q_W$ imply irrotational dissipation of kinetic energy. The joint PDF calculated from the whole channel domain shows a broad distribution across both invariants, but with a bias towards high $Q_W$ values. This is consistent with previous numerical studies [22, 23] suggesting that turbulent flows tend to arrange as worm–like or sheet–like vortices at the small scales, choosing such topology preferentially over irrotational dissipation.

The topology within vortex clusters is significantly more rotational than the flow as a whole, with the entire joint PDF shifted to high $Q_W$ values. The $Q_S - Q_W$ distribution peaks at a non–negligible value of $Q_W$, and the probability density drops to very small values near the $Q_S$ axis. Wall–detached vortex clusters appear to be more rotational than the wall-attached counterparts. This behaviour is consistent with previous observations in a turbulent boundary layer of greater sheet–like structure in the near–wall region [24]. Unlike the joint PDF for the whole channel (and for the high Reynolds stress structures), the joint PDFs for the vortex clusters display an increasing probability of high dissipation ($Q_S$) with increasing $Q_W$.

The shape of the $Q_S - Q_W$ joint PDFs for intense Reynolds stress structures more closely resembles that for the whole channel. A preference for rotational topology (high $Q_W$) over high negative $Q_S$ magnitudes may be observed in a broadly distributed PDF. As was the case in the $Q_A - R_A$ plane, $Q_4$ structures display a greater tendency towards small invariant magnitudes. The most striking feature here is shape of the $Q_S - Q_W$ joint PDF for wall-attached $Q_4$'s. Although overall there is a broad probability distribution, there is a significant proportion of these structures within which $Q_S$ and $Q_W$ balance one another manifesting in the highly correlated JPDF along the 45° line. This behaviour is associated with the existence of vortex sheet–like topology near the wall [24] within the lower parts of the attached $Q_4$'s and, since $Q_A = Q_W + Q_S$, explains the low invariant magnitudes within the $Q_4$ structures in the $Q_A - R_A$ plane.

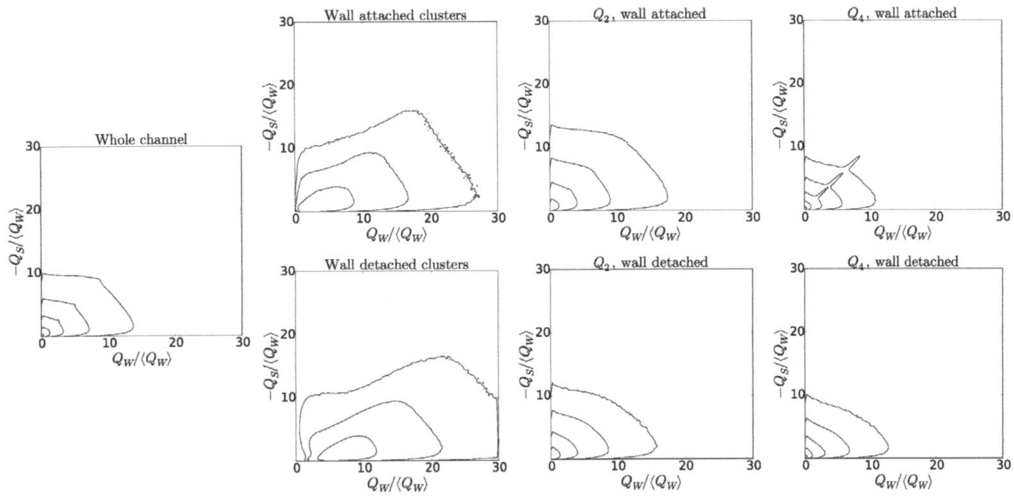


**Figure 7.** Joint PDFs of $Q_S$ vs $Q_W$ for each structure type. Decade–spaced contour intervals, with outermost contour at $10^{-3}$.

When the joint PDF of the invariant $Q_W$ and the vortex stretching, $\Sigma$, is plotted (figure 8) it becomes apparent that low values of $Q_W$ are associated with a wide range of vortex stretching rates, while the spread of vortex stretching rates diminishes with increasingly enstrophic flow. The probabilistic topology distribution is slightly biased towards positive vortex stretching rates. Similarly shaped distributions occur when the joint PDF is calculated conditional on structure type. For all structure types, the spread of vortex stretching diminishes with increasing $Q_W$, and in all cases there is a bias towards positive $\Sigma$. Additionally, vortex clusters have a greater bias towards positive vortex stretching, and this bias increases with increasing $Q_W$, indicating strong self–stretching behaviour of these structures. Consistent with figures 7, $Q_W$ values are generally higher in vortex clusters than in intense Reynolds stress structures, and $Q_4$ structures show less enstrophic behaviour than do the $Q_2$ structures.

## 6. Alignment of principal strains

The equation governing the evolution of enstrophy in the flow is

$$\frac{1}{2}\frac{D}{Dt}\left(\boldsymbol{\omega}^2\right) = \omega_i S_{ij}\omega_j + \nu\omega_i\nabla^2\omega_i, \tag{10}$$

where two terms on the right hand side of this equation relate to the production and dissipation of enstrophy. It is clear from the first term on the right hand side that the production of enstrophy is dependent on the strain field. This term may be rewritten in terms of the principal strains, $\alpha_i$, and their directions, $\hat{\mathbf{e}}_\mathbf{i}$, as

$$\omega_i S_{ij}\omega_j = \omega^2\alpha_i\left(\hat{\mathbf{e}}_\mathbf{i}\cdot\hat{\boldsymbol{\omega}}\right)^2. \tag{11}$$

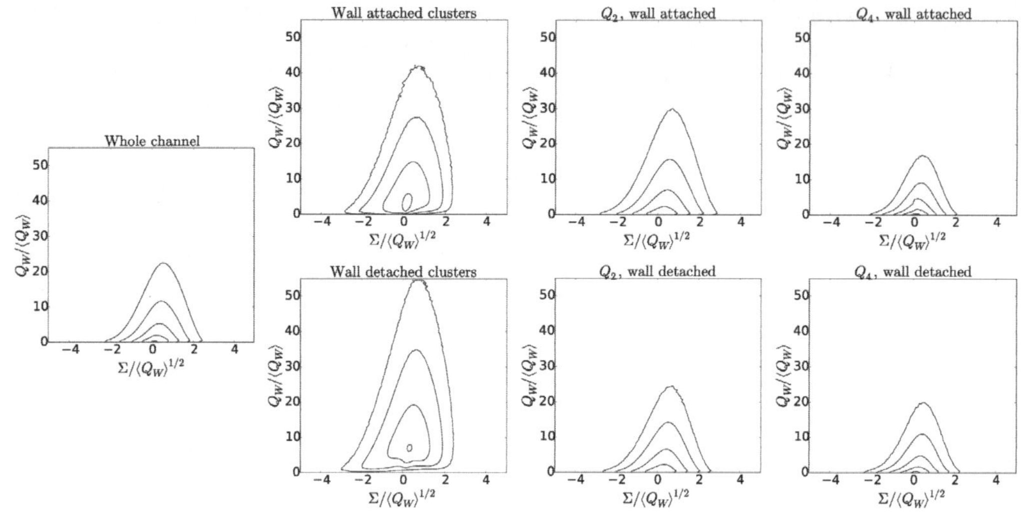


**Figure 8.** Joint PDFs of $Q_W$ vs $\Sigma$ for each structure type. Decade–spaced contour intervals, with outermost contour at $10^{-3}$.

The angle of alignment $\theta_i = \cos^{-1}(\hat{\mathbf{e}}_\mathbf{i} \cdot \hat{\boldsymbol{\omega}})$ between the local vorticity vector, $\boldsymbol{\omega}$, and the eigenvectors $\mathbf{e_i}$ of the rate-of-strain tensor thus directly drives the enstrophy production rate.

The cosine of the angle between $\mathbf{e_i}$ and $\boldsymbol{\omega}$ is calculated locally for each point in the flow as $\cos(\theta_i) = \hat{\mathbf{e}}_\mathbf{i} \cdot \hat{\boldsymbol{\omega}}$, and probability density functions conditional on structure type are given in figure 9. $\Phi$ represents the value of the PDF. In each set of axes the black lines represent the PDF taken over the entire computational domain, and PDFs conditional on each classification of turbulent structure are compared directly. 120 evenly spaced bins are used in the approximation of each PDF. Only every second symbol is shown, for clarity.

It is clear that the second, or intermediate, strain eigenvector aligns strongly with the the local vorticity direction, while the compressive eigenvector, $\hat{\mathbf{e}}_3$, predominantly aligns orthogonally with the vorticity vector. The first, most extensional eigenvector, $\hat{\mathbf{e}}_1$, appears mostly randomly oriented. This alignment has been noted previously by Ashurst *et al.* [25], Sondergaard *et al.* [26], and Tsinober *et al.* [27], and appears to be a property universal to turbulent flows.

**Table 2.** Ratios of probability of $\hat{\mathbf{e}}_2 - \boldsymbol{\omega}$ alignment to orthogonality.

| | Attached | Detached |
|---|---|---|
| Whole channel | 10.3:1 | |
| Vortex clusters | 17.4:1 | 21.6:1 |
| $Q_2$ | 10.3:1 | 7.5:1 |
| $Q_4$ | 7.2:1 | 5.1:1 |

Figure 9(b) shows that these alignment preferences persist within both wall-attached and wall-detached $Q_2$ structures, where all three eigenvectors appear to have very similar alignment to the alignment over the entire domain. In comparison, within vortex clusters the intermediate eigenvector is more strongly aligned with the vorticity field. The alignment is weaker within

$Q_4$ structures. Ashurst *et al.* [25] reports a ratio of the probability of perfect alignment of $\hat{\mathbf{e}}_2$ and $\boldsymbol{\omega}$ and the probability of these vectors orienting orthogonally as 15:1 in a free shear flow. The value here for the whole channel is approximately 10.3:1, indicating a less strongly aligned nature. The alignment ratios for each classification of coherent structure are listed in table 2.

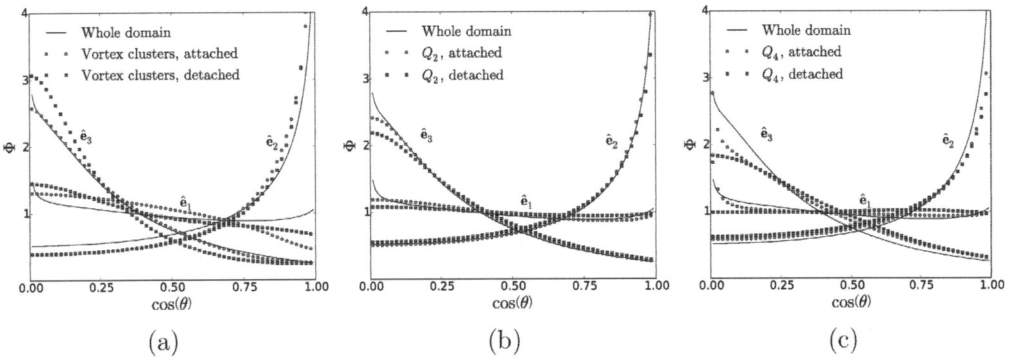


**Figure 9.** PDF of the alignment of the principal strain directions with the local vorticity vector. (a) Within vortex clusters, (b) Q2, and (c) Q4 structures, each compared with the distribution for the whole computational domain.

Within vortex clusters, the first eigenvector is no longer randomly oriented, but aligns transversely to the vorticity field. No such variation from the behaviour of the whole flowfield is observed in intense Reynolds stress structures of both the $Q_2$ and $Q_4$ type. It appears that the strong increase in $\hat{e}_1 - \boldsymbol{\omega}$ alignment close to $\cos(\theta) = 0$ is due primarily to the behaviour of the flow within wall-attached $Q_4$ structures. This is also the case for the similar feature seen in the alignment PDF relating to the $\hat{\mathbf{e}}_3$ eigenvector. For most of the structures for which alignment distributions are plotted, the third eigenvector is equivalently, or less strongly, transversally aligned with the vorticity field. The exception is the alignment distribution calculated within detached vortex clusters wherein the transversal alignment is stronger than that exhibited by the flow overall.

The Euclidean norm, $\|\boldsymbol{\alpha}\|$, of the principal strain eigenvalues is used to normalise the three eigenvalues of the rate-of-strain tensor, which are ordered by magnitude as

$$\frac{\alpha_1}{\|\boldsymbol{\alpha}\|} > \frac{\alpha_2}{\|\boldsymbol{\alpha}\|} > \frac{\alpha_3}{\|\boldsymbol{\alpha}\|} \tag{12}$$

Due to the incompressible nature of this flow, the largest strain rate $\alpha_1$ must be positive definite and the smallest, $\alpha_3$, must be always negative. The intermediate eigenvalue however may take either sign and, since the second eigenvector, $\hat{\mathbf{e}}_2$, is preferentially aligned with the vorticity within the vortex clusters and the intense Reynolds stress structures, its relative extensional or compressive nature strongly affects the behaviour of these structures [28]. Incompressibility also dictates that the maximum value that any one strain rate may take is $\alpha_i/\|\boldsymbol{\alpha}\| = \pm 2/\sqrt{6}$. Normalised probability density functions of the principal strain rates are computed and shown in figure 10. PDFs of $\alpha_i/\|\boldsymbol{\alpha}\|$ calculated over the entire computational domain are compared with conditional PDFs computed from vortex clusters, $Q_2$, and $Q_4$ structures in figure 10(a), (b), and (c), respectively.

The intermediate strain rate distribution is clearly skewed towards positive values in all cases, indicating a predominantly vortex stretching behaviour and thus the amplification of enstrophy. The distribution of the strain eigenvalues within $Q_2$ structures varies negligibly from the distribution calculated over the whole domain. Within $Q_4$ structures the first eigenvalue

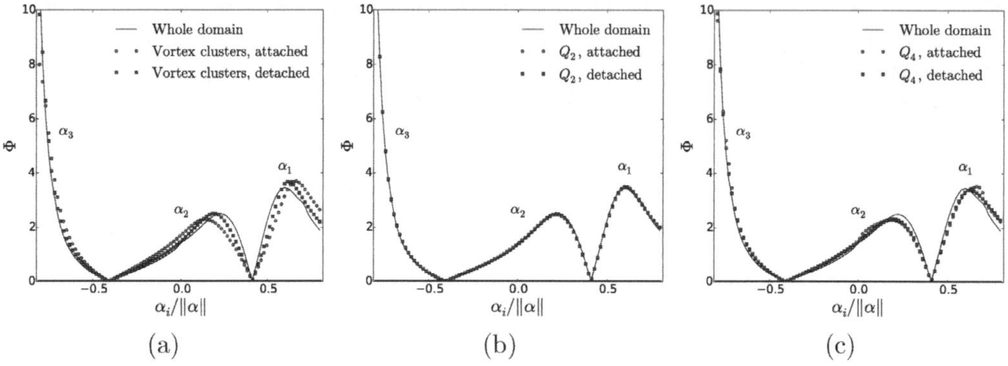


**Figure 10.** PDF of principal strain rates. (a) Within vortex clusters, (b) Q2, and (c) Q4 structures, each compared with the distribution for the whole computational domain.

tends to be more extensional, while the distribution of the second eigenvalue is less strongly skewed towards positive values. Very slight, raised bumps in the strain rate distributions are observed near the origin and at $\alpha_1/\|\boldsymbol{\alpha}\| = 1/\sqrt{2}$. These appear only in the whole domain PDFs and, more clearly, in the attached $Q_4$ structure PDFs, and are presumably driven by two–dimensional flow near the wall within the $Q_4$ structures. The distribution of principal strain rates within the vortex clusters displays similarly increased likelihood of highly extensional first principal strain, but to a greater extent than the $Q_4$ structures. This is especially true for vortex clusters originating within 20 wall–units of the wall.

It is notable that the PDFs of $\alpha_1$, $\alpha_2$, and $\alpha_3$ do not overlap. This implies that the most extensional strain rate eigenvalue at each point in the flow is larger than the intermediate eigenvalue at every other point in the flow, and that the same follows for the relationship between the intermediate and most compressional strain rate eigenvalues also. This is the case since, due to incompressibility, the values of neighbouring eigenvalues coincide when they are equal to $\alpha_i/\|\boldsymbol{\alpha}\| = \pm 1/\sqrt{6}$, and this represents a bound on each normalised principal strain rate.

## 7. Remarks

In comparing the topological distribution within intense Reynolds stress structures to that within vortex clusters, it was found that joint PDFs of the VGT invariants conditionally calculated within intense Reynolds stress structures show similar overall topology due to their non–mutually exclusive nature, yet still exhibit multiple distinct differences both with each other and with the overall channel flow itself. Wall–attached $Q_4$ structures originating at the lower bound within twenty wall units of the wall ($y^+ \leq 20$) show the greatest difference in topology from the other structures, with greater correlation between $Q_W$ (enstrophy) and $Q_S$ (dissipation) exhibited within these structures. This is likely due to attached $Q_4$ structures' closer proximity to the wall, which is reflected also in the distribution of principal strain rates, where the presence of two–dimensional flow is evident. Vortex clusters by definition include topology with only positive discriminant, and this is reflected in the shape of joint PDFs of the invariants $Q_A$ and $R_A$ of the velocity gradient tensor within these vortex clusters. They also show a different joint PDF structure to the intense Reynolds stress structures in the $Q_S - Q_W$ plane $Q_W - \Sigma$ plane, with stronger vortex stretching, and greater likelihood of strong rotational motion.

Alignment of the principal strain directions with the vorticity field is qualitatively similar to that observed in other turbulent flows, with strong alignment of the intermediate strain direction within all coherent structures examined here. The intermediate principal strain rate is skewed towards positive values, implying predominance of vorticity amplification within these structures, but both the likelihood of close alignment and the relative extensional nature of the intermediate strain rate vary depending on which type of coherent structure is considered. As with the joint PDFs of the VGT invariants, the differences between the $Q_2$ structures and the whole flow domain are less distinct when compared to the flow within $Q_4$ structures.

## Acknowledgements

This research was funded in part by the Multiflow program of the European Research Council (ERC) and was undertaken with the assistance of resources from the National Computational Infrastructure (NCI), which is supported by the Australian Government. Further computational resources were provided by the Partnership for Advance Computing in Europe (PRACE). The support of the Australian Research Council (ARC) is also gratefully acknowledged. Julio Soria gratefully acknowledges the support of an Australian Research Council Discovery Outstanding Researcher Award fellowship.

## References

[1] Lozano-Durán A, Flores O and Jiménez J 2012 The three-dimensional structure of momentum transfer in turbulent channels *J. Fluid Mech.* **694** 100–130

[2] Robinson S K 1991 Coherent structures in the turbulent boundary layer *Ann. Rev. Fluid Mech.* **23** 601–639

[3] Holmes P, Lumley J L and Berkooz G 1996 *Turbulence, coherent structures, dynamical systems and symmetry. Cambridge Monographs in Mechanics.* 1st ed (Cambridge Univ. Press)

[4] Hunt J C R, Wray A A and Moin P 1988 Eddies, streams, and convergence zones in turbulent flows *Proc. CTR Summer Prog., Center Turb. Res., NASA Ames/Stanford University*

[5] Chong M S, Perry A E and Cantwell B J 1990 A general classification of three-dimensional flow fields *Phys. Fluids* **2** 765–777

[6] del Álamo J C, Jiménez J, Zandonade P and Moser R D 2006 Self-similar vortex clusters in the turbulent logarithmic region *J. Fluid Mech.* **561** 329–358

[7] Haller G 2002 Lagrangian coherent structures from approximate velocity data *Phys. Fluids* **14** 1851–1861

[8] Shadden S C, Lekien F and Marsden J E 2005 Definition and properties of Lagrangian coherent structures from finite-time Lyapunov exponents in two-dimensional aperiodic flows *Physica D* **212** 271–304

[9] Cucitore R, Quadrio M and Baron A 1999 On the effectiveness and limitations of local criteria for the identification of a vortex *Europ. J. Mech. B: Fluids* **18** 261–282

[10] Chakraborty P, Balachandar S and Adrian R J 2005 Relationships between local vortex identification schemes *J. Fluid Mech.* **535** 189–214

[11] Soria J, Sondergard R, Cantwell B, Chong M and Perry A 1994 A study of the fine-scale motions of incompressible time-developing mixing layers *Phys. Fluids* **6** 871–884

[12] Perry A and Chong M 1994 Topology of flow patterns in vortex motions and turbulence *App. Sci. Res.* **53** 357–374

[13] Blackburn H, Mansour N and Cantwell B 1996 Topology of fine-scale motions in turbulent channel flow *J. Fluid Mech.* **310** 269–292

[14] Ooi A, Martin J, Soria J and Chong M 1999 A study of the evolution and characteristics of the invariants of the velocity-gradient tensor in isotropic turbulence *J. Fluid Mech.* **381** 141–174

[15] Wallace J M, Eckelman H and Brodkey R S 1972 The wall region in turbulent shear flow *J. Fluid Mech.* **54** 39–48

[16] Willmarth W W and Lu S S 1972 Structure of the Reynolds stress near the wall *J. Fluid Mech.* **55** 65–92

[17] Lu S S and Willmarth W W 1973 Measurements of the structure of the Reynolds stress in a turbulent boundary layer *J. Fluid Mech.* **60** 481–511

[18] del Álamo J C, Jiménez J, Zandonade P and Moser R D 2004 Scaling of the energy spectra of turbulent channels *J. Fluid Mech.* **500** 135–144

[19] Kim J, Moin P and Moser R D 1987 Turbulence statistics in fully developed channel flow at low Reynolds number *J. Fluid Mech.* **177** 133–166

[20] Moser R D, Kim J and Mansour N N 1999 Direct numerical simulation of turbulent channel flow up to $Re_\tau = 590$ *Phys. Fluids* **11** 943–945

[21] Atkinson C, Chumakov S, Bermejo–Moreno I and Soria J 2012 Lagrangian evolution of the invariants of the velocity gradient tensor in a turbulent boundary layer *Phys. Fluids* **24** 105104

[22] Siggia E D 1981 Numerical study of small-scale intermittency in three-dimensional turbulence *J. Fluid Mech.* **107** 375–406

[23] Jiménez J, Wray A A, Saffman P G and Rogallo R S 1993 The structure of intense vorticity in isotropic turbulence *J. Fluid Mech.* **255** 65–90

[24] Bermejo-Moreno I, Atkinson C, Chumakov S, Soria J and Wu X 2010 Flow topology and non–local geometry of structures in a flat–plate turbulent boundary layer *Proc. CTR Summer Prog., Center Turb. Res., NASA Ames/Stanford University*

[25] Ashurst W T, Kerstein A R, Kerr R M and Gibson C H 1987 Alignment of vorticity and scalar gradient with strain rate in simulated Navier–Stokes turbulence *Phys. Fluids* **30** 2343–2353

[26] Sondergaard R, Chen J H, Soria J and Cantwell B J 1991 Local topology of small scale motions in turbulent shear flows *8th Symp. Turbulent Shear Flows, Munich Germany*

[27] Tsinober A, Kit E and Dracos T 1992 Experimental investigation of the field of velocity gradients in turbulent flows *J. Fluid Mech.* **242** 169–192

[28] Vincent A and Meneguzzi M 1994 The dynamics of vorticity tubes in homogeneous turbulence *J. Fluid Mech.* **258** 245–254

Journal of Physics: Conference Series **708** (2016) 012006 doi:10.1088/1742-6596/708/1/012006

# Identifying coherent structures and vortex clusters in Taylor-Couette turbulence

**Vamsi Spandan, Rodolfo Ostilla-Monico, Detlef Lohse, Roberto Verzicco**

Physics of Fluids, Mesa+ Institute, Faculty of Science and Technology, University of Twente, Postbus 217, 7500AE Enschede

E-mail: vamsispandan@gmail.com

**Abstract.** The nature of the underlying structures in Taylor-Couette (TC) flow, the flow between two co-axial and independently rotating cylinders is investigated by two methods. First, the quadrant analysis technique for identifying structures with intense radial-azimuthal stresses (also referred to as 'Q's) of Lozano-Durán *et al.*, (*J. Fluid Mech.* 694, 100-130) is used to identify the main structures responsible for the transport of angular velocity. Second, the vortex clusters are identified based on the analysis by del Álamo *et al.*, (*J. Fluid. Mech.*, 561, 329-358). In order to test these criteria, two different radius ratios $\eta = r_i/r_o$ are considered, where $r_i$ and $r_o$ are the radii of inner and outer cylinder, respectively: (i) $\eta = 0.5$ and (ii) $\eta = 0.909$, which correspond to high and low curvature geometries, respectively and have different underlying structures. The Taylor rolls, i.e. the large-scale coherent structures, are effectively captured as 'Q's for the low curvature setup and it is observed that curvature plays a dominant role in influencing the size and volumes of these 'Q's. On the other hand, the vortex clusters are smaller in size when compared to the 'Q' structures. These vortex clusters are found to be taller in the case of $\eta = 0.909$, while the distribution of the lengths of these clusters is almost homogenous for both radius ratios.

## 1. Introduction

The ubiquitous nature of wall bounded turbulent flows both in nature and in process technology makes them a subject of interest for fundamental scientific research as well as for industrial applications. Since its earliest observation, turbulence has been always described as a process of random and chaotic mixing [1]. Turbulent wall flows [2] are dominated by complex multi-scale processes in which momentum (and other quantities such as mass or heat) are transported between the wall and the bulk of the flow and vice versa. However, even in the apparent disorder of turbulence, researchers have found the existence of some 'order' or, in other words, coherence in turbulent flows. In particular, for wall-bounded turbulent flows, the existence of various types of coherent structures such as hairpins, horse-shoe vortices, herringbone-like streaks, and many others has been the subject of research in several pioneering studies [3, 4, 5]. With consistent increase in computational capabilities, many advances have been made in understanding the role the underlying structures play in the turbulent cascades where energy and momentum are transferred across length and time scales [6].

Most studies of wall-bounded turbulence have been limited to the canonical case of pressure-driven flow between either two parallel plates or inside a pipe (hereafter referred to as a channel-

and pipe-flow, respectively) due to their simplicity and ease to conduct numerical simulations and high precision experiments. While much insight has been gained from these systems, it is not yet clear if the coherent structures and cascade processes present in these flows are universal. Simulations and experiments of other canonical flows, such as the zero-pressure-gradient boundary layer and the flow between two parallel-plates driven by shear (hereafter referred to as plane Couette (PC)) have taken off in the last years, and are raising interesting questions on the universality of wall-bounded turbulence. In particular, recent numerical simulations of the Taylor-Couette (TC) problem, i.e. the flow between two independently rotating co-axial cylinders have shown that some of the features occuring in channel flow are not present in TC flow. For example, in TC flow in many areas of the parameter space large-scale structures appear. These are known as Taylor rolls, and are fixed in space, persistent in time, and are responsible for majority of the angular momentum transport [7]. This is in contrast to channel flows, where analysis based on cospectra show that the large-scale wall-attached structures are inactive [6]. Even for PC flows, this cospectra analysis has shown that large-scale wall-attached structures are inactive [8].

TC flow can be described using four non-dimensional parameters: two for the geometry, i.e. the radius ratio $\eta = r_i/r_o$, where $r_i$ and $r_o$ are the radii of the inner and outer cylinders, respectively and the aspect ratio $\Gamma = L_z/d$, where $L_z$ is the axial length of the cylinders while $d = r_o - r_i$ is the gap-width, and two other parameters which relate to the strength of the driving, and are usually taken in terms of the inner- and outer cylinder Reynolds numbers as $Re_i = U_i d/\nu$ and $Re_o = U_o d/\nu$ where $U_i$ and $U_o$ are inner and outer cylinder velocities, respectively, and $\nu$ the kinematic viscosity of the fluid. We use a cylindrical coordinate system $(r, \theta, z)$ to mathematically tract TC flow, with $u_r$, $u_\theta$ and $u_z$ are the three velocities: radial (corresponding to wall-normal in channel flow), azimuthal (streamwise) and axial (spanwise). The angular velocity $\omega$ is simply $\omega = u_\theta/r$. From a geometrical point of view, the curvature of the wall present in TC flow is one of the major factors responsible for the differences between the dynamics of TC flow and other canonical flows. Studies of boundary layers over curved surfaces [9, 10, 11] have shown that even a small curvature added to the channel flow can be responsible for major changes in the flow dynamics. From the point of view of the driving there is another difference: channel flow is pressure driven while TC flow (and PC flow) are shear driven. Also, unlike the two other flows, TC flow is linearly unstable if $d|\mathcal{L}|/dr^2 < 0$, where $r$ is the radial coordinate and $\mathcal{L}$ is the angular momentum. When the driving exceeds a critical value, the instability gives rise to the formation of laminar Taylor rolls which fill the gap between both cylinders. With further increase in driving the laminar Taylor rolls transition into wavy Taylor rolls, and eventually into fully turbulent Taylor rolls. As mentioned previously, these rolls are persistent in both space and time in some regions of the ($\eta$, $Re_i$, $Re_o$) parameter space, and have been observed in both experiments and numerical simulations. Even at $Re_i \sim 10^6$, they can remain coherent at large-scales, and are responsible for majority of the angular momentum transport [7, 12, 13].

The geometry of the setup, and especially the radius ratio dictates the strength and coherence of the Taylor rolls. For pure inner cylinder rotation, it has been observed that when the curvature is low (i.e. $\eta$=0.909), strong signatures of the Taylor rolls exist not only in spectra and correlations, but also in the azimuthally- and time-averaged fields. The strength of the rolls are diminished in systems with higher curvature (i.e. $\eta$=0.5), an effect which can be attributed to the strong asymmetry between the inner and outer cylinders when $\eta = 0.5$ [14, 13, 7]. Adding a slight counter-rotation can make the large-scale structures reappear, as was seen in the TC experiments by van der Veen *et al.* [15], who for $\eta = 0.5$ found some signatures of large-scale coherent structures when the cylinders were counter-rotating but none for pure inner-cylinder rotation. The aspect ratio $\Gamma$ controlls the wavelength of the rolls, as only a even amount of rolls of the domain can fit. However, different roll wavelengths do not give very different flow physics

**Table 1.** Details of the numerical simulations. The first column is the radius ratio $\eta$. The second to fourth columns represent the amount of grid points in the azimuthal, radial and axial directions, the sixth and seventh columns show the length of the domain in the azimuthal (streamwise) and axial (spanwise) directions at the mid-gap, where $L_x = \pi(r_i + r_o)/n_s$. The seventh column shows the frictional Reynolds number at the inner cylinder, while the ninth and tenth columns show the resolution in inner wall-units at the mid-gap.

| $\eta$ | $N_\theta$ | $N_r$ | $N_z$ | $L_x/(2d)$ | $L_z/(2d)$ | $Re_{\tau,i}$ | $\Delta x^+$ | $\Delta z^+$ |
|---|---|---|---|---|---|---|---|---|
| 0.5 | 768 | 768 | 512 | 6.2 | 4.2 | 200 | 1.6 | 1.6 |
| 0.909 | 1024 | 768 | 512 | 6.5 | 4.2 | 380 | 2.6 | 3.1 |

if the driving is large enough [13].

The main focus of this study is to find more advanced ways of quantifying the large-scale coherent structures in TC, and to verify that they can be captured not only in time averaged flow fields and velocity spectra, but also through other methods by considering a single snapshot of the flow field, or a time series of snapshots sufficiently separated in time. A number of studies exist in the literature which elucidate the various techniques that can be used to identify organized 'structures' or 'clusters' in a turbulent flow field; in particular here we focus on the 'quadrant' analysis technique [16, 17, 18] and the identification of vortex clusters [19]. del Álamo *et al.* [20] adopted the approach of Chong *et al.* [19] based on the velocity gradient discriminant to identify clusters of vortical structures in a turbulent channel flow. For a similar flow configuration, Lozano-Durán *et al.* [21] studied the properties of three dimensional intense Reynolds stress structures (hereafter referred to as 'Q' structures) in line with the one-dimensional quadrant analysis of Lu and Willmarth [17]. In this work we follow the approach of Lozano-Durán *et al.* [21] and of del Álamo *et al.* [20] to identify both 'Q' structures and vortex clusters in a turbulent TC flow. In particular, the analysis of Lozano-Durán *et al.* is interesting for TC because it comes to different conclusions about the large-scale wall-attached structures in channels, namely that they actively transport Reynolds stresses, similar to the Taylor-rolls.

In order to obtain the flow field snapshots, direct numerical simulations of TC flow were performed using a second-order central finite-difference scheme with fractional time stepping [22, 23]. This scheme has already been tested and used extensively to study the dynamics of TC flow over a wide parameter range [24, 25, 26, 14]. As mentioned previously, for the purpose of this study two different radius ratios of the TC setup were considered, namely (i) $\eta = 0.5$ (high curvature) and (ii) $\eta = 0.909$ (low curvature). The aspect ratio for both setups is fixed to $\Gamma = 2.09$ so the domain can accommodate a single Taylor roll pair. To reduce computational costs, for $\eta = 0.5$, a rotational symmetry of $n_s = 3$ was used, while for $\eta = 0.909$ we took $n_s = 20$. These two rotational symmetries ensure that the azimuthal length of the domain is of the order of $2\pi$ half-gaps. Full details of the numerical resolution used is presented in Table 1. The outer cylinder is kept at rest (i.e. $Re_o = 0$) and the inner cylinder is rotated to reach a Reynolds number of $Re_i = 2 \times 10^4$, which corresponds to a frictional Reynolds number at the inner cylinder of approximately $Re_{\tau,i} = 200$ and $Re_{\tau,i} = 380$ for $\eta = 0.5$ and $\eta = 0.909$, respectively. For $\eta = 0.5$, the dynamics of the TC flow is dominated by the asymmetry between both the cylinders while for $\eta = 0.909$, although the curvature is much smaller it is sufficient enough for the linear instability to give rise to Taylor rolls. Three snapshots of the flow field separated by at least ten large-eddy turnover times $d/U_i$ are selected for each radius ratio, to ensure that the statistics are not dependent on the selected snapshot.

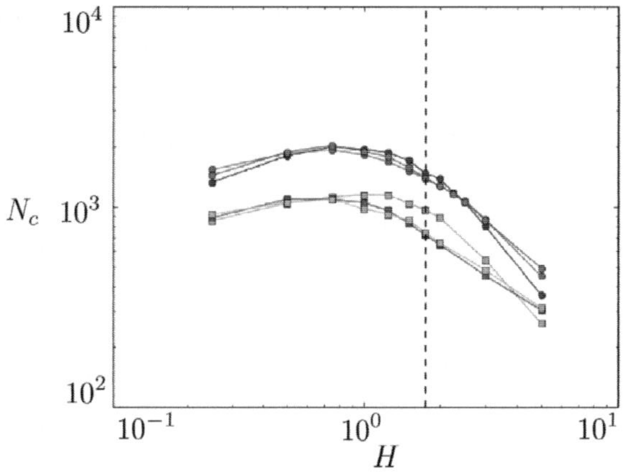


**Figure 1.** Total number of identified 'Q' structures for $\eta = 0.909$ (squares) and $\eta = 0.5$ (circles); three different snapshots are considered. The vertical line marks the chosen threshold ($H = 1.75$)

The manuscript is organized as follows: section 2 deals with the procedure for identifying 'Q's, which are responsible for transport of tangential Reynolds stresses. These structures are identified with the Taylor-roll for the low curvature case. Some aspect of these 'Q's are also quantified. Section 3 is about the identification and statistics of vortex clusters, with a focus on the geometry and dissipation. Section 4 is a summary of the present work and an outlook for the future.

## 2. 'Q' structures

### *2.1. Identification and percolation analysis*

In this section we focus on the 'Q's, i.e. the structures responsible for the transport of Reynolds stresses. In TC flow the transported (and conserved) quantity is the angular velocity flux $J^\omega$ which is defined as $J^\omega = r^3(\langle u_r\omega\rangle_{A,t} - \nu\partial_r\langle\omega\rangle_{A,t})$, where $\langle ...\rangle_{A,t}$ represents averaging in the two homogenous (azimuthal and axial) directions and in time and $u_r$ and $\omega$ are the radial velocity and angular velocity, respectively [27]. In the definition of $J^\omega$, the first term represents the Reynolds stress contribution while the second term is the viscous contribution to the angular velocity transport. We note that this equation is analogous to the definition of the total stress in channels, with the difference that the torque (stress) is constant throughout the domain, as the flow is shear driven.

Following the approach of Lozano-Durán *et al.* [21] the 'Q' structures in the case of TC flow are defined as connected regions satisfying the following criterion:

$$|u_r u_\theta| > H u_r' u_\theta'. \tag{1}$$

where $H$ is the hyperbolic hole size and can be thought of as a threshold parameter for the 'Q' structures, $u_r$, $u_\theta$ are the instantaneous radial and azimuthal velocities while $u_r'$, $u_\theta'$ are the root mean square fluctuations respectively. Connectivity of the vortex clusters is defined in terms of its six neighbouring grid points and any object smaller than $30^3$ wall units in volume is filtered out for the analysis.

The first step in identifying the relevant 'Q' structures and vortex clusters is to perform a percolation analysis to calculate the best threshold parameters, i.e. $H$. Thresholding parameters have to be chosen carefully, especially in an inhomogenous flow such as wall bounded turbulence.

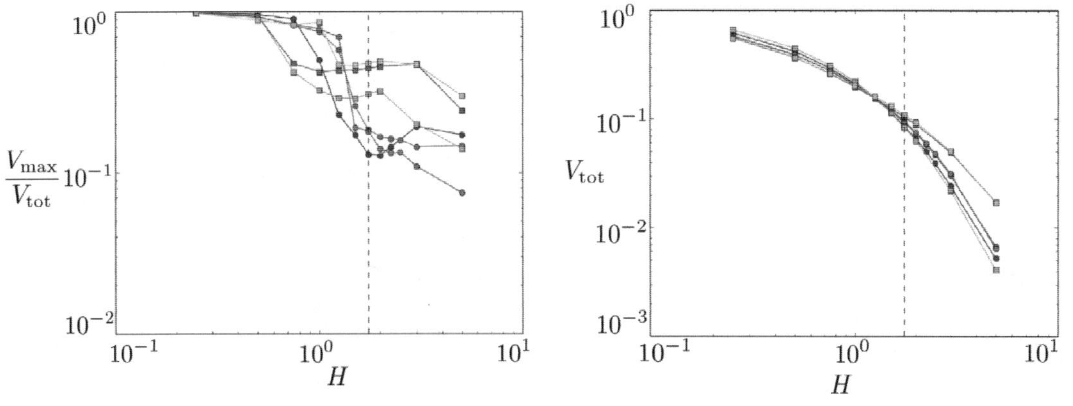


**Figure 2.** Left panel: Maximum volume of a 'Q' normalised by the total volume of all the identified 'Q' s in the domain. Right panel: Total volume of all the identified 'Q' s normalised by the volume of the TC domain. Square symbols refer to $\eta = 0.909$, while the circles refer to $\eta = 0.5$. Three different snapshots for each radius ratio are considered.

In figure 1 we plot the total number $N_c$ of identified 'Q' structures for both radius ratios and the three different snapshots. At a high value of ($H \sim 3$) the number of structures identified are of the order $N_c \sim \mathcal{O}(10^2)$. On decreasing the value of $H$, more structures arise while the structures grow in size (this can also be seen in the left panel of figure 2). Further decreasing $H$ just results in the structures merging together without increasing in size until a single structure fills up the entire volume. Another interesting aspect is that the total number of identified 'Q's in the TC flow ($N_c \sim \mathcal{O}(10^3)$) is much smaller than what was previously seen for channel flow (i.e. $N_c \sim \mathcal{O}(10^6)$) by Lozano-Durán *et al.* [21]. Since this cannot be accounted for only due to computational box-size difference, it indicates either the existence of larger 'Q' s in TC flow, or that the 'Q' s fill a smaller part of the domain.

In the left panel of figure 2 we plot the volume of the largest 'Q' structure normalized by the total volume of all identified 'Q's for different values of $H$. We observe that the percolation behaviour of the 'Q's in TC flow is very similar to that seen in channel flow by Lozano-Durán *et al.* [21]. When $H > 2$, there is no significant change in the volume of the largest structures computed, while a percolation crisis (recognized by the steepest gradient in the plot of the normalized maximum volume of the structures) occurs for $0.7 < H < 2$ (c.f. left panel of figure 2). The steep fall in the maximum volume is not very obvious for $\eta = 0.909$ and it occurs at a smaller value of $H$ when compared to the cases with $\eta = 0.5$. As noted before, on further decreasing $H$ all the 'Q's merge into a single structure. A percolation threshold of $H = 1.75$ is thus chosen for identifying the 'Q' structures (shown as a vertical dotted line in figures 2 and 1). The right panel of figure 2 shows the total volume of all the identified 'Q' structures normalized by the total volume of the TC domain. We can observe that while there is some difference at high values of $H$, the curves overlap for most of the range of $H$ where the percolation crisis occurs. In this way, we ensure that the selected snapshot does not have an influence on subsequent analysis. However, we note that quite some variation exists across snapshots, and by using only three snapshots we might be incurring in some statistical errors.

### *2.2. Visualization of structures*

In figure 3 we show a visualisation of the 'Q' structures for two different radius ratios and also the averaged contour plots of the azimuthal velocity. When the curvature of the setup

Journal of Physics: Conference Series **708** (2016) 012006 doi:10.1088/1742-6596/708/1/012006

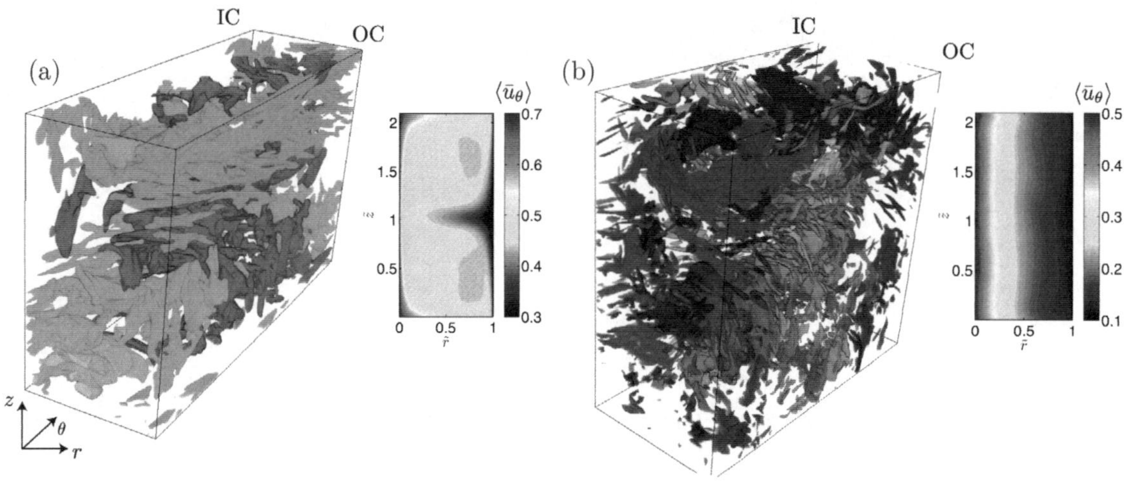


**Figure 3.** 'Q' structures identified for two different radius ratios (a) $\eta = 0.909$ (b) $\eta = 0.5$. The right panels show azimuthally and time averaged contour plots of azimuthal velocity. For both cases $H = 1.75$

is low (figure 3$a$), the averaged contours of azimuthal velocity show a strong signature of the coherent Taylor rolls (one-pair here). The Taylor rolls can also be visually seen through a three dimensional volume plot of the 'Q' structures calculated using the criterion introduced in equation 1. Indeed, in figure 3$a$ we can observe large scale structures (identified by different colours) which occupy most of the volume. These large scale structures, co-exist with other small scale structures such as near wall streaks. However, most of these small structures, including 'Q 's from every quadrant, eventually merge into a large-scale Taylor roll. For very low values of $H$, the large 'Q 's merge with the smaller ones to fill the TC domain completely as seen in the right panel of figure 2.

Figure 3$b$ shows the case of a high curvature setup, where the rolls seem to break up, and there is no clear signature of a large scale roll both in the averaged contour plot and also from the visualisation of the 'Q' structures. Nonetheless, some large scale structures still exist which may correspond to the plume ejections from the inner cylinder. We study the nature of these structures while analysing the distribution of 'Q' s later. It seems that while some small scale structures are preserved in both systems, the roll structure observed in the case of $\eta = 0.909$ breaks up into multiple 'Q's when the curvature is high i.e. $\eta = 0.5$. This is also reflected in figure 1 where the total number of 'Q's identified is higher for $\eta = 0.5$ when compared to the case with $\eta = 0.909$, again a difference higher than what can be accounted for by the difference in computational box size. Using the quadrant analysis technique allows us to classify the identified structures into various categories such as $Q_1$ (outward interactions), $Q_2$ (ejections), $Q_3$ (inward interactions) and $Q_4$ (sweeps) based on the volume integral of the stream-wise and wall-normal fluctuations over a single structure. The mean stream-wise and wall-normal fluctuation in a structure is computed as $u_m = \frac{\int u dV}{\int dV}$ and $v_m = \frac{\int v dV}{\int dV}$; the structure is termed $Q_1$ if $u_m > 0$ and $v_m > 0$, $Q_2$ if $u_m < 0$ and $v_m > 0$, $Q_3$ if $u_m < 0$ and $v_m < 0$, and $Q_4$ when $u_m > 0$ and $v_m < 0$. In order to study how the distribution of these structures depends on the curvature of the setup we categorise the various 'Q's in table 2 for $\eta = 0.5$ and $\eta = 0.909$. For channel

**Table 2.** Distribution of the 'Q' structures computed over three different snapshots for the low-curvature ($\eta = 0.5$) and high-curvature ($\eta = 0.909$) setup respectively.

| $\eta$ | $Q_1$ (outward interactions) | $Q_2$ (ejections) | $Q_3$ (inward interactions) | $Q_4$ (sweeps) |
|---|---|---|---|---|
| 0.5 | 23.9±1.2% | 29.6±0.8% | 22.8±0.9% | 23.7±1.4% |
| 0.909 | 17.2±0.9% | 29.0±1.8% | 21.8±% | 32.0±1.5% |

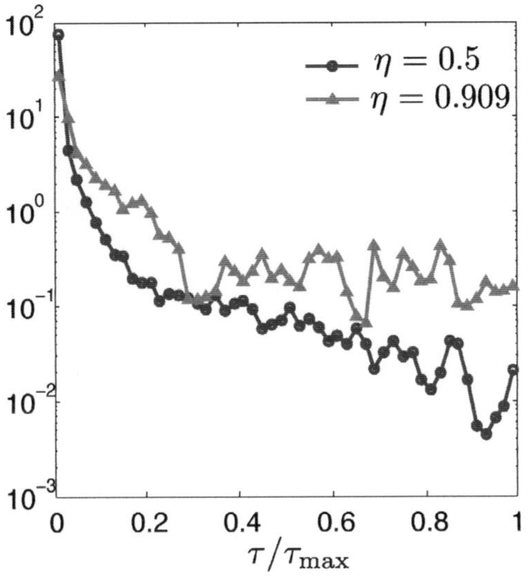


**Figure 4.** Normalised probability distribution function of the absolute value of the Reynolds stress (c.f. equation 1) in the 'Q' structures for $\eta = 0.5$ (circles) and $\eta = 0.909$ (triangles). For both cases $H = 1.75$; $\tau_{\max} = \max(\tau_i)$, where $\tau_i$ is the total Reynolds stress in an individual 'Q' structure

flow Lozano-Durán *et al.* [21] found that at least 60% of the total number of 'Q' structures were composed of $Q_2$ and $Q_4$'s independent of the Reynolds number ($Re_\tau = 934, 2003$). In TC flow, $Q_2$ and $Q_4$'s arise from the events leading to the formation of Taylor rolls i.e. the plume ejections from the inner cylinder which impact onto the outer cylinder and the plume ejections from the outer cylinder which are a result of the impact events. Both these events are strong markers of $Q_2$ and $Q_4$'s respectively and from table 2 we find that when the curvature is low ($\eta = 0.909$) they account for almost 60% of the total 'Q's. In the case of a high curvature setup ($\eta = 0.5$) the percentage of $Q_2$'s does not change significantly and the large scale structures observed in figure 3*b* belong to this quadrant. However, we find that the sweep events are drastically reduced. This is a consequence of less intense ejection events from the outer cylinder when $\eta = 0.5$ (c.f. figure 3*b*) which are crucial for the formation of Taylor rolls.

### *2.3. Reynolds stress distribution*

In this subsection we look at how the tangential stresses are distributed among the 'Q' structures. We compute the total Reynolds stress in each 'Q' structure as the volume integral of the stress (c.f. equation 1) at each computational node enclosed in an individual structure. As seen

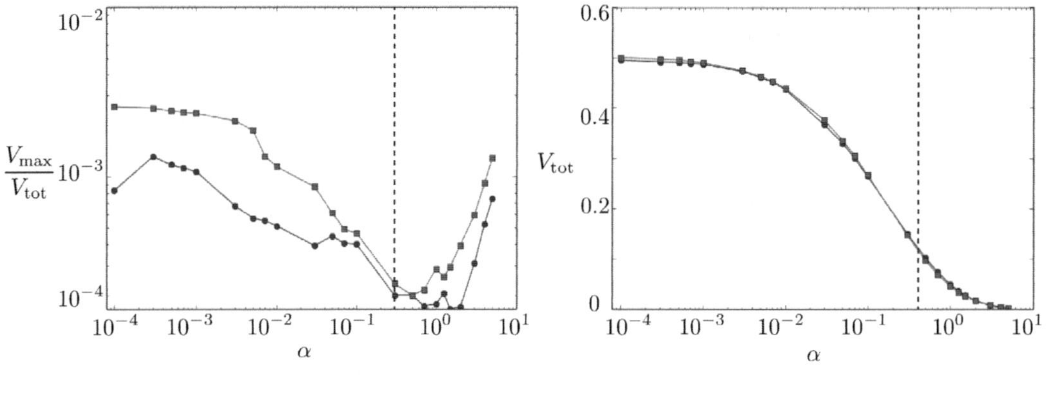


**Figure 5.** Left panel: Maximum volume of a vortex cluster normalised by the total volume of all the identified clusters in the domain. Right panel: Total volume of all the identified clusters in the domain. Square symbols refer to $\eta = 0.909$, while the circles refer to $\eta = 0.5$; dashed vertical line represents the chosen percolation parameter $\alpha$

previously, structures in all four quadrants are present. Thus, structures which contain either positive or negative stresses are present, indicating the existence of both regions of ejecting herring-bone streaks, which detach from the boundary layer and form regions with on average, net positive transport, and impacting regions in the flow, where the structures originating from the other boundary layer impact the cylinder's boundary layer, and form regions of net negative transport. In figure 4 we plot histograms of the absolute value of normalised Reynolds stresses contained in the 'Q' structures for two different radius ratios to study the distribution of the Reynolds stress in the 'Q' structures. When the curvature is low ($\eta = 0.909$), we observe a uniform distribution of structures with high Reynolds stress ($0.6 < \tau/\tau_{\text{max}} < 1$). These packets of large scale structures are responsible for majority of the angular velocity transport in a low-curvature setup. From the visualisation shown in figure 3 we observed that although small scale structures exist in both low and high curvature system, large scale rolls are more apparent when the curvature is low. On reducing $H$ the large scale rolls merge forming large scale rolls which fill up the entire volume (c.f. figure 2). However when the curvature is high ($\eta = 0.5$), the probability to find structures with intense Reynolds stresses are much lower as compared to the case where $\eta = 0.909$.

## 3. Vortex clusters

### *3.1. Identification and percolation analysis*

In this section we focus on the vortex clusters. A similar approach as in the previous section is adopted, where each vortex cluster is identified as a region satisfying the following criterion:

$$D(\mathbf{x}) > \alpha \overline{D'^2(r)}^{1/2}, \tag{2}$$

where $D(\mathbf{x})$ is the discriminant of the velocity gradient tensor at a single point, $\overline{D'^2}^{1/2}$ is the standard deviation of $D$ over wall-normal planes while $\alpha$ is a thresholding parameter. Again, the thresholding parameter, in this case $\alpha$ has to be chosen carefully. A percolation analysis is done for vortex clusters as done for 'Q' structures in the previous sub-section, following the procedure by del Álamo *et al.* [20] for channel flows. Again, the connectivity of the vortex

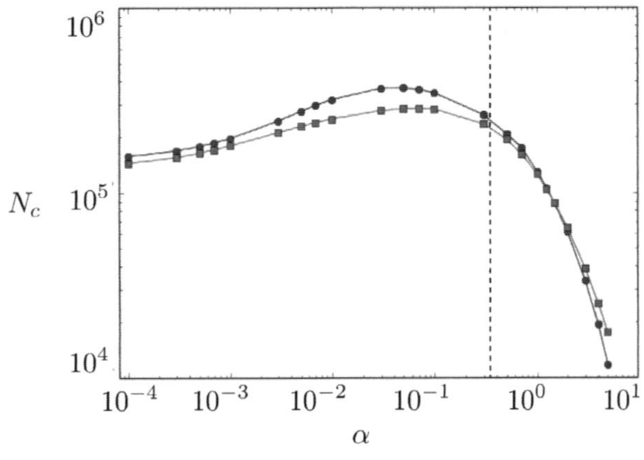


**Figure 6.** Total number of identified vortex clusters in the domain for two different radius ratios $\eta = 0.5$ (circles), $\eta = 0.909$ (squares); dashed vertical line represents the chosen percolation parameter $\alpha$

clusters is defined in terms of its six neighbouring grid points and any object smaller than $30^3$ wall units in volume is filtered out for the analysis.

In figure 5 we show the normalized maximum volume of a vortex cluster and the total volume of all the clusters for different values of $\alpha$. The trend of $V_{\text{max}}/V_{\text{tot}}$ for vortex clusters is similar to the one observed for 'Q' structures and a region of percolation crisis of the vortex clusters can also be identified around $\alpha_c \sim 0.08 - 0.1$. However, the maximum volume of any individual vortex cluster on reducing $\alpha$ does not go beyond $10^{-2}$ and the total volume occupied by these objects is roughly half of the total domain volume. This is an indication that the vortex clusters are much smaller objects as compared to the 'Q' structures. Following del Álamo *et al.* [20], we chose a percolation parameter of approximately $\alpha = 2.5\alpha_c$ i.e. $\alpha = 0.25$.

Figure 6 shows the total number of identified vortex clusters for different $\alpha$. With increase in $\alpha$, the total number of vortex clusters falls more steeply than the identified 'Q' structures with increase in $H$. An interesting observation is that the number of vortex clusters ($N_c \sim 10^5$) is much higher than that of 'Q' structures ($N_c \sim 10^3$; c.f. figure 1*a*) which shows that while the volume of individual vortex clusters is very small they are in much higher number as compared to the 'Q' structures.

### *3.2. Cluster geometries*

As the volume of individual vortex cluster is extremely small compared to the total volume occupied by all the vortex clusters, it is challenging to analyze these clusters visually as done for the 'Q' structures. Instead, in this subsection we focus on the geometrical features of these vortex clusters. In the case of a turbulent channel flow, del Álamo *et al.* [20] found that the logarithmic region consists of small packs of clusters which are nearly homogenous and isotropic and another group of tall structures with their base in the near wall region ($y^+ \sim 20$). By circumscribing each vortex cluster with a parallelepipedal box, they found that the tall attached clusters are geometrically self-similar. We follow a similar approach and characterize the radial, axial and azimuthal lengths of each vortex cluster.

To characterize the geometry of the vortex clusters in the stream-wise ($\hat{\theta}$), wall-normal ($\hat{r}$) and span-wise ($\hat{z}$) directions, we compute the length of each cluster in three directions as

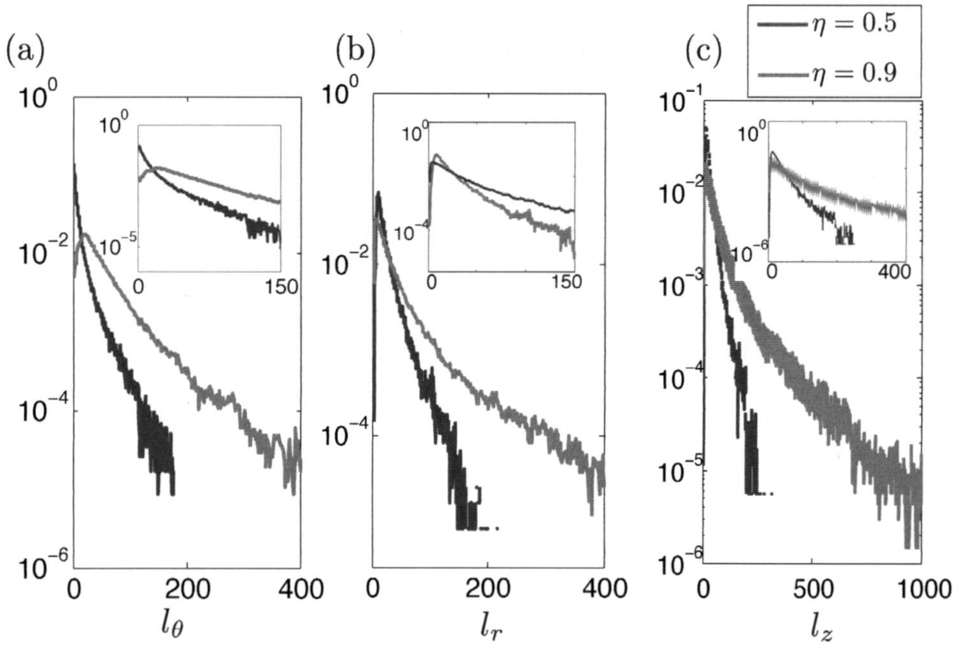


**Figure 7.** Probability distribution functions of the (a) stream-wise ($l_\theta$) (b) wall-normal ($l_r$) and (c) span-wise ($l_z$) lengths of the vortex clusters. All lengths are computed in viscous wall units.

$l_i = \langle l_i \rangle_{\text{max}} - \langle l_i \rangle_{\text{min}}$. In figure 7, we plot the probability distribution functions of the lengths of these clusters for $\eta = 0.5$ and $\eta = 0.909$. Figure 7*a* shows that the stream-wise extent of the clusters is different for setups with different curvature. From the inset of figure 7*a* we find that it is more probable to find small scale stream-wise structures in a low-curvature setup ($\eta = 0.5$) as compared to a high curvature setup ($\eta = 0.909$). The radial and axial extents of the vortex clusters are much higher in a low-curvature setup as compared to a high-curvature setup. The stronger shear owing to the higher shear Reynolds number when $\eta = 0.909$ may be responsible for the vortex clusters to be larger as compared to when $\eta = 0.5$. We do not find any packets of self-similar clusters in the near wall region as observed by del Álamo *et al.* [20] in the case of a turbulent channel flow.

In order to study if these clusters have a preferential site of origin, we plot the joint probability distribution functions of the minimum and maximum radial and axial positions of the vortex clusters for both radius ratios in figure 8. While del Álamo *et al.* [20] found groups of tall structures originating from the near wall region in case of TC flow we find that the clusters are of uniform wall-normal lengths throughout the domain in case of both $\eta = 0.5$ and $\eta = 0.909$. The origin of this effect is unclear and further analysis is required in understanding the formation of the vortex clusters in TC flow. Additionally, the clusters in $\eta = 0.909$ case are at least twice the length of the clusters when $\eta = 0.5$.

### *3.3. Dissipation in vortex clusters*

Similar to the analysis done in section 2.3, here we compute the total dissipation in each vortex cluster as a volume integral of the net dissipation at every computational grid node enclosed in each vortex cluster i.e $\epsilon = \frac{\int \epsilon_i dV}{\int dV}$, where $\epsilon_i$ is the dissipation computed in a computational box

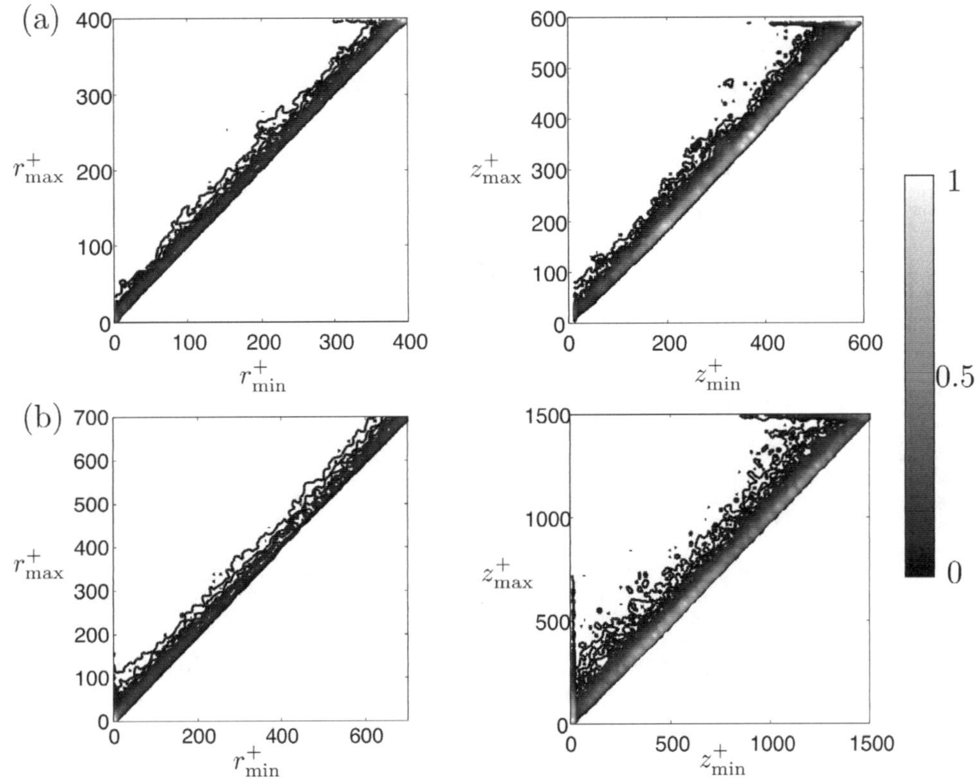


**Figure 8.** Joint probability distribution functions to describe the geometry of the vortex clusters (a) $\eta = 0.5$ (b) $\eta = 0.909$. The plot is in greyscale and the color scale varies from 0 to 1 in each plot; dark colour indicates lower probability while a lighter colour refers to high probability.

of volume $dV$. The result is shown in figure 9. Contrary to figure 4, in the case of vortex clusters we observe almost no differences in the net dissipation contained in each vortex cluster between both cases. Although the low curvature setups have taller/longer vortex cluster extending in the wall-normal direction, as observed from figure 8, the distribution of the dissipation in these clusters is almost homogenous, and very few clusters have extreme dissipation. Additionally the difference in the p.d.f of the dissipation in the vortex clusters for three different snapshots of the flow is less than 5 %.

## 4. Summary and outlook

The study of formation, evolution and decay of coherent structures in wall bounded turbulent flows is crucial for our understanding of the complex multi-scale process of turbulence. While many previous studies have focussed on gaining insight on the role of these coherent structures in canonical flow setups such as pipe flow or channel flows, other flows where the curvature of the wall is along the direction of the streamlines have been hardly investigated. In this work we consider such a system, i.e. Taylor-Couette (TC) flow, where the flow is driven between two independently rotating co-axial cylinders. By following the approach of Lozano-Durán *et al.* [21] and del Álamo *et al.* [20] we have developed suitable criteria and identified Reynolds stress structures ('Q's) and vortex clusters in a turbulent Taylor-Couette flow.

Through the analysis of single flow-field snapshots of two different radius ratios ($\eta = 0.5$, and

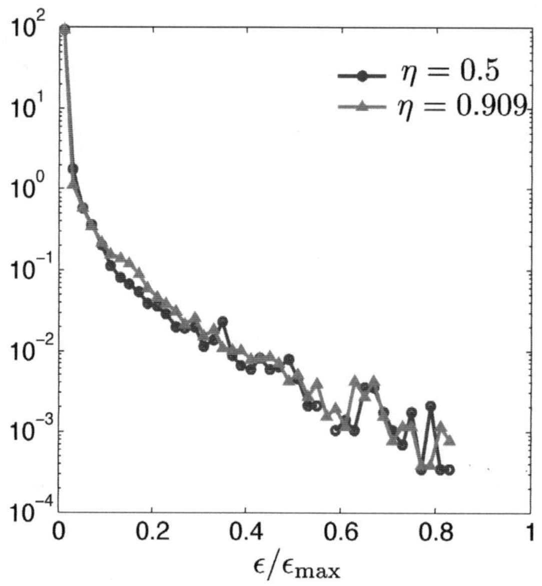


**Figure 9.** Normalised probability distribution function of the total dissipation in the vortex structures for $\eta = 0.5$ (circles) and $\eta = 0.909$ (triangles).

$\eta = 0.909$) we find, consistent with previous analysis, that low-curvature TC systems have wall-attached, large-scale Taylor rolls which fill up the entire volume, and transport the majority of Reynolds stresses. This is a further way of quantifying the existence of these structures, which have also been observed in the time averaged contours, axial autocorrelations and velocity spectra. Unlike the case of channels, where different results were found for the transport of the large-scales when applying the 'Q 'analysis [21], here, the results obtained are consistent with previous results from co-spectra [13].

When the curvature is high, the large-scale rolls break up into smaller structures, herringbone-like streaks and the flow structure is more similar to what is seen in other wall-bounded turbulent flows. Contrary to the 'Q's, we find that vortex clusters are much higher in number. While the 'Q's fill up the entire volume of the domain, the vortex clusters occupy a much smaller volume, and, in general are smaller and thus come in higher numbers. The distribution of the sizes of the vortex clusters is mostly homogenous and no groups of wall-attached and wall-detached clusters are found in TC flow for both the low-curvature and high-curvature systems. The clusters are found to be larger for the $\eta = 0.909$ case, however, dissipation statistics did not change between both cases.

During the Multiflow Summer workshop, due to time constraints and due to limited computational resources this analysis was applied to relatively small boxes of $(6.2 \times 2 \times 4.2)d/2$ at moderate frictional Reynolds numbers $Re_\tau \approx 400$. We plan on extending this analysis to small boxes at frictional Reynolds numbers up to $Re_\tau \approx 3000$, and also to large computational boxes of up to $(24 \times 2 \times 16)d/2$, and to a larger number of flowfield napshots to reduce the statistical errors, and to be able to separate the Qs into their quadrants while obtaining useful information about their physics. Further analysis at other radius ratio or for counter-rotating cylinders will be useful to determine the validity of the obtained insights.

**Acknowledgments**
We thank Miguel P. Encinar for being our host during this workshop, funded by the Multiflow programme of the European Research Council, and J. Jiménez for the invitation. Furthermore, we acknowledge all participants for the fruitful discussions, and computing time from SurfSARA through NWO for the simulations from which the flowfields used in this analysis come. We also acknowledge financial support by FOM.

## References

[1] Pope S B 2000 *Turbulent Flow* (Cambridge U. Press)
[2] Smits A J, McKeon B J and Marusic I 2011 *Annu. Rev. Fluid Mech.* **43** 353–375
[3] Townsend A E 1980 *The structure of turbulent shear flow* (Cambridge U. Press)
[4] Perry A E and Chong M S 1982 *J. Fluid Mech.* **119** 173–217
[5] Perry A E, Henbest S M and Chong M S 1986 *J. Fluid Mech.* **165** 163–199
[6] Jiménez J 2011 *Annu. Rev. Fluid Mech.* **44** 27
[7] Grossmann S, Lohse D and Sun C 2016 *Annu. Rev. Fluid Mech.* **48**
[8] Avsarkisov V, Hoyas S, Oberlack M and García-Galache J P 2014 *J. Fluid Mech.* **751** R1–8
[9] Bradshaw P 1973 Effects of streamline curvature on turbulent flow. *Agardograph* AG-169 (AGARD)
[10] Hunt I A and Joubert P N 1979 *J. Fluid Mech.* **91** 633–659
[11] Hoffmann P H, Muck K C and Bradshaw P 1985 *J. Fluid Mech.* **161** 371–403
[12] Huisman S G, van der Veen R C, Sun C and Lohse D 2014 *Nature Communications* **5** 3820
[13] Ostilla-Mónico R, Verzicco R, Grossmann S and Lohse D 2015 *J. Fluid Mech.* **788** 95–117
[14] Ostilla-Mónico R, van der Poel E, Verzicco R, Grossmann S and Lohse D 2014 *J. Fluid Mech.* **761** 1–26
[15] van der Veen R, Huisman S, Merbold S, Harlander U, Egbers C, Lohse D and Sun C 2015 *arXiv:1508.05802*
[16] Wallace J M, Eckelmann H and Brodkey R S 1972 *J. Fluid Mech.* **54** 39–48
[17] Lu S S and Willmarth W W 1973 *J. Fluid Mech.* **60** 481–511
[18] Willmarth W W and Lu S S 1972 *J. Fluid Mech.* **55** 65–92
[19] Chong M S, Perry A E and Cantwell B J 1990 *Phys. Fluids* **2** 765–777
[20] del Álamo J, Jiménez J, Zandonade P and Moser R D 2006 *J. Fluid Mech.* **561** 329–358
[21] Lozano-Durán A, Flores O and Jiménez J 2012 *J. Fluid Mech.* **694** 100–130
[22] Verzicco R and Orlandi P 1996 *J. Comput. Phys.* **123** 402–414
[23] van der Poel E P, Ostilla–Mónico R, Donners J and Verzicco R 2015 *Comput. Fluids* **116** 10–16
[24] Ostilla–Mónico R, Stevens R, Grossmann S, Verzicco R and Lohse D 2013 *J. Fluid Mech.* **719** 14–46
[25] Ostilla–Mónico R, Huisman S G, Jannink T, Van Gils D, Verzicco R, Grossmann S, Sun C and Lohse D 2014 *J. Fluid Mech.* **747** 1–29
[26] Ostilla–Mónico R, van der Poel E, Verzicco R, Grossmann S and Lohse D 2014 *Phys. Fluids* **26** 015114
[27] Eckhardt B, Grossmann S and Lohse D 2007 *J. Fluid Mech.* **581** 221–250

Journal of Physics: Conference Series **708** (2016) 012007 doi:10.1088/1742-6596/708/1/012007

# Space and time behaviour of the temperature second-order structure function in Rayleigh-Bénard convection

**Riccardo Togni**[1], **Andrea Cimarelli**[1,2], **Adrián Lozano-Durán**[3] **and Elisabetta De Angelis**[4]

[1]DIN, Università di Bologna, Via Fontanelle 40, 47121 Forlì, Italy
[2]DISMI, Università degli Studi di Modena e Reggio Emilia, Via Giovanni Amendola 2, 42122 Reggio Emilia, Italy
[3]School of Aeronautics, Universidad Politécnica de Madrid, 28040 Madrid, Spain
[4]School of Engineering - Cardiff University, Queen's Buildings, The Parade, Cardiff, CF24 3AA, United Kingdom

E-mail: DeAngelisE@cardiff.ac.uk

**Abstract.** One of the most peculiar aspects of turbulence in wall bounded-flows is the ability of the turbulent fluctuations to regenerate themselves through self-sustained processes. The dynamics of these self-sustaining mechanisms has been extensively investigated in the past via two complementary approaches. From one side, the possibility to identify very robust kinematic features within the flow feeds the hope of the scientific community to obtain a complete and consistent dynamical description of the physics of the turbulent regeneration cycles in terms of the so-called coherent structures. From the other side, the multi-scale and inhomogeneous features of the self-sustaining mechanisms of turbulence have been addressed by means of global statistical quantities based on two-point averages such as second-order structure functions. The present work attempts to link these two approaches, by identifying how turbulent cycle mechanisms and turbulent structures reflect on the global statistical properties of second-order structure function. To this aim we use Direct Numerical Simulation data of thermally driven turbulence in the Rayleigh-Bénard convection and we analyse for the first time the behaviour of the second-order structure function of temperature in the complete four-dimensional space of spatio-temporal scales and wall-distances. The observed behaviour is then interpreted in terms of the dynamics of coherent thermal structures and of their commonly accepted model of life-cycle.

## 1. Introduction

A fundamental concept of turbulence theory is represented by the multi-scale scenario introduced by Richardson [1] and further developed by Kolmogorov [2]. At sufficiently high Reynolds numbers and away from the boundaries or other singularities, a clear distinction between large and small scales can be made. Large eddies are directly influenced by the geometry of the flow, whereas the behaviour of the small ones is almost universal as it depends only on the rate at which they receive energy and on viscosity. The latter represents one of the few exact and nontrivial results in the field of turbulence, nevertheless the limitations of the leading hypotheses makes this framework a particular case, although conceptually of invaluable importance [3].

In the next few decades after Kolmogorov [4, 5] until this day [6, 7, 8, 9], many research groups have spent a lot of efforts trying to understand and model turbulence. However, even if the same goal is shared by all these groups, the methodologies adopted are very different, which lead sometimes to contradicting results. Historically, turbulence has been approached in a *statistical* and in a *structural* sense. The decomposition introduced by Reynolds represents the first attempt to describe turbulence in a statistical way by splitting the field into an average component and a fluctuating one [10]. On the other side, Richardson was the pioneer of the structural approach by portraying turbulence in terms of eddies of different sizes [1].

Loosely speaking, the statistical point of view deals with the global statistics of the turbulent flow field whereas the structural one considers turbulence as a property of deterministic solutions of the governing equations rather than as a stochastic process [11]. While in the structural approach the results obtained are mainly influenced by the choice of the eduction technique (e.g. quadrant analysis [12], VITA [13], POD [14], based on invariants of the velocity-gradient tensor [15], etc. ), in the statistical one the results depend only on the choice of the observable (e.g. mean velocity, turbulent intensities, etc. ). Quantitatively the results of these two approaches are not comparable but they can complement each other and the results obtained by one method can support or clarify the evidences obtained by the other.

The aim of the present work is to try to bridge the gap between these two approaches by searching possible explanations of the behaviour shown by the second-order structure function in the different features shown by turbulent structures. To this purpose we use the Rayleigh-Bénard convection, see Ref. [16] for a recent review. In such a flow, thermally driven turbulence shows a much more coherent pattern with respect to more classical pressure-driven wall-turbulence. Indeed, the turbulent cycle is characterized by one single coherent structure, the so-called thermal plume. The prominent role played by thermal plumes is then expected to leave a marked footprint on global statistical quantities. Indeed, we here propose the analysis of the second-order structure function for the temperature field. To further increase the possibility of recognize features related to turbulent structures on the behaviour of second-order structure function, we address here for the first time, the entire augmented-space of wall-turbulence, i.e. the compound space of spatial and temporal scales and wall-distances. Needless to say this study represents only a first step toward a more complete analysis represented by the Yaglom equation, i.e. the balance for the second order structure function, where the study of the different ranges of scales and positions encountered by the related fluxes endows the second-order moment with its physical interpretation.

## 2. Direct Numerical Simulation

The Rayleigh-Bénard convection (RBC) consists of a fluid layer heated from below and cooled from above by two horizonatal plates in a laterally unbounded domain. The governing equations for RBC are the continuity, Navier-Stokes and energy equations under the Boussineq approximation

$$\begin{aligned}
&\frac{\partial u_i}{\partial x_i} = 0, \\
&\frac{\partial u_i}{\partial t} + \frac{\partial u_i u_j}{\partial x_j} = -\frac{\partial p}{\partial x_i} + \sqrt{\frac{Pr}{Ra}}\frac{\partial^2 u_i}{\partial x_j \partial x_j} + \theta\delta_{i2}, \\
&\frac{\partial \theta}{\partial t} + \frac{\partial \theta u_j}{\partial x_j} = \frac{1}{\sqrt{Pr Ra}}\frac{\partial^2 \theta}{\partial x_j \partial x_j},
\end{aligned} \tag{1}$$

where $i, j = 1, 2, 3$, $\delta_{ij}$ is the Kronecker delta and the variables $u_i$, $p$, $\theta$ are respectively the velocity, pressure and temperature fields. We introduce the commonly-used spatial coordinates $x = x_1$, $y = x_2$, $z = x_3$ and velocity components $u = u_1$, $v = u_2$, $w = u_3$. The Cartesian coordinate system is cell-centered, with the $xz$-plane parallel to the horizontal plates and the $y$-axis pointing in the direction opposite to that of gravity acceleration, see figure 1. Equations

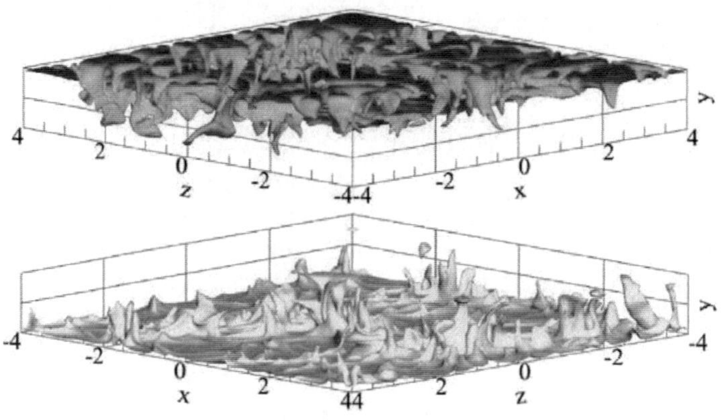


**Figure 1.** Isosurfaces of temperature ($\theta = 0.3$ down and $\theta = -0.3$ up) for $Ra = 1.0 \times 10^6$.

(1) are written in a dimensionless form using the distance between the plates $H$, the temperature difference between the lower and the upper plates $\Delta\Theta$ and the free-fall velocity $U_f = \sqrt{g\alpha\Delta\Theta H}$ as characteristic scales, where $g$ and $\alpha$ are the gravity acceleration and the thermal expansion coefficient respectively. In what follows, all the results will be reported in a non-dimensional form by using $H$, $\Delta\Theta$ and $U_f$ as characteristic quantities. The resulting non-dimensional groups are the Prandtl $Pr = \nu/\kappa$ and Rayleigh numbers $Ra = g\alpha\Delta\Theta H^3/\nu\kappa$, where $\nu$ the kinematic viscosity and $\kappa$ the thermal diffusivity. The Boussinesq equations (1) are solved using a pseudospectral method which discretizes space with Chebychev polynomials in the $y$ direction and with Fourier modes in the $x$ and $z$ directions. Time integration is performed with a fourth-order Runge-Kutta scheme for the nonlinear terms and a Crank-Nicholson scheme for the linear ones. Two DNS are performed for $Pr = 0.7$ at $Ra = 1.7 \times 10^5$ and $1.0 \times 10^6$ in a rectangular box of sizes $L_x = 8$, $L_y = 1$, $L_z = 8$. Periodic boundary conditions are imposed at the lateral sidewalls whereas isothermal and no-slip boundary conditions are used on the top and bottom plates. This rather large computational domain has been considered, at the expense of the $Ra$ values reachable, in order to analyze the second-order structure function without being influenced by the horizontal periodicity. The DNS at $Ra = 1.7 \times 10^5$ and at $Ra = 1.0 \times 10^6$ employ $128 \times 129 \times 128$ and $256 \times 129 \times 256$ dealiased modes and polynomials respectively, see Ref. [17] for further details about the simulations and for the validation of the dataset.

## 3. Flow structures in thermally driven turbulence

Let us now briefly describe the topology of the flow. The turbulence regeneration cycle of RBC is composed by coherent thermal and velocity structures which emerge from the chaotic regime, combine themselves and create a sort of persistent machinery. The most relevant coherent structure that can be identified in this cycle is the so-called *thermal plume*, which can be loosely defined as a portion of fluid having a temperature difference with respect to the background [16] (see figure 2($a$)). These structures follow a precise life-cycle [18], in particular hot and cold plumes detach from the bottom and the upper plate respectively, move throughout the fluid layer driven by buoyancy force and finally impinge on the opposite wall triggering the ejection of new plumes [17]. In the core, thermal plumes have a mushroom-like shape whereas close to the walls they have a sheet-like one, see figure 1. Generally speaking, the sheet-like plumes create a fine network across the plates and the mushroom-like plumes are emitted from the intersection spots [19]. Both the ejection and impingement events of plumes take place not at the wall, but approximately on the edge of a conductive layer, made of quiescent fluid, called *thermal*

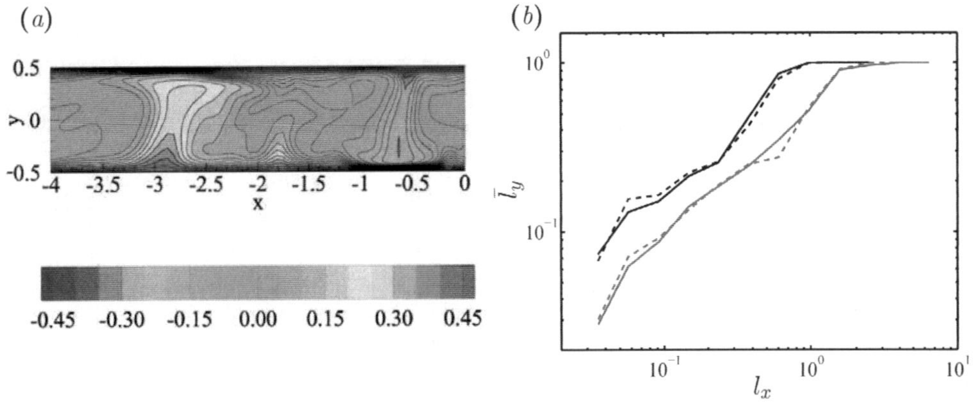


**Figure 2.** ($a$) Vertical section of the fluid domain colored with the temperature at $Ra = 1.7 \times 10^5$. ($b$) Mean wall-normal length, $\bar{l}_y$, as a function of the streamwise lengths of the structures $l_x$. Solid and dashed lines for high- and low-temperature structures respectively. Colors are black for $Ra = 1.7 \times 10^5$ and red for $Ra = 1.0 \times 10^6$.

*boundary layer.* The thermal boundary layer thickness $\delta_\theta$ is usually defined by considering a thickness based on the slope of the mean temperature profile

$$\delta_\theta = \frac{1}{2}\left(\left|\frac{d\Theta}{dy}\right|_{y=y_{wall}}\right)^{-1} = \frac{1}{2Nu}, \tag{2}$$

where $\Theta$ is the mean temperature and $Nu$ is the Nusselt number. The Nusselt number for $Ra = 1.7\times10^5$ is equal to 5.0 whereas for $Ra = 1.0\times10^6$ is equal to 8.2 [17]. It is worth noting that also large-scale circulations are found to characterize thermally driven turbulence with different features according to the geometry of the domain [20, 21, 22]. In doubly-periodic rectangular geometries, the large-scale circulation is found to be composed by ascending and descending plumes locking into a roll-like structure; the latter undergoes a continuous displacement along the horizontal directions $x$, $z$ and exhibits a quite persistence in time [21].

To further address the topology of the flow we use now a thresholding technique similar to the one adopted in [23] and [9] in order to identify scalings of properly defined three-dimensional high- and low-temperature structures. We consider temperature structures the connected region of the flow such that $|\theta'(x, y, z)| > \alpha\theta'_{rms}(y)$, where $\theta'$ is the fluctuating temperature, $\alpha$ a thresholding parameter, and $\theta'_{rms}(y)$ the root-mean-square of the fluctuating temperature as a function of the wall-normal coordinate. As apparent, the resulting set of structures depends on the value of the threshold. By means of a percolation analysis, we select a threshold, $\alpha = 1.5$, which is found to maximize the number of identified structure. Every single object is also classified as a high- or low-temperature structure according to the sign of $\theta'_m$ computed over the domain $\Omega$ of all its constituent points as,

$$\theta'_m = \frac{\int_\Omega \theta'(x, y, z)dV}{\int_\Omega dV}. \tag{3}$$

The above procedure yields about $10^3$ objects for $Ra = 1.7 \times 10^5$ and $4 \times 10^3$ for $Ra = 1.0 \times 10^6$. In order to characterize this population, we circumscribe each structure in a box aligned to the Cartesian axes, whose streamwise, wall-normal and spanwise sizes are denoted by $l_x$, $l_y$ and $l_z$. This procedure allows us to characterize in a more quantitative way the structural properties of the temperature field. Indeed, as expected, we found that these objects are characterized by having roughly the same horizontal lengths, $\bar{l}_x \sim l_z$ and $\bar{l}_z \sim l_x$ where $\bar{\cdot}$ denotes the ensemble average over the population of identified thermal structures. The analysis of the $(l_x, \bar{l}_y)$-space shown in figure 2($b$) reveals the presence of a hierarchy of thermal structures whose sizes satisfy $\bar{l}_y \sim l_x$. The results also show that there are many structures spanning the full height, $\bar{l}_y \sim 1$. It is worth underlining the fact that this specific result might reasonably change when higher Rayleigh numbers are addressed. This analysis can be also extended by considering a tracking of the time evolution of these objects thus fully exploiting the potentiality of the approach.

## 4. The four-dimensional augmented-space of thermally driven turbulence

It is well-known that turbulence is a phenomenon characterized by fluctuations occurring at many space and time scales [24]. Beyond the intuitive idea which arises from observation, a *scale* of turbulence is quite an elusive quantity which is not univocally defined. From a statistical point of view, the scale of a turbulent field can be defined by using second-order structure function as the separation vector between two points and two times. The second-order structure function for a generic quantity $\beta$ can be written as

$$\left\langle \delta\beta^2 \right\rangle = \left\langle \delta\beta\delta\beta \right\rangle , \tag{4}$$

where

$$\delta\beta = \beta'\left(X_{cj} + r_j/2, T_c + \tau/2\right) - \beta'\left(X_{cj} - r_j/2, T_c - \tau/2\right) \tag{5}$$

denotes the increment of the fluctuations $\beta'$ between the point $X_{cj} - r_j/2$ at time $T_c - \tau/2$ and the point $X_{cj} + r_j/2$ at time $T_c + \tau/2$. Thus, the second-order structure function (4) depends on the spatial separation vector $r_i$, on the spatial location of the mid-point $X_{ci}$, on the temporal separation $\tau$ and on the temporal location of the mid-time $T_c$. Due to the statistical symmetries of RBC in both space and time, the dependence on the mid-point $X_{ci}$ reduces simply to the distance from the wall $Y_c = X_{c2}$ whereas the dependence on the mid-time $T_c$ vanishes. Furthermore, by accounting for the statistical isotropy in the horizontal planes, the functional dependence of $\left\langle \delta\beta^2 \right\rangle$ on $(r_x, r_y, r_z, \tau, Y_c)$ reduces to $(r_\pi, r_y, \tau, Y_c)$, where $r_\pi = \sqrt{r_x^2 + r_z^2}$. It is worth noting that, the $r_y$ direction is limited by the presence of the wall, i.e. wall-normal scales cannot extend for a distance greater than twice the distance of the mid-point from the wall, $r_y \leq 2Y_c$ [25].

As previously mentioned, the most relevant coherent structure that can be identified in RBC, for Rayleigh numbers sufficiently above the onset of convection [26], is the so-called thermal plume. In accordance with the definition outlined above, plumes are coherent structures of temperature rather than being structures of momentum, therefore their statistical footprint is mostly marked on observables built with the temperature field. For this reason, we analyze the behaviour of the second-order structure function of temperature $\left\langle \delta\theta^2 \right\rangle$ as a function of $(r_\pi, r_y, \tau, Y_c)$. This statistical observable can be considered a measure of the temperature variance of turbulence at a scale $(r_\pi, r_y, \tau)$ and at a position $Y_c$, thus it will be hereafter called also *scale variance*. For the sake of statistical convergence, two sets of uncorrelated time-series are collected, one for $Ra = 1.7 \times 10^5$ (14 time-series) and another for $Ra = 1.0 \times 10^6$ (7 time-series). Being $2H/U_f$ the dimensional large eddy turnover time and $\tau_\ell = 2$ its non-dimensional value, these time series are $2.5\tau_\ell$ long within which 51 samples are collected to compute two-points and two-times statistics.

## 5. Results

Figure 3($a$) and ($b$) show the isocontours of the scale variance $\left\langle \delta\theta^2 \right\rangle$ in the reduced $(r_\pi, r_y, \tau)$-space for $Y_c = 0.1$, $Y_c = 0.3$ at $Ra = 1.7 \times 10^5$. The planes colored with $\left\langle \delta\theta^2 \right\rangle$ are orthogonal and chosen in such a way that they cross the maxima achieved by $\left\langle \delta\theta^2 \right\rangle$ at $Y_c = 0.1$ and $Y_c = 0.3$

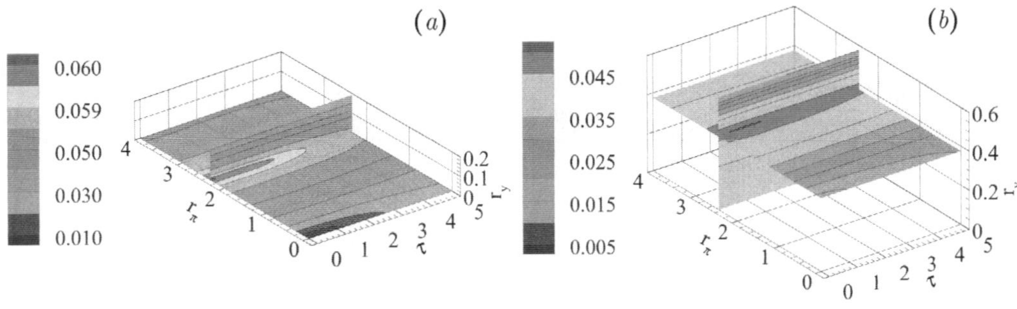

**Figure 3.** Isocontours of $\langle \delta\theta^2 \rangle$ in the reduced $(r_\pi, r_y, \tau)$-space at $(a)$ $Y_c = 0.1$ and $(b)$ $Y_c = 0.3$ for $Ra = 1.7 \times 10^5$.

respectively. Around the origin and for both wall-distances, scale variance starts from zero at $r_\pi = r_y = \tau = 0$ and then increases when augmenting the separations in space and time. The location of the maximum, on the other hand, changes depending on the wall-distance; close to the plate for $Y_c = 0.1$, the scale variance is peaked at $r_\pi = 2.4$ and $r_y = \tau = 0$. On the contrary, further away from the wall at $Y_c = 0.3$, the maximum occurs at well-defined scales $r_\pi = 2.4$, $r_y = 0.4$ and $\tau = 1$. The same qualitative behaviour is observed also for the larger Rayleigh number considered.

In order to set quantitatively the behaviour of scale variance, let us consider the $Y_c$-behaviour of the *characteristic scales* $(\tilde{r}_\pi, \tilde{r}_y, \tilde{\tau})$ defined as the locus of the scale variance maxima at different distances from the plates, i.e.

$$\langle \delta\theta^2 \rangle_{max} (Y_c) = \langle \delta\theta^2 \rangle (Y_c | \tilde{r}_\pi, \tilde{r}_y, \tilde{\tau}) \, .$$

The analysis of $(\tilde{r}_\pi, \tilde{r}_y, \tilde{\tau})$ as a function of $Y_c$ reveals that the horizontal scale $\tilde{r}_\pi$ is almost constant with $Y_c$. In particular, $\tilde{r}_\pi = 2.4$ for $Ra = 1.7 \times 10^5$ and $\tilde{r}_\pi = 2.7$ for $Ra = 1.0 \times 10^6$. The fact that the characteristic horizontal length remains constant with the distance from the wall means that the most intense temperature fluctuations occur at a specific horizontal scale. Furthermore, it sounds reasonable that $\tilde{r}_\pi$ could be indicative of the average distance between an ascending and a descending hot and cold temperature structure. In support of this conjecture, we can see in figure 2$(a)$ that the distance between the hot and the cold plumes for the DNS at $Ra = 1.7 \times 10^5$ is around 2.5, which is quite similar to the correspondent $\tilde{r}_\pi$. An analogous result is reported in Ref. [21], where a symmetry-accounting ensemble-averaging method is employed to identify large-scale rolls in an unbounded RBC at $Ra = 1.0 \times 10^7$ and $Pr = 1$; they found that the distance between ascending and descending fluid is, on average, almost equal to two-times the height of the fluid domain.

Contrary to the horizontal scale, the wall-normal scale is found to be zero close to the plates and to increase linearly with the wall-distance in the core of the flow as shown in figure 4$(a)$ and $(b)$. Let us analyse more in detail the linear behaviour of the wall-normal scale in the core of the flow. The maxima of scale variance for each wall-distance are located near the $(r_y = 2Y_c)$-plane. In particular, we measure that the exact plane where the maxima occur is

$$\tilde{r}_y = 2 \left[ Y_c - \frac{1}{2Nu} \right] ,$$

for both the Rayleigh numbers considered. Hence, these wall-normal scales are representative of thermal fluctuations occurring on scales extending down to a distance from the wall $1/(2Nu)$

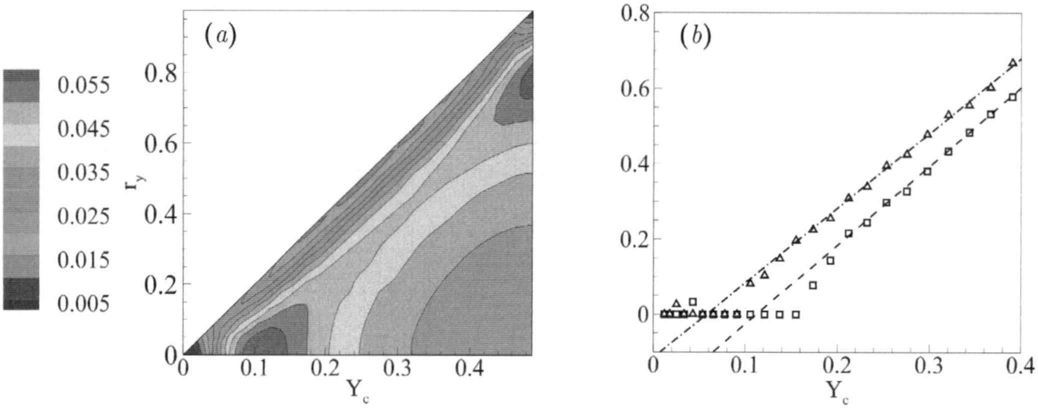


**Figure 4.** ($a$) Section of the space $(r_\pi, r_y, \tau, Y_c)$ in the $(r_\pi = 2.4, \tau = 0)$-plane colored with $\langle \delta\theta^2 \rangle$ for $Ra = 1.7 \times 10^5$. ($b$) Characteristic wall-normal scale $\tilde{r}_y(Y_c)$ as a function of $Y_c$ for $Ra = 1.7 \times 10^5$ (squares) and $Ra = 1.0 \times 10^6$ (triangles). The equation of the dashed line is $2.1(Y_c - 0.12)$ whereas the equation of the dashed-dot line is $2.0(Y_c - 0.06)$.

which is exactly the non dimensional thermal boundary layer thickness defined in (2). Hence, we can claim that the most intense thermal fluctuations are *attached* to the thermal boundary layer. This behaviour may be explained in terms of dynamics of turbulent structures in the following way. Thermal plumes detach from the near-wall region and span the entire domain impinging on the opposite wall but retaining their roots attached to the original wall. On the other hand, both the ejection and the impingement of plumes take place not at the wall, but approximately on the edge of a conductive layer, made of quiescent fluid, the thermal boundary layer. Thus, the overall picture is consistent with attached thermal structure spanning the entire convection cell from one thermal boundary layer to the opposite one.

It is finally interesting to note that, as shown in figure 4, two well-defined peaks of scale variance (colored in red) are present around $Y_c = 1/(2Nu) = 0.1$ for $r_y = 0$, $r_\pi = 2.4$ and $Y_c = 0.5$ for $r_y = 2\,[Y_c - 1/\,(2Nu)]$, $r_\pi = 2.4$, respectively on top of the thermal boundary layer and at the center of fluid layer. Both these two peaks appear to be a footprint of the same temperature structure and to be strongly related with the inhomogeneity of the flow in the vertical direction. In particular, the temperature structure is again a couple of ascending (hot) and descending (cold) fluid statistically separated in the horizontal direction by $r_\pi = 2.4$ and these two peaks are a statistical evidence of the impinging/ejection events. Indeed, as sketched in figure 5, on top of the thermal boundary layer, $Y_c = 1/(2Nu) = 0.1$, the horizontal scale of the maximum is $r_\pi = 2.4$ while the wall-normal scale is zero because at this position non-zero increments, $r_y \neq 0$, would involve the temperature field closer to the plate and, hence, outside the thermal plume. On the contrary, at the channel center, $Y_c = 0.5$, the statistical occurrence of impinging and ejection events of couples of hot/cold plumes appear again for $r_\pi = 2.4$ but for $r_y = 2\,[Y_c - 1/\,(2Nu)]$ which involve exactly the two points lying on top of the two opposite thermal boundary layer where the impinging/ejection events occur, see again sketch 5. As previously mentioned, these aspects are strongly related with the inhomogeneity of the flow. Indeed, by considering increments only in the wall-normal scales, the second-order structure function of temperature can be written as

$$\langle \delta\theta^2 \rangle (r_y, Y_c) = \langle \theta'^2 \rangle (Y_c + r_y/2) + \langle \theta'^2 \rangle (Y_c - r_y/2) - 2 \langle \theta' (Y_c + r_y/2)\, \theta' (Y_c - r_y/2) \rangle\,, \quad (6)$$

thus highlighting the different contributions to the $r_y$ behaviour. It is clear that for large

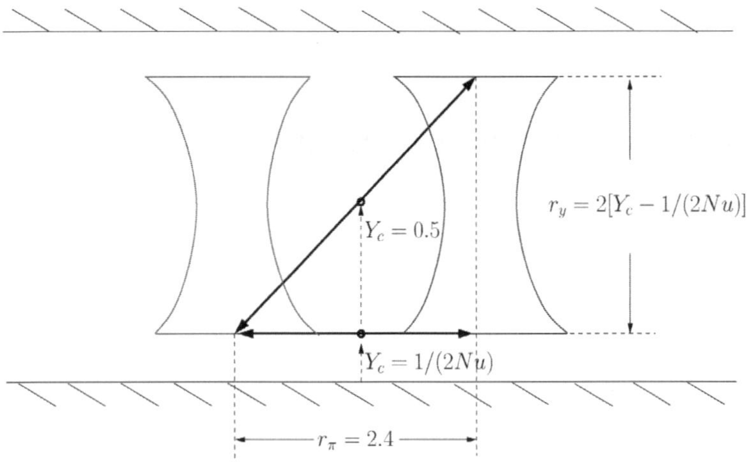


**Figure 5.** Sketch of the statistical occurrence of ascending (hot) and descending (cold) fluid and of the increments involved in the two maxima of the second-order structure function of temperature occurring on top of the thermal boundary layer $Y_c = 1/(2Nu) = 0.1$ and at the cell center $Y_c = 0.5$. The value $r_\pi = 2.4$ is relative to the $Ra = 1.7 \times 10^5$ case.

separation, the inhomogeneity strongly influences the space of wall-normal scales by setting the locus of the maxima of $\langle \delta\theta^2 \rangle$ in the plane of attached scales, $r_y = 2\left[Y_c - 1/(2Nu)\right]$, via the the spatial distribution of temperature variance given by the first two terms of (6). Since, the maxima of $\langle \delta\theta^2 \rangle$ in the $r_y = 2\left[Y_c - 1/(2Nu)\right]$-plane are here linked with the statistical occurrence of ascending (hot) and descending (cold) plumes being occurring for $r_\pi = 2.4$, we can conjecture that the location of the maximum of temperature variance $\langle \theta^2 \rangle$ at $Y_c = 1/(2Nu)$ and the thickness of the thermal boundary layer $\delta_\theta$ are settled by the impinging/ejection events of thermal plumes.

Let us now focus on the time scale behaviour of scale variance. In figure 6($a$) and ($b$) the isocontours of $\langle \delta\theta^2 \rangle$ in the $(r_\pi = 2.4, Y_c = 0.1)$-plane and $(r_\pi = 2.4, Y_c = 0.3)$-plane are shown respectively. As previously noted, close to the plates, the maximum of scale variance occurs for $r_\pi = 2.4$ and for $r_y = \tau = 0$. Increasing both $r_y$ and $\tau$ a decrease of $\langle \delta\theta^2 \rangle$ is observed. On the contrary, further away from the walls, characteristic scales are observed also in the wall-normal and time scales being present a maximum at attached wall-normal scales and for $\tau > 0$. However, let us notice that while the shape of this maximum is peaked in the wall-normal scales, the temporal behaviour is very smooth thus leading to significant uncertainties on the measurement of $\tilde{\tau}$. The behaviour of the characteristic time scales $\tilde{\tau}(Y_c)$ as a function of the wall-distance is shown in figure 6($c$). In the near-wall region no characteristic time scale is defined being $\tilde{\tau}(Y_c) = 0$ in this region of the flow. This aspect actually suggests that the impinging events on top of the thermal boundary layer statistically causes ejection events at the same time, $\tau = 0$, at an horizontal distance $r_\pi = 2.4$, see again sketch 5. Then, as expected, a characteristic time scale, $\tilde{\tau} > 0$, is observed further away from the wall where the maxima of $\langle \delta\theta^2 \rangle$ show an attached behaviour in the wall-normal scales. This time scale is a statistical measure of the time needed by the ejection event created on top of the thermal boundary layer by an impinging event, to reach the wall-distance $y = Y_c + r_y/2$, see sketch 5. In particular, we measure that to reach the cell center $y = Y_c + r_y/2 \sim 0.5$, the thermal structure statistically spends 2 large eddy turnover time $\tau_\ell$. Then, for $Y_c > 0.33$, the time scale $\tilde{\tau}$ drops to zero. This aspects can be associated to the fact that the impinging event that created the ejection statistically vanishes

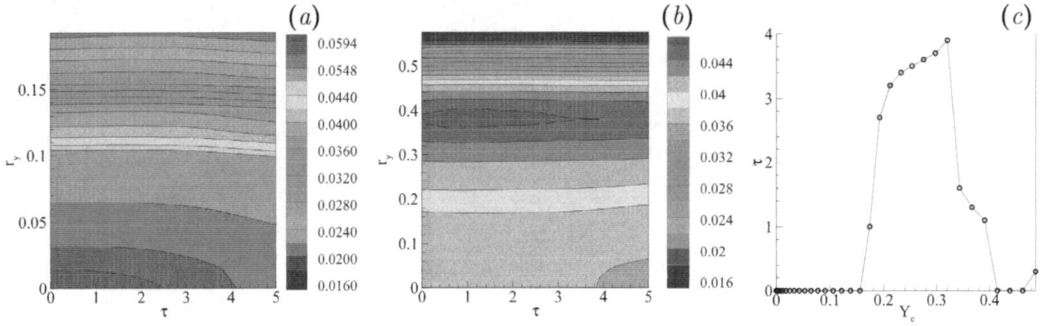

**Figure 6.** Section of the space $(r_\pi, r_y, \tau, Y_c)$ in the $(a)$ $(r_\pi = 2.4, Y_c = 0.1)$-plane and $(b)$ $(r_{pi} = 2.4, Y_c = 0.3)$-plane colored with $\langle \delta\theta^2 \rangle$ for $Ra = 1.7 \times 10^5$. $(c)$ Characteristic time scale $\tilde{\tau}(Y_c)$ as a function of $Y_c$ for $Ra = 1.7 \times 10^5$.

after 2 large eddy turnover times which is the time employed by the ascending plume to reach the wall-distance $y = Y_c + r_y/2 \sim 0.5$. However, further investigation accounting for a longer range of time scales is mandatory to draw a conclusion with any certainty. Indeed, it would be also possible that, at these wall-distances, the characteristic time scale exceeds the maximum time scale here analysed.

## 6. Conclusions

Direct Numerical Simulation data of turbulent Rayleigh-Bénard convection are analysed for the first time in terms of second-order structure function of the temperature field in the complete four-dimensional augmented space of spatial and temporal scales and wall-distances. The observed behaviour for the second-order structure function is then clarified from a dynamical point of view by means of the commonly accepted model of life-cycle for the thermal structures. The second-order structure function shows marked maxima for a fixed horizontal length scale at different wall-distances that is found to be related with the average distance between a couple of ascending (hot) and descending (cold) thermal structures. This length is Rayleigh dependent and is found to be $r_\pi = 2.4$ and $2.7$ for $Ra = 1.7 \times 10^5$ and $Ra = 1.0 \times 10^6$, respectively. The life-cycle of thermal structures is found to leave a clear footprint also in the wall-normal scales. Indeed, we found that $\tilde{r}_y = 2\,[Y_c - 1/(2Nu)]$ highlighting the fact that the most intense thermal fluctuations occur at scales extending down to the edge of the thermal boundary layer in agreement with the commonly accepted model for the dynamics of plumes [18]. We also found that the statistical occurrence of an impinging event on top of the thermal boundary layer is related with the simultaneous appearance of an ejection event at an average horizontal distance $r_\pi = 2.4$ and $2.7$ respectively for the $Ra = 1.7 \times 10^5$ and $Ra = 1.0 \times 10^6$ cases. The resulting ejection of fluid is found to spend 2 large eddy turnover time $\tau_\ell$ to reach the cell center. Afterwards, the impinging event which created it vanishes thus highlighting that the life-time of impinging is 2 large eddy turnover time $\tau_\ell$. The results presented in this work surely justify a deeper analysis of the problem and the very next step will be investigating more turbulent systems at higher Rayleigh numbers.

### Acknowledgments

The authors would like to acknowledge the support of the European Research Council, which funded the Multiflow summer program within which this work was developed.

## References

[1] Richardson L F 1922 Weather prediction by numerical process (Cambridge U. Press)
[2] Kolmogorov A N 1941 Dissipation of energy in locally isotropic turbulence *Dokl. Akad. Nauk SSSR* **32** 19–21
[3] Pope S B 2000 Turbulent flows (Cambridge U. Press)
[4] Monin A S and Yaglom A M 1971 Statistical Fluid Mechanics: Mechanics of Turbulence, Volume 1 (MIT Press, Cambridge, MA)
[5] Townsend A A 1976 The structure of Turbulent Shear Flow (Cambridge U. Press)
[6] Chong M S, Perry A E and Cantwell B J 1990 A general classification of three-dimensional flow fields *Phys. Fluids A* **2** 765–777
[7] Domaradzki J A and Rogallo R S 1990 Local energy transfer and nonlocal interactions in homogeneous, isotropic turbulence *Phys. Fluids A* **2** 413–426
[8] Cimarelli A and De Angelis E 2014 The physics of energy transfer toward improved subgrid-scale models *Phys. Fluids* **26** 055103
[9] Lozano-Durán A and Jiménez J 2014 Time-resolved evolution of coherent structures in turbulent channels: characterization of eddies and cascades *J. Fluid Mech.* **759** 432–471
[10] Reynolds, O. 1894 On the dynamical theory of incompressible viscous fluids and the determination of the criterion *Proc. R. Soc. Lond.* **56** 40–45
[11] Jiménez J 2013 Near-wall turbulence *Phys. Fluids* **25**, 101302.
[12] Willmarth W W and Lu S S 1972 Structure of the Reynolds stress near the wall *J. Fluid Mech.* **55** 65–92
[13] Blackwelder R F and Kaplan R E 1976 On the wall structure of the turbulent boundary layer *J. Fluid Mech.* **76** 89–112
[14] Berkooz G, Holmes P and Lumley J L 1993 The proper orthogonal decomposition in the analysis of turbulent flows *Annu. Rev. Fluid Mech.* **25** 539–575
[15] Jeong J, Hussain F, Schoppa W and Kim J 1997 Coherent structures near the wall in a turbulent channel flow *J. Fluid Mech.* **332** 185–214
[16] Chillà F and Schumacher J 2012 New perspectives in turbulent Rayleigh-Bénard convection *Eur. Phys. J. E* **35** 1–25
[17] Togni R, Cimarelli A and De Angelis E 2015 Physical and scale-by-scale analysis of Rayleigh-Bénard convection *J. Fluid Mech.* **782** 380–404
[18] Kadanoff L P 2001 Turbulent heat flow: structures and scaling *Phys. Today* **54** 34–39
[19] Zhou Q, Sun C and Xia K Q 2007 Morphological evolution of thermal plumes in turbulent Rayleigh-Bénard convection *Phys. Rev. Lett.* **98** 074501
[20] Niemela J J, Skrbek L, Sreenivasan K R and Donnelly R J 2001 The wind in confined thermal convection *J. Fluid Mech.* **449** 169–178
[21] Van Reeuwijk M, Jonker H J J and Hanjalić K 2005 Identification of the wind in Rayleigh-Bénard convection *Phys. Fluids* **17** 051704
[22] Bailon-Cuba J, Emran M S and Schumacher J 2010 Aspect ratio dependence of heat transfer and large-scale flow in turbulent convection *J. Fluid Mech.* **655** 152–173
[23] Lozano-Durán A, Flores O and Jiménez J 2012 The three-dimensional structure of momentum transfer in turbulent channels *J. Fluid Mech.* **694** 100–130
[24] Frish U 1995 Turbulence: the legacy of A. N. Kolmogorov (Cambridge U. Press)
[25] Cimarelli A, De Angelis E, Talamelli A, Casciola C M and Jiménez J 2014 The attached reverse and detached forward cascades in wall-turbulent flows *J. Phys. Conf. Seri.* **506** 012005
[26] Ahlers G 2009 Trend: Turbulent convection *Physics* **2** 74 (Am. Phys. Soc.)

Journal of Physics: Conference Series **708** (2016) 012008 doi:10.1088/1742-6596/708/1/012008

# Reynolds Stress Structures in the Hybrid RANS/LES of a Planar Channel

**Jack Weatheritt**

Post-doctoral Researcher, Department of Mechanical Engineering, University of Melbourne, Australia

E-mail: jack.weatheritt@unimelb.edu.au

**Richard Sandberg**

Professor, Department of Mechanical Engineering, University of Melbourne, Australia

**Adrian Lozano-Durán**

Post-doctoral Researcher, School of Aeronautics, Universidad Politécnica de Madrid, Spain

**Abstract.** The near-wall cycle of hybrid RANS/LES is studied by calculating the flow through a planar channel. Statistical results are commented on and related to instantaneous structures which are extracted from the flow field. The problematic structures in the artificial near-wall cycle, well known to be super-streaks, are identified and quantified. The calibration of such closures provides a correct mixing length argument in the logarithmic layer. However because of these overly intense streamwise streaks, it is impossible to simultaneously predict the Reynolds streamwise normal and shear stress components correctly. Further, because the location of the RANS-LES interface changes spatially and temporally, we see these structures are more free to move vertically and this further worsens statistical results.

## 1. Introduction

Wall-bounded flows are of great engineering importance. Unfortunately, near-wall regions are the most expensive to explicitly resolve in numerical simulations. The original concept behind combining Reynolds-Averaged Navier-Stokes (RANS) modelling with Large Eddy Simulation (LES) is to ease the near-wall grid requirements of the latter. The paradigm is often stated as, 'perform RANS where required and LES where possible.' This is often understood, with respect to simple wall-bounded flows, as treating the boundary layer with RANS and switching to LES in the outer layer. However, as shall be seen, this is not the case. The near-wall is in reality solved by some kind of unsteady RANS, which contains resolved fluctuations, operating in conditions much different from the original purpose. These fluctuations are well known to be large streaks that are non-physical in nature, however are necessary as boundary conditions to the LES region. Work by Jiménez *et al.* [1, 2] shows that, in simulations, the near-wall cycle of the small scales is reasonably localised and unaffected by the outer layer. This at least provides some hope for hybrid RANS/LES as a philosophy. Resolving large streaks whilst modelling this near-wall cycle is a possibility, although currently far from reality. This paper uses arguably the most popular near-wall approach [3] to show that current mechanisms do not properly account for the break up of these streaks, in particular, the artificial buffer layer formed by the delay in recovering LES seriously affects statistical results. These findings are reasonably well known, however this paper provides more quantative information regarding these structures.

Chapman [4] estimated that an LES of a flat plate boundary layer requires a total number of grid points $N$ proportional to the Reynolds number $\mathrm{Re}_{L_x}$ via the relation

$$N \sim \mathrm{Re}_{L_x}^{9/5}, \tag{1}$$

which is only marginally better than Direct Numerical Simulation (DNS). The exponent was later revised by Choi and Moin [5] to be 13/7. $L_x$ is the length of the flat plate. Strong dependence on and poor scaling with Reynolds number puts the majority of industrially relevant geometries out of reach for LES.

Work on wall-modelled LES started with the seminal work of Schumann [6] and Deardorff [7] who derived equilibrium laws to relate the wall shear stress with the velocity in the core, assuming a universal law of the wall. These laws made it possible to relax the near-wall grid requirements and good results [6] showed the feasibility of wall modelled LES. Note that Deardorff [7] used a resolution much too low in the outer layer, reducing the solution quality.

This kind of wall model is only possible when a universal law of the wall can be assumed such that the only feasible applications are simple geometries like pipes and channels. From this seminal work, varied hybrid RANS/LES methodologies have been developed that extend the wall-modelled LES philosophy. Broadly speaking, efforts can be categorised as: wall function approaches (above paragraph), two-layer models and manual/implicit controlling of the RANS-LES interface by modifying the length scale [8], subgrid viscosity [9] or turbulent kinetic energy [10]. This control of the RANS/LES switch can be achieved either by blending between turbulence and subgrid models [11, 12]; or by a unified model, where only one turbulence model is used and its contribution has to be damped to mimic a subgrid closure [8, 13, 14, 15]. Note, for completeness we mention the zonal method [16, 17] which employs two distinct regions, one LES and one RANS. The interface between the two zones is prescribed by the user and turbulent fluctuations must be generated as a boundary condition for the LES. This interface can be either perpendicular to the wall [18] or parallel [19]. The former is used when near-wall interactions are deemed crucial to the overall flow prediction and the latter when computational power cannot be spared to resolve the boundary layer.

Two-layer models, first proposed by Balaras and Benocci [20], use a coarse grid LES up to the wall, but embed a second finer grid for solving the near-wall region with the boundary layer equations, closed with an algebraic eddy viscosity. The Poisson equation is not solved near the wall and the wall normal velocity is obtained by enforcing mass conservation across the interface. The wall stress components obtained from this inner layer are then applied as a boundary condition to the LES. This additional inner layer only marginally increases the cost of the computation and obtains accurate results in planar channel flow.

Spalart *et al.* [21] and Speziale [8] essentially devised the same philosophy of damping the length scales in a RANS model, although their approaches were not formally described as such. These two approaches became known as Detached Eddy Simulation (DES) and Flow Simulation Methodology (FSM). DES contains an explicit switch between the RANS and LES modes,

$$\ell^{\mathrm{des}} = \min(y, C_{\mathrm{des}}\Delta) \tag{2}$$

where $\ell^{\mathrm{des}}$ replaces the dissipation length scale in the Spalart-Allmaras turbulence model [22], $\Delta$ is the grid length scale, $y$ is the wall distance and $C_{\mathrm{des}} = 0.65$. The model becomes an 'LES' when $C_{\mathrm{des}}\Delta < y$, increasing the model dissipation and thus lowering the effective viscosity to a level appropriate for the subgrid scales. This idea is easily generalisable to other turbulence models,

$$\ell^{\text{hyb}} = \min(\ell_t, \ell^{\text{sgs}}), \tag{3}$$

where $\ell_t$ is the RANS length scale and $\ell^{\text{sgs}}$ is the subgrid length scale, which in traditional DES is defined as $\ell^{\text{sgs}} = C_{\text{des}}\Delta$. The length $\ell^{\text{hyb}}$ is general notation for the modified scale, denoting the hybrid length scale which replaces all occurrences of $\ell_t$ in a given RANS model. FSM on the other hand dos not have such a treatment and the switch is made continuously by damping the RANS length scale. This concept is detailed in Section 2.

Equation 3 works well when the streamwise $\Delta_x$ and spanwise $\Delta_z$ grid spacings are much bigger than the boundary layer thickness. Such a constraint forces the closure to act in RANS mode near the wall, through the sensitivity in $\Delta$. Even with a $\Delta_x^+ > 8000$, Nikitin *et al.* [23] showed that turbulence is sustained in the bulk of a channel flow, lending support for the method. Despite this, hybrid RANS/LES has been shown by numerous authors to be sensitive to the streamwise grid spacing [3, 21, 23, 24, 25]. It should be noted that most of the analysis has been performed using DES; this does not reflect its shortcomings relative to other hybrid methodologies, rather the opposite. Early success of DES saw it become the industry standard in the 2000s and so the deficiencies of hybrid RANS/LES were discovered with this technique first.

When an aggressive streamwise spacing is not used, the switch between RANS-LES is pushed deep into the inner layer. This is because the cell Reynolds number is reduced and resolved eddies are allowed to form. This lowers the modelled stress contribution which is not compensated by resolved turbulence. This phenomena has become known as Model Stress Depletion (MSD) and in some cases has been shown to cause premature separation [25]. Spalart *et al.* [3] argue that this was slightly overstated but the sensitivity to grid spacing was acknowledged. Further, the near-wall, being a RANS solution, contains its own log-layer several wall units shy of the correct value, due to the model being artificially suppressed. Once LES takes over, the resolved stress dominates and a new log-layer is formed several wall units too big. This discrepancy, also called the log-layer mismatch, is amplified as the Reynolds number is increased [26]. Baggett [24] argues from channel flow results that both MSD and the log-layer mismatch are produced by the 'artificial' near-wall turbulence cycle characterised by large streamwise streaks, which is entirely correct, although he argues that this is resulting from resolved structures being entirely parametrised by the grid. This conclusion however was the result of a low resolution in the spanwise direction, and this author has numerous experiences of observing large-scale streaks on much finer meshes.

In such unified modelling, methods have surfaced that more directly control the RANS-LES interface to reduce the grid sensitivity of the switch. Piomelli *et al.* [27] introduced a backscatter method in the inner layer of DES, which helped in the breakdown of the large streamwise streaks. Introduced as a proof of concept only, with no physical justification to the forcing term, the modelled log-layer was removed somewhat.

In recent advances of the Partially-Averaged Navier-Stokes (PANS) method [28], wall functions have been revisited in a VLES approach [29]. The method blends integration to the wall with equilibrium wall functions [30]. The planar channel flow is very well predicted for $Re_\tau = 395$ on a fine mesh, with no evidence of a log-layer mismatch. This is unsurprising however, as the model contribution was very low and as such the modelled log-layer is absorbed fully into the buffer layer.

Spalart *et al.* [3] detailed the Delayed Detached Eddy Simulation (DDES) formulation which further controls the length scale, expanding on Eq. 3. Weinmann *et al.* [13] used a similar function that effectively acts on the model dissipation rate $\varepsilon$. The former is the subject of this paper, in an effort to assess the near-wall behaviour of unified models.

The DDES method uses the equation,

$$\ell^{\text{hyb}} = \ell_t - f_d \min\left(0, \ell_t - \ell^{\text{sgs}}\right), \tag{4}$$

where for the standard DDES approach, $\ell^{\text{sgs}} = C_{\text{des}}\Delta$. $f_d$ is given as,

$$f_d = 1 - \tanh(24 r_d^3), \tag{5}$$

with

$$r_d = \frac{\nu_{\text{sgs}} + \nu}{\sqrt{\partial_{x_j} u_i \partial_{x_j} u_i \kappa^2 y^2}}. \tag{6}$$

$\kappa = 0.41$ is the von Kármán constant and $r_d$ controls the switch from RANS to LES, based on the effective viscosity. The length scale is then adjusted using Eq. 4.

This report is as follows. Section 2 details the hybrid RANS/LES methodology. This is formally an FSM approach, however the near-wall switching mechanism of DDES has been adopted. Consequently the methodology contain within this report is referred to as FSM. In Section 3, this modified FSM is applied to turbulent channel flow. Whilst results are reasonable, the discrepancies in the artificial buffer layer are linked to resolved structures originating in this region. Section 3.2 quantatively looks at these structures. Finally Section 4 is a summary that briefly mentions one possible remedy.

## 2. Hybrid RANS/LES Methodology

The unified framework adopted in this work is the latest incarnation of the FSM blending philosophy [8, 13]. The grid-filtered momentum and continuity equations for an incompressible fluid,

$$\partial_t \widetilde{u}_i + \widetilde{u}_j \partial_{x_j} \widetilde{u}_i = -\partial_{x_i} \widetilde{p} + \nu \partial_{x_j}^2 \widetilde{u}_i - \partial_{x_j} \tau_{ij}^{\text{hyb}}, \tag{7}$$

$$\partial_{x_i} \widetilde{u}_i = 0, \tag{8}$$

are solved. $\widetilde{(\cdot)}$ is the grid-filter and $\tau_{ij}^{\text{hyb}}$ is the model term that must pick up the subgrid scales caused by this cut-off. In the coarse grid limit, $\tau_{ij}^{\text{hyb}} \to \tau_{ij}^{\text{rans}}$, no turbulence may be explicitly resolved and the grid-filter satisfies the definition of a Reynolds operator. As the grid scale $\Delta$ becomes sufficiently small, say $\Delta \sim \ell_k$ ($\ell_k$ being the Kolmogorov length scale), then $\tau_{ij}^{\text{hyb}} \to 0$ and Eq. 8 reduces to the Navier-Stokes. Thus it is the philosophy of unified hybrid RANS/LES to operate as a DNS, RANS or anywhere in between; depending on the local values of $\Delta$ and $\ell_k$.

To respect the coarse grid RANS limit, $\tau_{ij}^{\text{hyb}}$ is a modified RANS turbulence model. Away from this limit, the RANS model acts as in subgrid mode and the formulation can be thought of as a non-classical LES. Consider the classical RANS stress-strain relationship,

$$\tau_{ij}^{\text{rans}} = \frac{2}{3} k \delta_{ij} - 2\nu_t S_{ij} + a_{ij}^x, \tag{9}$$

where $k$ is the turbulent kinetic energy, $\nu_t$ the eddy viscosity and $S_{ij}$ the mean strain rate. By setting the extra anisotropy $a_{ij}^x$ to 0, one recovers the linear Boussinesq approximation. In general $a_{ij}^x$ is a non-linear polynomial in the velocity gradient tensor. For more detail see the work of Pope [31] or Speziale *et al.* [32].

Equation 9 can be coerced into a hybrid formulation by,

$$\tau_{ij}^{\text{hyb}} = \frac{2}{3} k \delta_{ij} - 2\nu_{\text{sgs}} S_{ij} + F a_{ij}^x, \tag{10}$$

where the subgrid viscosity is given as,

$$\nu_{\text{sgs}} = F\nu_t \sim F \cdot (u_t \ell_t) = u_t \cdot (F\ell_t) = u_t \ell^{\text{hyb}}, \tag{11}$$

where $\ell_t$ and $u_t$ are characteristic velocity and length scales of the largest unresolved motions. Note, $k$ is now the turbulent kinetic energy of the unresolved scales. $F$ is a damping function that is responsible for reducing $\ell_t$ to be characteristic of the smallest resolved eddies. Scales below $\ell^{\text{hyb}}$ are therefore not resolved and $\nu_{\text{sgs}}$ is appropriately defined as a subgrid viscosity. Note, $u_t$ is not damped by F because the unsteady flow field is already a characteristic velocity scale of the resolved motions.

By similar arguments, consider the dissipation rate $\varepsilon$ in first a RANS framework,

$$\varepsilon_{\text{rans}} \sim \frac{u_t^3}{\ell_t}. \tag{12}$$

Then, for the hybrid formulation,

$$\varepsilon \sim \frac{u_t^3}{F\ell_t}. \tag{13}$$

Therefore to formulate a unified hybrid model from RANS, one may simply define

$$\ell^{\text{hyb}} = F\ell_t \tag{14}$$

in place of all instances of $\ell_t$ throughout a RANS closure. Note that this is a key difference between the DES and FSM philosophies, the former only damps $\ell_t$ in the dissipation term.

The damping function $F$ is the crucial element for success. Apart from being bound between 0 and 1, the DNS and RANS limits, there is very little one can write down. We assume that $F$ is a function of the turbulent scales $\ell_k$ and $\ell_t$ and also $\Delta$. This is intuitive; through this functional dependancy, $F$ can assess the local and instantaneous turbulence and decide if the grid resolution is sufficient. The grid-filter is mesh dependent and a formal definition does not exist, therefore no derivation exists of a fully analytical $F = F(\Delta, \ell_t, \ell_k, \dots)$. One must always begin a derivation from assumptions, which do not hold for every situation. For example, by considering Eq. 3 one can define the (D)DES damping function away from solid boundaries as $F = 0.65\Delta/\ell_t$. Clearly if $\Delta \approx \ell_t$, one desires the model in RANS mode, however the constant prevents this — $\Delta \approx \ell_t \implies F \approx 0.65 < 1$. Therefore on coarse meshes DES acts as a poorly calibrated RANS model.

To alleviate this, Weatheritt and Sandberg [15] used an evolutionary algorithm to create a functional form not from assumptions, but from physical flow data. This bases the damping function on scenarios that actually transpired, not on ones assumed to occur. This damping function is used throughout this work and is given as,

$$F = \min\left[\log\left(1 + c_1\frac{\Delta}{\ell_k}\right), 1.0\right] \min\left[c_2\frac{\Delta}{\ell_t}, 1.0\right], \tag{15}$$

with the constants $c_1 = 0.75$ and $c_2 = 2.1$. In order to make use of the DDES near-wall switch, Eq. 4 is used as the definition of $\ell^{\text{hyb}}$, in favour of Eq. 14, where we define $\ell^{\text{sgs}} = F\ell_t$. This implies that as $y \to \infty$, $\ell^{\text{hyb}} = F\ell_t$ and as $y \to 0$, $\ell^{\text{hyb}}$ is found by Eq. 4 which includes the delayed switch to LES feature discussed in Section 1.

The rest of this section details the specific RANS model used in this work. The authors choose the pressure-strain correlation of Speziale, Sarkar and Gatski [32], as the starting point for the turbulence model, because it is derived for homogeneous turbulence. In other words an assumption of slowly changing mean shear seems reasonable for the subgrid scales. This model is then inserted into an explicit algebraic equation for the Reynolds stress. The model is defined by Eq. 10. The extra anisotropy is found from,

$$a_{ij}^x = \beta_2\tau^2\left(\widetilde{S}_{ik}\widetilde{S}_{kj} - \frac{1}{3}\widetilde{S}_{mn}\widetilde{S}_{nm}\delta_{ij}\right) + \beta_4\tau^2\left(\widetilde{S}_{ik}\widetilde{\Omega}_{kj} - \widetilde{\Omega}_{ik}\widetilde{S}_{kj}\right) \tag{16}$$

where $\Omega_{ij}$ is the rotation rate tensor and $\tau = (1/\beta^*\omega)$ is the turbulent time scale. The coefficients $\beta$ and the specific dissipation rate $\omega$ will be defined below. The subgrid viscosity required for Eq. 10 is defined as,

$$\nu_{\text{sgs}} = -\frac{1}{2}F\beta_1 k\tau. \tag{17}$$

The dissipation rate $\varepsilon$ is evaluated from

$$\varepsilon = \frac{k^{3/2}}{F\ell_t}, \tag{18}$$

where the characteristic length scale is $\ell_t = k^{1/2}/\beta^*\omega$.

The coefficients $\beta$ are respectively,

$$\beta^* = 0.09, \quad \beta_1 = -A_1N/Q, \quad \beta_2 = 2A_1A_2/Q, \quad \beta_4 = -A_1/Q. \tag{19}$$

Where the constants $A$ are,

$$A_1 = 1.22, \quad A_2 = 0.47, \quad A_3 = 0.88, \quad A_4 = 2.37. \tag{20}$$

$Q$ is found from,

$$Q = N^2 - 2\tau^2\widetilde{\Omega}_{mn}\widetilde{\Omega}_{nm} - \frac{2}{3}\tau^2A_2^2\widetilde{S}_{mn}\widetilde{S}_{nm}, \tag{21}$$

where $N$ is,

$$N = \begin{cases} \frac{A_3}{3} + (P_1 + \sqrt{P_2})^{1/3} + (P_1 - \sqrt{P_2})^{1/3}, & P_2 \geq 0 \\ \frac{A_3}{3} + 2(P_1^2 - P_2)^{1/6}\cos\left[\frac{1}{3}\arccos\left(P_1(P_1^2 - P_2)^{-1/2}\right)\right], & P_2 < 0, \end{cases} \tag{22}$$

and the $P$,

$$P_1 = \left[\frac{A_3^2}{27} + \left(\frac{A_1A_4}{6} - \frac{2}{9}A_2^2\right)\tau^2\widetilde{S}_{mn}\widetilde{S}_{nm} - \frac{2}{3}\tau^2\widetilde{\Omega}_{mn}\widetilde{\Omega}_{nm}\right]A_3, \tag{23}$$

$$P_2 = P_1^2 - \left[\frac{A_3^2}{9} + \left(\frac{A_1A_4}{6} + \frac{2}{9}A_2^2\right)\tau^2\widetilde{S}_{mn}\widetilde{S}_{nm} + \frac{2}{3}\tau^2\widetilde{\Omega}_{mn}\widetilde{\Omega}_{nm}\right]^3. \tag{24}$$

The relations for $N$, $P$ and $Q$ are the result of assuming the anisotropy tensor $a_{ij} = \tau_{ij}^{\text{hyb}} - 2/3k\delta_{ij}$ is a function of $S_{ij}$, $\Omega_{ij}$ and $P_k/\varepsilon$. $P_k = -\tau_{ij}^{\text{hyb}}\partial_{x_j}\widetilde{u}_i$ is the production of turbulent kinetic energy. By further assuming two dimensional mean flow, the tensor basis required to span $a_{ij}$ reduces to just three components. Finally, because $a_{ii} = 0$, there are only two non-zero scalar invariants. $N$ is then an algebraic approximation of the ratio $P_k/\varepsilon$. For a full and guided derivation see the detailed work of Wallin and Johansen [33].

Finally, to close the equation set, transport equations for $k$ and $\omega$ are required. These are written down in the formulation of the Menter Shear Stress Transport (SST) [34] model, modified by Hellsten [35]

$$\partial_t k + \widetilde{u}_j\partial_{x_j}k = P_k - \varepsilon + \partial_{x_j}\left[(\nu + \sigma_k\nu_t)\partial_{x_j}k\right] \tag{25}$$

$$\partial_t\omega + \widetilde{u}_j\partial_{x_j}\omega = \gamma\frac{\omega}{k}P_k - \zeta\omega^2 + \partial_{x_j}\left[(\nu + \sigma_\omega\nu_t)\partial_{x_j}\omega\right] + \sigma_d CD_{k\omega}^+. \tag{26}$$

Left to be defined is,

$$
\begin{aligned}
CD_{k\omega} &= \partial_{x_j} k \partial_{x_j} \omega, & CD^{+}_{k\omega} &= \max\left[CD_{k\omega}, 0\right], \\
\psi &= \psi_2 + F_1(\psi_1 - \psi_2), & F_1 &= \tanh\left(1.5\Gamma^4\right) \\
\Gamma &= \min\left[\max\left[\frac{k^{\frac{1}{2}}}{\beta^* \omega y}, \frac{500\nu}{y^2\omega}\right], \varphi\right], & \varphi &= \frac{20k}{\max\left[y^2 CD_{k\omega}, 200k_\infty\right]}.
\end{aligned}
\tag{27}
$$

$\psi$ is a coefficient in Eqs. 25-26 that has two definitions, defined by a subscript 1 or 2. Each $\psi$ is blended between two values $\psi_1$ and $\psi_2$, such that $\psi_1$ is active near the wall and $\psi_2$ far from it. These coefficients, along with $\sigma_k$, are given as,

$$
\begin{aligned}
&\sigma_{\omega 1} = 0.53, \quad \sigma_{\omega 2} = 1.00, \quad \sigma_{d1} = 1.0, \quad \sigma_{d2} = 0.4, \quad \sigma_k = 1.1, \\
&\gamma_1 = 0.518, \quad \gamma_2 = 0.44, \quad \zeta_1 = 0.0747, \quad \zeta_2 = 0.0828.
\end{aligned}
\tag{28}
$$

This hybrid closure is implemented into OpenFOAM [36], this limits numerical schemes to $2^{\text{nd}}$ order. Further for the convection of $\widetilde{u}_i$, a blending of 95% central differencing and 5% upwinding is performed, sacrificing this second order accuracy.[1] Being an unstructured code, $\Delta$ is defined as the cube root volume of each cell,

$$
\Delta = (\Delta_x \Delta_y \Delta_z)^{1/3}. \tag{29}
$$

## 3. Planar Channel

The turbulent channel simulation used in this work is described as follows. The flow is driven at a constant mass rate between two parallel zero slip flat plates of infinite extent. In practice this is achieved by periodic boundary conditions in the streamwise and spanwise directions and a body force added to the momentum equation. The streamwise, wall normal and spanwise directions are respectively parallel to the $x$, $y$ and $z$ coordinates. The Reynolds number based on friction velocity $u_\tau$ and channel half height $h$ is $\mathrm{Re}_\tau = 590$. The grid is deliberately prescribed such that the RANS-LES switch is 'sensitive' to the streamwise spacing — see Section 1. To achieve this $\Delta_x = h/20$ and $\Delta_z = \Delta_x/2$. The streamwise spacing is set to match the original diagnosis by Spalart *et al.* [3]. Normally $\Delta_x \geq \Delta_z$, which assumes that the smallest structures are streaks. Note, the grid here is set such that near-wall super-streaks of domain length are not governed by an overly coarse spanwise spacing. Because the hybrid RANS/LES formulation does not utilise wall functions, $y_1^+$ is set to 0.7. These grid requirements are satisfied by a computational domain of $2\pi \times 2h \times \pi$ spanned by $108 \times 108 \times 108$ cells.

The reference DNS data for statistical comparison is that of Moser, Kim and Mansour [37], whilst the reference structures are from the DNS data of Lozano-Durán and Jiménez [38]. This latter data set was analysed with the method outlined by Lozano-Durán *et al.* [39] and is followed within this report. Both reference sets [37, 38] are denoted purely by DNS without ambiguity.

### *3.1. Statistical Results*

Presented in this section are the time-, span- and stream- wise averaged quantities of typical interest in a planar channel flow. Table 1 is a list of global results for the hybrid case. Because of the low Reynolds number, the buffer layer is defined by the interval bounded by the modelled and resolved peaks $(y^+(\overline{\tau_{12}}^{\max}_{\text{sgs}}),\ y^+(\overline{\tau_{12}}^{\max}_{\text{res}}))$. From Fig. 1(a), it is evident the buffer layer is

[1] This is not required for the channel simulation here, but merely an artefact of the calibration considerations of more complex geometries.

**Table 1.** Location of the peak modelled, resolved and total stresses, the average RANS-LES interface and the size of the buffer layer. The RANS-LES interface is defined by the smallest value of $y^+$ such that $F < 1$.

| | $y^+(\overline{\tau_{12}}_{\rm sgs}^{\rm max})$ | $y^+(\overline{\tau_{12}}_{\rm res}^{\rm max})$ | $y^+(\overline{\tau_{12}}_{\rm tot}^{\rm max})$ | $F^+_{\rm switch}$ | $\delta^+_{\rm buff}$ |
|---|---|---|---|---|---|
| FSM | 15.69 | 84.27 | 53.89 | 17.67 | 68.58 |

incorrect with a modelled log-layer partially developing from the near-wall RANS model. It is this region that is known as the *artificial* buffer layer. The resolved log-layer intercept is correctly predicted, as is the shear stress, and consequently the skin friction is under predicted due to this buffer layer discrepancy.

The shear stress, with the modelled and resolved component dissection, is seen correctly predicted in Fig. 1(c). This is unsurprising, the methodology was originally calibrated for this very quantity. The normal stress components are also included, $u_i^{\rm rms}$ denotes the root mean square of the velocity fluctuations about the mean. The added turning points at $y^+ \approx 80$ in the $v^{\rm rms}$ and $w^{\rm rms}$ hint at destruction of the model component and the delayed growth of resolved normal stress. This delay is because initially, at the RANS-LES interface $y^+ = F^+_{\rm switch}$, resolved motion is produced in the streamwise direction. Consider the transport equations for the fluctuations about the effective filtered velocity $u'_i = u_i - \widetilde{u}_i$. In the limit $F \to 1$, the production terms reduce to,

$$\begin{aligned} P_{u'} &= -v'\partial_y \overline{u} \\ P_{v'} &= -v'\partial_y v' \\ P_{w'} &= -v'\partial_y w'. \end{aligned} \tag{30}$$

Note that in the RANS limit, $\overline{(\cdot)}$ satisfies the properties of a Reynolds operator and $\widetilde{u}_i \to \overline{u}_i$. $\overline{u}_i$ is the velocity of the RANS part of the flow field. As $F \to 0.5$, see Fig. 1(d), the effective viscosity is not low enough for proper structures to form, instead $\overline{u} \to \widetilde{u}$ guides more energy into $u'$. It is not until $F$ is a function of $\Delta$ and not the consequence of the DDES wall switching mechanism (Eq. 4), that the anisotropy levels recover somewhat. Clearly, near the wall at least, an isotropic damping of the length scale $\ell_t$ is insufficient.

The anisotropy is shown best through the invariant map, displayed in Fig. 1(b). The bounding lines, known as the Lumley triangle [40], enclose all realisable states of turbulence. $II$ and $III$ are the second and third invariants of $a_{ij}/2k$ given as $a_{ij}a_{ji}/8k$ and $\det(a_{ij}/2k)$ respectively. For more information, see for example the paper by Simonsen and Krogstad [41]. The right-hand edge corresponds to a state where one component of $a_{ii}$ (no summation) dominates (cigar shaped stress tensor) and the left-hand edge is where one component is negligible (pancake shaped stress tensor). The upper boundary is all other possible states where only two components exist and $(0, 0)$ is the point of perfect isotropy. Walking the DNS profile, from the origin, we see that we begin with an isotropic state (on average) in the bulk of the channel, moving to the small kink at $y^+ = 100$ and then as we move closer to the wall we finally see two-component turbulence. For the FSM, we see the largest discrepancy in the artificial buffer layer. Further, the profile is translated to the right of the DNS because of a global over prediction of $u^{\rm rms}$ and isotropy is never fully recovered in the bulk. Finally, the two-component state of turbulence is never achieved; the underlying RANS model, see Section 2, was derived assuming homogeneous turbulence, a condition that is strongly violated near the wall.

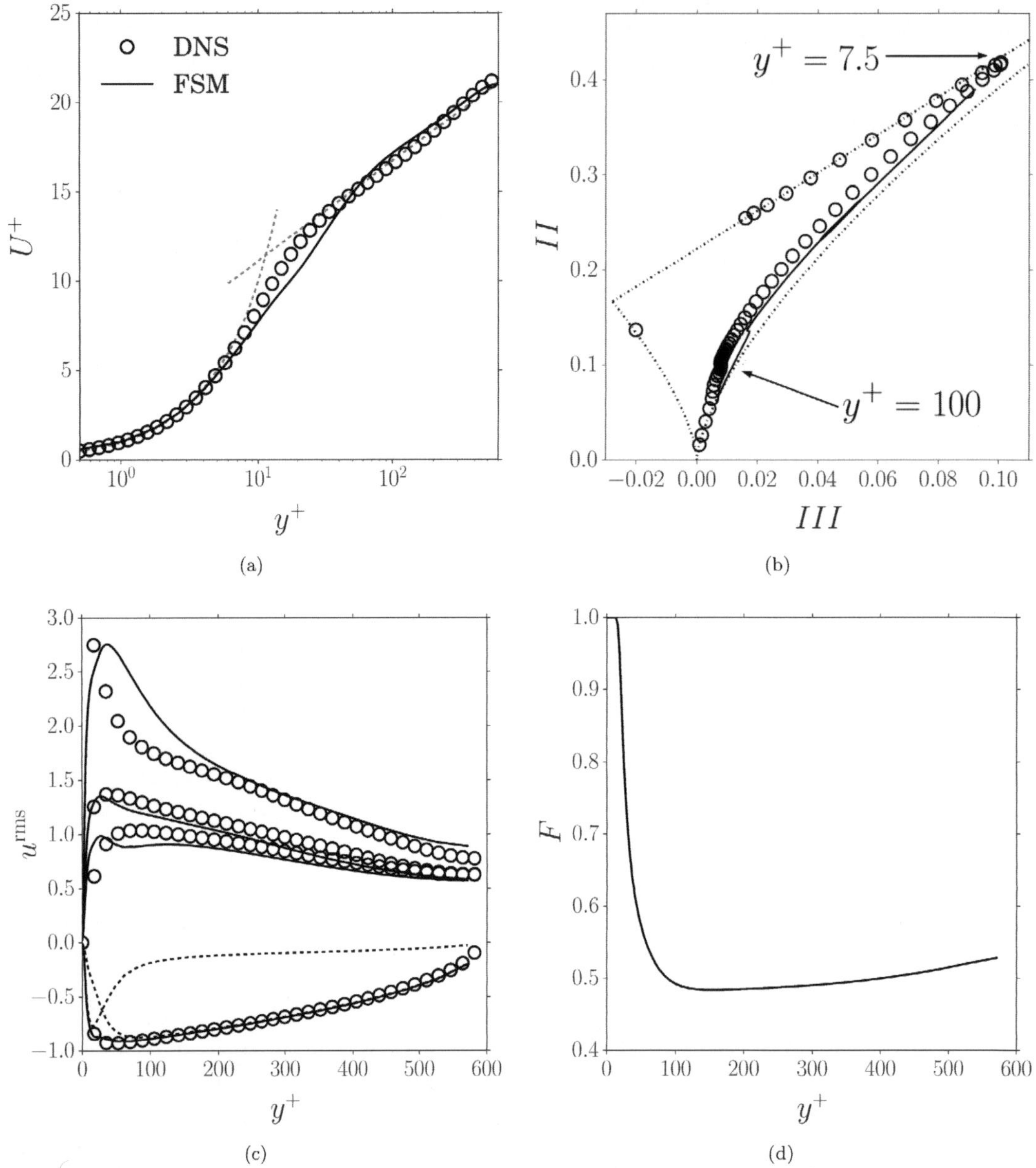


**Figure 1.** Statistical results: (a) Streamwise velocity, (b) Invariant Map, (c) Root mean square of velocity flucuations, (d) Average $F$. Symbols: ○ DNS, —— FSM, - - - - modelled and resolved break down of $\overline{\tau}_{12}^{1/2}$, — · — Lumley triangle.

### *3.2. The Instantaneous Field*

To better understand the shape of the statistics profiles, the instantaneous fields are now inspected. Figure 2 are profiles of $\widetilde{u}$ and $\widetilde{v}$ in the $x$-$z$ planes. Note initially that $y^+ = 10$ is deep within the 'RANS' zone ($F^+_{\mathrm{switch}} = 17.67$) and the fluctuations one observes provide the boundary condition for the LES region.

The streamwise fluctuations are heavily correlated. The familiar super-streaks can be seen in the artificial buffer layer, which we define to be the region between the peaks of modelled

$y^+(\overline{\tau_{12}}_{\text{sgs}}^{\max}) = 15.69$ and resolved $y^+(\overline{\tau_{12}}_{\text{res}}^{\max}) = 84.27$ stress — see Table 1. Further, the streaks at $y^+ = 10$ are visually correlated with those at $y^+ = 50$; implying the presence of structures as tall as this region.

Within the artificial buffer layer, one can see an anticorrelation between $u'$ and $v'$ — red $v'$ values indicates the lift up of these streaks and blue is the downwash of more homogeneous shaped structures. In order to analyse these apparent structures further, momentum transport structures are extracted from the flow and compared with DNS simulations [42], following the method of Lozano-Durán *et al.* [39]. The DNS Reynolds number is $\text{Re}_\tau = 950$. Despite being significantly higher than the hybrid case, it was shown in the original analysis [39] that channel flows at $\text{Re}_\tau = 950$ and $\text{Re}_\tau = 2000$ yielded similar results. As such, the FSM is contemplated alongside the $\text{Re}_\tau = 950$ DNS with some confidence of drawing meaningful comparisons.

A structure $\mathcal{S}$ is defined as a connected region of space that satisfies,

$$|u'v'(x, y, z)| \geq H \cdot u^{\text{rms}}(y) v^{\text{rms}}(y). \tag{31}$$

Connectivity is defined by considering the six adjacent points in the mesh including the periodic boundaries. Percolation analysis (initially developed for isotropic turbulence [43] and later adapted to channels [39]) showed that a hyperbolic hole size $H = 1.75$ yielded the threshold value of the ratio of the volume of the largest structure and the total volume of all structures. For the FSM, percolation analysis revealed $H \approx 1.25 \sim 1.75$ yielded similar behaviour and the upper bound is taken in this report to match the reference data.

Once structures have been extracted from uncorrelated snapshots, properties that define each $\mathcal{S}$ can be calculated. The structure velocity is defined as,

$$\langle u \rangle_i = \frac{\int_{\mathcal{S}} u_i' dV}{\int_{\mathcal{S}} dV}. \tag{32}$$

Quadrant analysis [44, 45] can be performed by considering the signs of $\langle u \rangle$ and $\langle v \rangle$. The most common types of structures are ejections (Q2's, $\langle u \rangle < 0, \langle v \rangle > 0$) and sweeps (Q4's, $\langle u \rangle > 0, \langle v \rangle < 0$), smaller interactions of outward (Q1, $\langle u \rangle > 0, \langle v \rangle > 0$) and inward (Q3, $\langle u \rangle < 0, \langle v \rangle < 0$) natures complete the quadrant. Also of particular interest in this work are the dimensions of the bounding box and the minimum and maximum wall distance of each structure. These are denoted respectively as: $(\ell_x, \ell_y, \ell_z)$, $y_{\min}$ and $y_{\max}$. Special care is taken to ensure these quantities are properly defined across periodic boundaries and that the normal velocity points away from the nearest wall.

A total of $10^6$ structures were extracted from 208 DNS snapshots, whilst 6000 structures were extracted from 71 FSM snapshots. Each structure can be reduced to a data point, such that it is described by its list of features $(\langle u \rangle, \langle v \rangle, \ell_x, \ell_z, y_{\min}, y_{\max}, \dots)$. This results in a high dimensionality data set.

The velocity quadrant of all structures is displayed in Fig. 3(a). The filled contours are the DNS and the red the FSM. From this point, DNS structures will be denoted DNS-$\mathcal{S}$ and similarly the FSM as FSM-$\mathcal{S}$. The inward and outward interactions are almost absent from the hybrid flow field. This is because they are much smaller in nature than Q2's and Q4's and are removed by the effective filter. The hybrid case has produced sweeps and ejections with much too great streamwise velocity. Generally the wall normal velocity is in much better agreement, although some sweeps are too intense in this direction. These erroneous structures are those closest to the wall. Knowing that the outer boundary of each histogram is made up of the near-wall structures, it is deduced that the velocity in the streamwise streaks of the hybrid RANS/LES is approximately 1.5-1.75 times too great. This is not immediately apparent in Fig 1(c) because the RANS-LES interface fluctuates vertically and RANS regions with no structures persist (making the resolved streaks rarer). Once deep in the artificial buffer layer, this no longer occurs and one

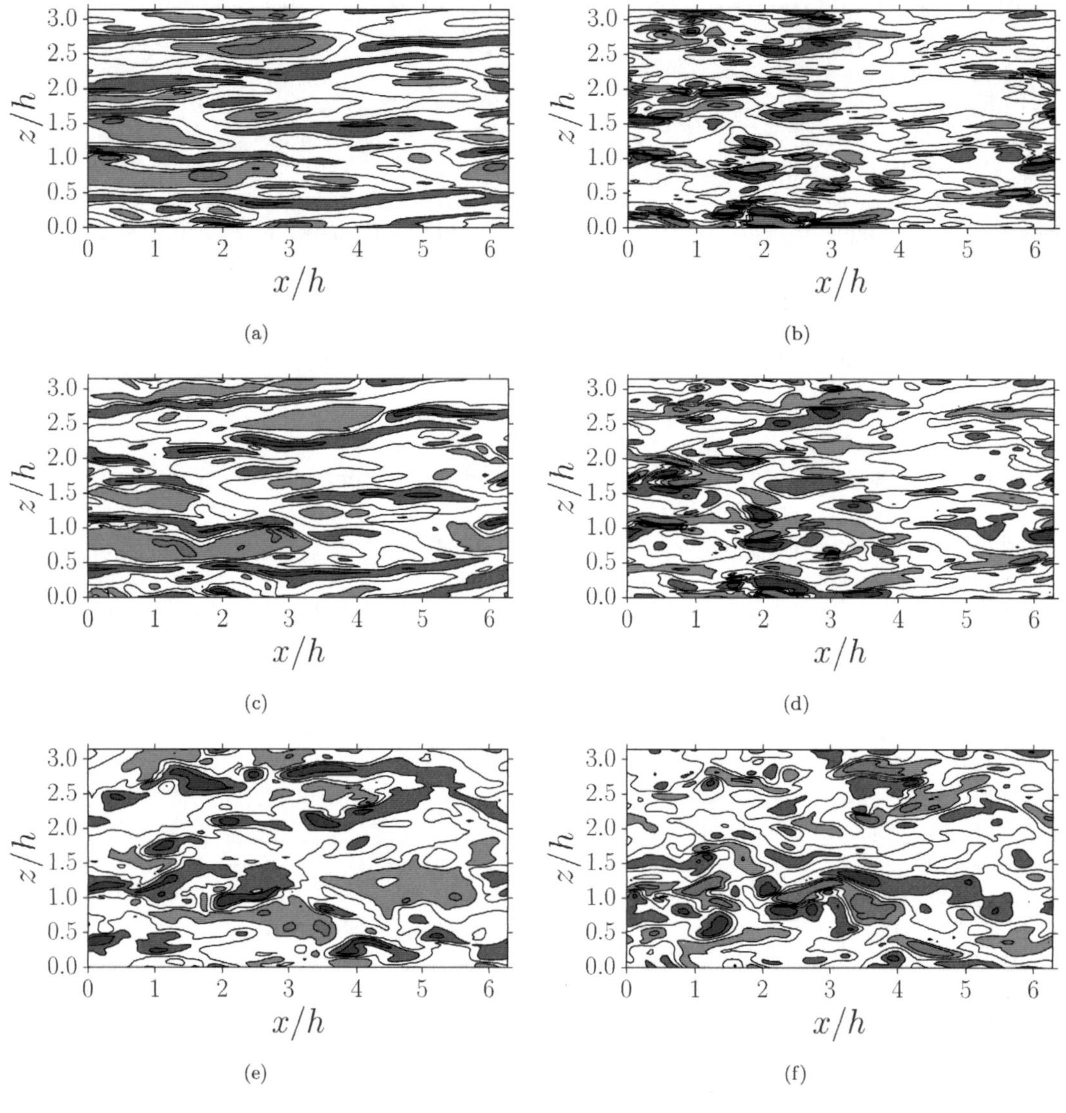


**Figure 2.** Instantaneous values of $u'$ (left) and $v'$ (right) at $y^+$ values of 10 (top), 50 (middle) and 150 (bottom). Red: $+\sigma$ and $+2\sigma$. Blue: $-\sigma$ and $-2\sigma$. $\sigma$ is the standard deviation for each plane. —— integer intervals of $\sigma$.

can see an overprediction in $u^{\mathrm{rms}}$. Currently this phenomena is mesh/methodology/Reynolds number dependent, however further simulations should make their discrepancy parametrisable.

Figure 3(b) is a plot of the structure averaged shear stress against its averaged wall normal velocity for FSM-$\mathcal{S}$ that live in the log-layer only. This is defined as $y < 0.2h$. To extract data for these log-layer plots, structures satisfying $y_{\max} > 0.2h$ are removed. We see the FSM structures closely follow the included lines. These lines are [39],

$$|\overline{uv}^+| = \pm 2.5\overline{v}^+. \tag{33}$$

This is based on the mixing length argument, that $\ell \propto y$ and the fluctuation (displacement of

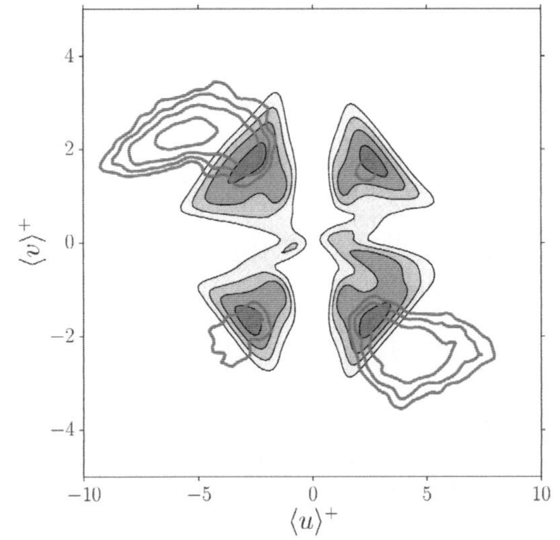


(a)

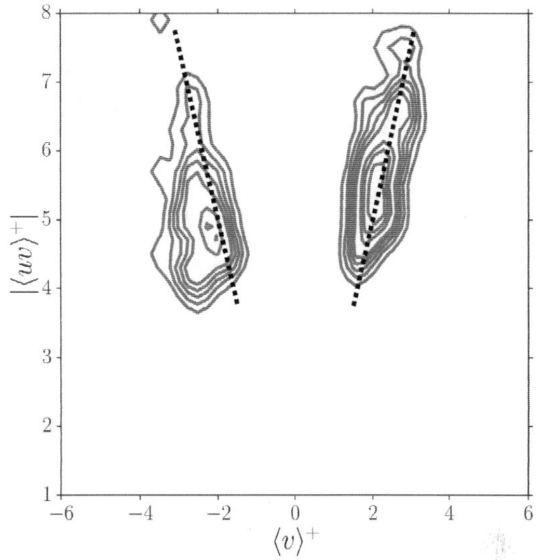


(b)

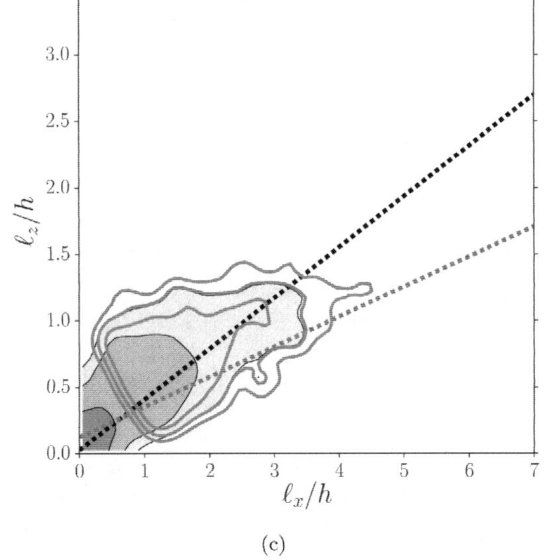


(c)

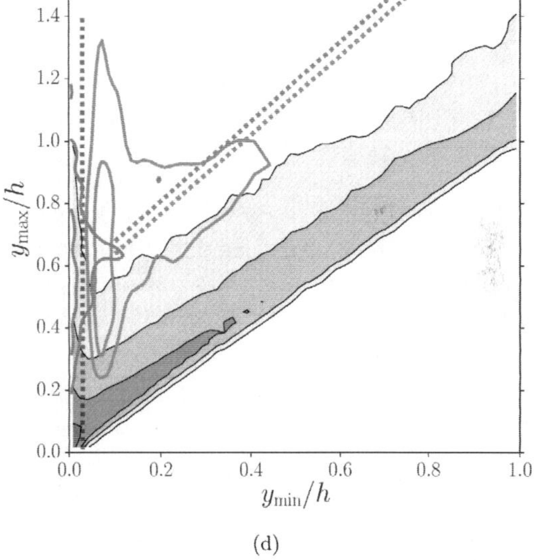


(d)

**Figure 3.** Histograms of structures, all contour levels are with respect to the maximum density on a logarithmic scale. Colours are consistent across all plots and plot type. Contours/lines: filled/black, DNS-$\mathcal{S}$; red, FSM-$\mathcal{S}$; blue, DNS-$\mathcal{S}^+$; green, FSM-$\mathcal{S}^+$. (a) $\langle v \rangle$ against $\langle u \rangle$. Contour levels: 30%, 50%, 70%, 90%. (b) $|\langle uv \rangle^+|$ against $\langle v \rangle^+$ for structures in the logarithmic layer only. Contour levels 10% increments. Lines: Eq. 33. (c) $\ell_z$ against $\ell_x$. Contour levels: 40%, 60%, 80%. Lines: least squares regression fits $\ell_z = m\ell_x + c$. (d) $y_{\max}$ against $y_{\min}$. Contour levels: filled, 30%, 50%, 70%, 90%; blue and green, 60%, 90%. Diagonal Lines: least squares regression for all structures above Eq. 36 such that $y_{\min} > 0.1h$; vertical line, $y_{\min} = F_{\text{switch}}$.

velocity mean) is proportional to this eddy size $\ell$,

$$\overline{uv}^+ \propto (\overline{v}\ell\partial_y\overline{u})^+ \propto \overline{v}^+. \tag{34}$$

The last simplification is based on the log-law equation,

$$\partial_y\overline{u} = \frac{u_\tau}{\kappa y}. \tag{35}$$

Therefore resolved structures in hybrid RANS/LES, once developed, do reproduce the correct mixing. The model has been calibrated such that this is reproduced, however because of the well documented increase in streamwise fluctuations, occurring from the RANS-LES interface, it becomes an impossibility to tune the model for both $u^{\mathrm{rms}}$ and $uv^{\mathrm{rms}}$ in its current state and we see an over prediction in the normal stress (Fig. 1(c)).

The black and red lines in Fig. 3(c) are linear regressions of FSM-$\mathcal{S}$ and DNS-$\mathcal{S}$ respectively. An approximate relationship between the spanwise and streamwise length, $\ell_x = m\ell_z$, can be formulated. For the DNS-$\mathcal{S}$ this gradient is $m \approx 2.6$, whilst for the FSM-$\mathcal{S}$ the value is considerably larger at $m \approx 4.5$. This indicates that FSM-$\mathcal{S}$ constitutes structures whose ratio $\ell_x/\ell_z$ is approximately 1.75 times greater than equivalent structures in DNS-$\mathcal{S}$. Although not shown in this preliminary report, the strong, small near-wall ejections in DNS-$\mathcal{S}$ satisfy $\ell_x/\ell_z \approx 5.3$ but the corresponding structures in FSM-$\mathcal{S}$ satisfy $\ell_x/\ell_z \approx 7.8$ — showing that the streakiness is a feature at the smallest scales also. This comparison is tentatively made however; the bounding boxes of FSM-$\mathcal{S}$ of this type are approximately 10 times their DNS-$\mathcal{S}$ counterparts.

In order to probe the DNS data further, a slightly modified version of an evolutionary clustering algorithm [46] is employed to divide the structures into groups. Evolutionary Algorithms (EAs) are non-deterministic optimisation tools that iterate towards a solution via survival of the fittest. The specifics and the optimality of solutions are not dwelt on here, rather a few comments about the structure groupings will be discussed below. Typically, DNS structures can be categorised as wall attached or wall detached. For example the filled contours in Fig. 3(c) show two arms; the thin, left arm are wall attached structures that have grown tall and the much larger, right arm are structures that are not as tall but exist away from $y = 0$. Instead, the EA finds the distinction,

$$\frac{4}{5}\ell_x + \ell_z = h, \tag{36}$$

which categorises the structure based implicitly on its volume. Figure 3(c) is the joint probability density function showing the distribution of the DNS structures with respect to $\ell_z$ and $\ell_x$. The blue contours are those that lie above Eq. 36. It is the latter grouping that is the primary focus of the rest of this report, used to contrast with equivalent FSM structures. The DNS structures of this sort shall now be denoted by DNS-$\mathcal{S}_+$ and equivalently the FSM by FSM-$\mathcal{S}_+$. Note that FSM-$\mathcal{S}_+$ contains 27% of the turbulent kinetic energy in the entire flow field and thus warrants a better understanding. Further, as will be shown this group of structures are non-physical and a major cause of error in the statistical profiles.

The blue contours of Fig. 3(d) are the DNS-$\mathcal{S}_+$. In other words, Eq. 36 describes a constraint that a structure must exceed to stay wall attached. Once a structure is $y_{\mathrm{max}} = \ell_y \approx 0.6h$ tall it detaches. The blue line of the same figure is a regression line of the detached DNS-$\mathcal{S}_+$, included as there are too few to show up in the histogram.

Equivalently, the red contours of Fig. 3(d) are the FSM-$\mathcal{S}_+$ and the red line is a linear regression of those detached. The term detached has a slightly different meaning for the FSM. Because of the RANS zone, structures cannot properly form near the wall that satisfy Eq. 31 and instead structures really take shape above the RANS-LES switch. The grey dashed line of the same figure highlights $y_{\min} = F_{\text{switch}}$. In this sense, attached means 'interface' attached. From Table 1 we can read that the artificial buffer layer begins at $y^+ = 15.69$ ($y/h = 0.027$) and ends at $y^+ = 84.27$ ($y/h = 0.14$). Figure 3(d) shows that between these bounds, interface attached structures grow much taller than their DNS counterparts, with many extending beyond the half height of the channel. Interestingly, their number is slightly fewer — 26% of DNS-$\mathcal{S}_+$ satisfy $y_{\max} > h$ but only 22% of FSM-$\mathcal{S}_+$. Crucially however, the differences lie in when they detach and also that the wall is a solid boundary whilst the interface is not. The tall FSM-$\mathcal{S}_+$ structures exist over a wider band of possible $y_{\min}$'s and become detached from the interface much more sporadically than the DNS-$\mathcal{S}_+$. At a height of $\ell_y/h \approx 0.6$, the DNS-$\mathcal{S}_+$ begin to detach and follow the regression line outlined in Fig. 3(d). Remarkably the FSM-$\mathcal{S}_+$ follow a similar line, the gradients differ by only $0.01h$. However, due to the freedom allowed by the interface over the wall, the standard deviation from this line in the artificial buffer layer is $0.27h$ for FSM-$\mathcal{S}_+$ and only $0.21h$ for the DNS-$\mathcal{S}_+$. This increase of 33% is indicative of the sporadic detachment, which pollutes the bulk flow with the elongated structures that contribute to the regression lines in Fig. 3(c).

## 4. Conclusions and Future Avenues

In short we are now in a better position to understand the discrepancies of Fig. 1. The correct shear stress comes at the price of an incorrectly predicted streamwise stress. The streamwise streaks that cause the issues are these very tall, thin structures that exist in the artificial buffer layer. These are malformed due to the delay in dropping the length scale to an LES value and are allowed more freedom to move around because they are only interface attached, not wall attached. This has the knock-on effect of never recovering isotropy in the bulk and a smeared $u^{\text{rms}}$ peak.

These structures (FSM-$\mathcal{S}_+$) characterise the incorrect buffer layer, which spans from $y^+(\overline{\tau_{12}}_{\text{sgs}}^{\max})$ to $y^+(\overline{\tau_{12}}_{\text{res}}^{\max})$. In this region, the incorrect velocity profile is caused by the set in of the underlying RANS model or in other words the delay in turbulence generation.

To improve the hybrid RANS/LES approach, the break-up of the interface attached FSM-$\mathcal{S}_+$ must be tackled. Piomelli *et al.* [26] have added white noise forcing to the interface with some success but without physical justification. This work is the very early beginnings of motivating a physical argument to such a backscatter method. Namely a forcing method that breakups up the interface attached FSM-$\mathcal{S}_+$ and reduces the variation in their height when they detach. Such a methodology could be incorporated as a body forcing term on the right hand side of Eq. 8 or alternatively each Reynolds stress component $\tau_{ij}^{\text{hyb}}$ could be damped individually to account for the anisotropy of the resolved structures at the RANS-LES interface.

## Acknowledgments

This work was funded in part by an EPSRC Doctoral Training Centre grant (EP/G03690X/1) and by the Multiflow program of the European Research Council.

## References

[1] Jiménez J and Pinelli A 1999 *J. Fluid Mech.* **389** 335–359
[2] Jiménez J, Del Álamo J C and Flores O 2004 *J. Fluid Mech.* **505** 179–199
[3] Spalart P R, Deck S, Shur M, Squires K, Strelets M K and Travin A 2006 *Theor. Comp. Fluid. Dyn.* **20** 181–195
[4] Chapman D K 1979 *AIAA J.* **17** 1293–1313
[5] Choi H and Moin P 2012 *Phys. Fluids* **24** 011702
[6] Schumann U 1975 *J. Comput. Phys.* **18** 376–404
[7] Deardorff J W 1970 *J. Fluid Mech.* **41** 453–480
[8] Speziale C G 1997 Computing non-equilibrium turbulent flows with time-dependent RANS and VLES *Fifteenth Int. Conf. Numerical Methods in Fluid Dyn.* (Springer) pp 123–129

[9] Rodi W, Mansour N N and Michelassi V 1993 *J. Fluid Eng.* **115** 196–205
[10] Breuer M, Jaffrézic B and Arora K 2008 *Theor. Comp. Fluid. Dyn.* **22** 157–187
[11] Fan T C, Tian M, Edwards J R, Hassan H A and Baurle R A 2001 *AIAA Pap.* **2001–2929**.
[12] Xiao X, Edwards J and Hassan H 2004 *AIAA J.* **42** 2508–2515
[13] Weinmann M, Sandberg R D and Doolan C 2014 *Int. J. Heat Fluid Fl.* **50** 263–278
[14] Weatheritt J and Sandberg R D 2014 A new Reynolds-stress damping function for Hybrid RANS/LES with an evolved functional form *IUTAM Symp. Adv. in Computation, Modeling and Control of Transitional and Turbulent Flows* (Goa, India)
[15] Weatheritt J and Sandberg R D 2015 *AIAA Pap.* **2015-0312**.
[16] Sergent E, Bertoglio J and Laurence D 2003 Coupling between large-eddy simulation and Reynolds-averaged Navier–Stokes *3rd Int. Worksh. Direct and Large-Eddy Simulation*, Munich
[17] Mathey F, Cokljat D, Bertoglio J P and Sergent E 2006 *Prog. Comp. Fluid Dy.* **6** 58–67
[18] Anupindi K and Sandberg R D 2015 *AIAA Pap.* **2015-2920**
[19] Davidson L and Peng S H 2003 *Int. J. Numer. Meth. Fl.* **43** 1003–1018
[20] Balaras E and Benocci C 1994 *AGARD Conf. Proc.* **551** 2.1–6.
[21] Spalart P R, Jou W H, Strelets M and Allmaras S R 1997 Comments on the feasibility of LES for wings, and on a Hybrid RANS/LES approach *Advances in DNS/LES* ed Liu C and Liu Z (Gayden Press) pp 137–147
[22] Spalart P and Allmaras S 1994 *AIAA Pap.* **92-0439**
[23] Nikitin N, Nicoud F, Wasistho B, Squires K and Spalart P 2000 *Phys. Fluids* **12** 1629–1632
[24] Baggett J S 1998 *Ann. Res. Briefs CTR* 267–277, Stanford U.
[25] Menter F R, Kuntz M and Bender R 2003 *AIAA Pap.* **2003-0767**
[26] Piomelli U, Balaras E, Pasinato H, Squires K D and Spalart P R 2003 *Int. J. Heat Fluid Fl.* **24** 538–550
[27] Piomelli U and Balaras E 2002 *Annu. Rev. Fluid Mech.* **34** 349–374
[28] Girimaji S S 2006 *J. Appl. Mech.* **73** 413–421
[29] Jakirlic S, Chang C Y, Kutej L and Tropea C 2014 *Bull. Am. Phys. Soc.* **59**(20) R2.2
[30] Popovac M and Hanjalic K 2007 *Flow Turbul. Combust.* **78** 177–202
[31] Pope S B 1975 *J. Fluid Mech.* **72** 331–340
[32] Speziale C G, Sarkar S and Gatski T B 1991 *J. Fluid Mech.* **227** 245–272
[33] Wallin S and Johansen A V 2000 *J. Fluid Mech.* **403** 89–132
[34] Menter F R 1994 *AIAA J.* **32** 1598–1605
[35] Hellsten A K 2005 *AIAA J.* **43** 1857–1869
[36] 2014 OpenFOAM: The Open Source CFD Toolbox. User Guide. Tech. Rep. SGI Corp
[37] Moser R, Kim J and Mansour N N 1999 *Phys. Fluids* **11** 943–945
[38] Lozano-Durán A and Jiménez J 2014 *Phys. Fluids* **26** 011702
[39] Lozano-Durán A, Flores O and Jiménez J 2012 *J. Fluid Mech.* **694** 100–130
[40] Lumley J L 1978 *Adv. Appl. Mech.* **18** 213
[41] Simonsen A J and Krogstad P A 2005 *Phys. Fluids* **17** 88103
[42] Del Álamo J C, Jiménez J, Zandonade P and Moser R D 2004 *J. Fluid Mech.* **500** 135–144
[43] Moisy F and Jiménez J 2004 *J. Fluid Mech.* **513** 111–133
[44] Wallace J M, Eckelmann H and Brodkey R S 1972 *J. Fluid Mech.* **54** 39–48
[45] Willmarth W and Lu S 1972 *J. Fluid Mech.* **55** 65–92
[46] Zheng Y, Jia L and Cao H 2012 *Inf. Technol. Control* **41** 283–294

# Geometry-induced fluctuations in the transitionally rough regime

**Nabil Abderrahaman-Elena and Ricardo García-Mayoral**
Engineering Dept. University of Cambridge. Trumpington Street. CB2 1PZ. UK
E-mail: r.gmayoral@eng.cam.ac.uk

**Abstract.** Direct numerical simulations of turbulent flows over rough surfaces are conducted to investigate the physics of the transitionally rough regime. Different roughness sizes are analysed within the transitional regime, while keeping the shape of the surface geometry constant. To study the effect of roughness on the flow field, a novel decomposition is used to divide the velocity into two components: a turbulent, geometry-independent contribution, and a geometry-induced contribution, whose intensity is modulated by the overlying turbulence. In the onset of the transitionally rough regime, the turbulent component remains essentially unmodified, and it is anticipated that all the roughness effects can be attributed entirely to the geometry-induced fluctuations. As the roughness size increases further, the turbulent component is also modified, and the fluid-surface interaction becomes more complex.

## 1. Introduction

Turbulent flows over rough surfaces have been studied by engineers and physicists for several decades due to their ubiquity. In industrial applications, roughness usually has an undesirable impact as it generates additional mixing near the wall, which results in an increase of drag. In other applications, however, this additional mixing can be beneficial, for instance to enhance heat transfer. Engineers like Darcy (1857), especially concerned about pressure losses in pipe flows, began studying turbulent flow over rough walls more than a century and a half ago. Since then, there has been a growing interest in rough-wall turbulence, with a large number of experimental, theoretical and numerical works being carried out. Extensive reviews of the subject can be found in Raupach et al. (1991); Jiménez (2004) and Flack and Schultz (2010).

The three regimes in which wall-bounded flows are traditionally classified are sketched in figure 1. For low Reynolds number, the flow is laminar but, as the Reynolds number increases beyond a certain critical value, the flow becomes turbulent. Provided that the characteristic size of roughness is still too small to affect the flow, rough surfaces behave as smooth walls, in what is known as the hydraulically smooth regime. However, for a fixed geometry of characteristic size $k$, as the Reynolds number increases the value of $k^+$ increases, where the superscript '+' denotes scaling in wall-units. For a sufficiently large Reynolds number, the roughness begins to affect the turbulent flow as $k^+$ is no longer negligible. Eventually, $k^+$ is large and inertial terms dominate even in the roughness sublayer, which is the region in which turbulent fluctuations depend predominantly on the roughness geometry (Schultz and Flack, 2007). The pressure drag is then dominant and the friction becomes independent of viscosity. The friction coefficient reaches an asymptotic state in which it becomes independent of the Reynolds number, as shown

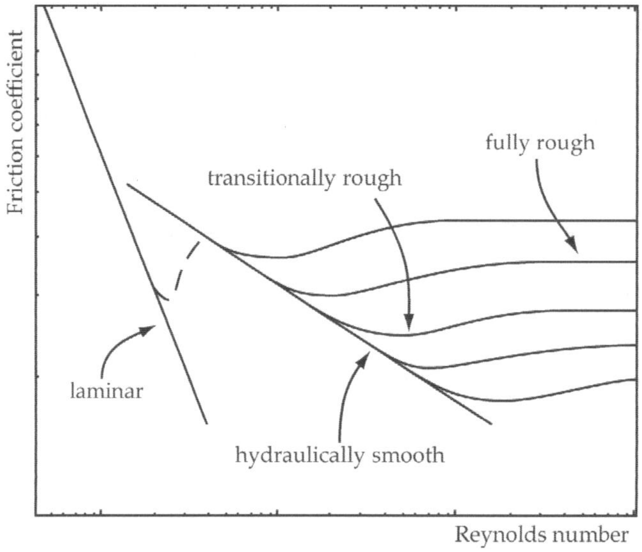


**Figure 1.** Friction coefficient against bulk-velocity Reynolds number.

in figure 1. This is known as the fully rough turbulent regime. For values of $k^+$ between the hydraulically smooth and the fully rough regimes, the flow is said to be transitionally rough.

Because of the complexity of any real rough geometry, it is useful to find a simple way to characterise roughness. Nikuradse (1933) conducted a series of experiments in pipe flows with rough walls where the inner walls were coated with packed sand grains of equal size, and measurements were taken for several sand grain sizes. The parameter chosen to characterise the different cases was the sand grain mean diameter, $k_s$. Far from the wall, Hama (1954) found that roughness only modifies the mean velocity profile by a shift $\Delta U^+$. In the logarithmic region, the shape of the profile and Kármán constant, $\kappa$, are otherwise unaffected,

$$U^+(y^+) = \kappa^{-1} \log y^+ + 5.1 - \Delta U^+. \quad (1)$$

Schlichting (1936) found that $k_s^+$ could be use to characterise any flow over roughness. The strategy is then not to consider the actual roughness size, $k^+$, but instead to find the equivalent sand grain roughness, $k_s^+$, which gives the same $\Delta U^+$ as the actual roughness. Any flow with an equivalent sand roughness of $k_s^+$ produces the same $\Delta U^+$ as a sand roughness of grain size $k_s^+$. In the fully rough regime, $k_s^+/k^+$ becomes constant and is only a function of the geometry. However, in the transitionally rough regime $k_s^+/k^+$ is variable and depends on the flow also. An alternative parameter is $k_{s_\infty}^+$, which is equal to $k_s^+$ in the fully rough regime, but maintains a constant ratio $k_{s_\infty}^+/k^+$ even in the transitional regime (Jiménez, 2004). As a result, the ratio $k_{s_\infty}^+/k^+$ depends only on the geometry in all the regimes, but different geometries exhibit a different $\Delta U^+$ for the same $k_{s_\infty}^+$ in the transitional regime. Still, $\Delta U^+$ curves eventually collapse in the fully rough regime, as illustrated in figure 3(b).

Starting from $k^+ \approx 0$, as $k^+$ increases roughness begins to affect the viscous sublayer, friction deviates from hydraulically smooth values, and the flow enters the transitionally rough regime. Some experimental works pay special attention to this regime (Flack et al., 2012; Flack and Schultz, 2014). However the key questions remain unsolved: what mechanism triggers the departure from the hydraulically smooth regime as $k^+$ increases; and how to define a threshold for this transition based on the roughness geometry only. One of the main issues is that, while $k_{s_\infty}^+$ characterises appropriately the fully rough regime for most rough surfaces, it seems not to be the correct parameter in the transitionally rough regime. This was already noticed by Colebrook (1939) as his results and Nikuradse's do not match in the mentioned transitional regime. In general, different geometries of roughness lead to different transitions. Comparing several sizes and combinations of sand grains, Colebrook and White (1937) proposed this transition is abrupt in organised geometries, while the presence of a wider range of roughness sizes makes the transition smoother. This assumption is also supported by more recent experiments, such as those using uniformly arranged spherical rough elements by Ligrani and Moffat (1986), in which the transition takes place for even higher values of $k^+_{s_\infty}$, compared to Nikuradse's and Colebrook's experiments, but also in a sharper manner. What is more, the discrepancy is such that Flack and Schultz (2010) report a relatively wide range of experimental values in which the transition takes place, spanning between $1.4$–$15 < k_s^+ < 18$–$70$.

Recent research by Chung et al. (2015) is aiming to find a methodology to capture the effect of roughness, particularly the increase in friction and $\Delta U^+$, at a minimum cost. They conduct fully turbulent simulations, but in minimal-span boxes (Jiménez and Moin, 1991) that reduce the computational requirements. The present project also aims to provide estimates for $\Delta U^+$ at a reduced cost, but removing the need to carry out turbulent simulations altogether, by capturing the effect of the geometry with a reduced order model, if possible.

In the present work we aim to gain insight into the physics triggering the transition from the hydraulically smooth regime. We carry out a series of direct numerical simulations (DNSs) of turbulent channels with rough walls. We study a roughness texture formed by equispaced square posts of the same size in a rectangular arrangement, as sketched in figure 2. Similar geometries have been studied by other authors (Leonardi et al., 2007; Leonardi and Castro, 2010) although their simulations mainly explore the fully rough regime. The purpose of our simulations is to investigate the evolution of turbulent flow fluctuations through the transitionally rough regime. The surface shape is kept constant throughout the simulations, but the size of the elements measured in wall units is varied from $k^+ \approx 6$ to $k^+ \approx 38$, aiming to span the whole transitionally rough regime. These flow fluctuations are split into turbulent and roughness-induced contributions. We restrict ourselves to the study of streamwise velocity fluctuations and leave the study of the other variables for future works. We aim to decompose the fluctuation into turbulent and roughness-induced contributions, with the purpose of isolating the effect of the roughness geometry.

The paper is organised as follows. In section 2 the numerical method is briefly described. We present statistics of the flow in section 3 and propose a decomposition of the flow in simpler components. Section 4 concludes this work.

## 2. The numerical method

The numerical experiments are conducted in a plane turbulent channel with rough texture on the top and bottom walls. The domain is periodic in the wall-parallel directions. The streamwise, wall-normal and spanwise coordinates are $x$, $y$ and $z$, with $u$, $v$ and $w$ the corresponding components of the velocity $\boldsymbol{u}$. The flow is assumed incompressible with a density $\rho = 1$. The numerical method, briefly outlined below, is adapted from that of García-Mayoral and Jiménez (2011*b*) for riblet channels, is modified to handle fully three-dimensional geometries. The temporal integrator is a fractional-step method combined with a three-substep Runge-Kutta, and with pressure correction at the final step only (Le and Moin, 1991). The channel half-height is $\delta$ measured from the roughness tips, the length is $L_x = 2\pi\delta$ and the width $L_z = \pi\delta$. The spatial discretisation is pseudo-spectral. The two periodic directions, $x$ and $z$, are discretised using Fourier series, while the wall-normal direction, $y$, is discretised using a second order finite difference scheme in a collocated grid. The integration of the pressure in these collocated grids commonly produces an undesired chequerboard effect (Ferziger and Perić, 2002). To avoid this issue, the code uses a quasi-divergence-free formulation based on the work by Nordström et al.

Journal of Physics: Conference Series **708** (2016) 012009 doi:10.1088/1742-6596/708/1/012009

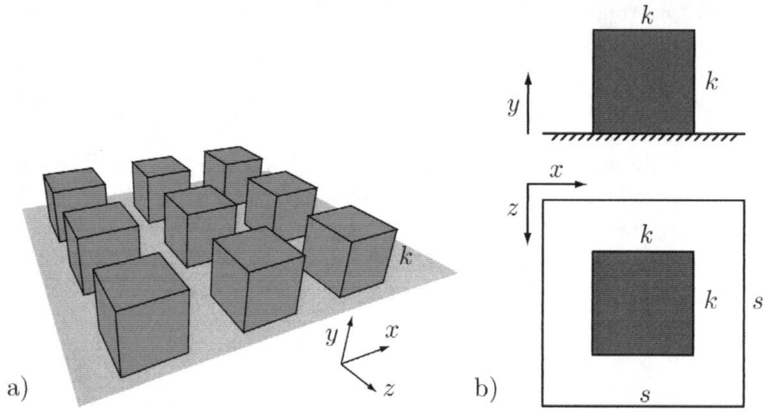


**Figure 2.** Sketch of the roughness geometry over the bottom wall. (a) Collocated arrangement of the elements. (b) Sketch of each periodic unit in the lattice shown in (a). In grey the post, formed by a cube of side $k^+$. The outer line represents the clearance between elements of size $s = 2k$.

(2007). This strategy consists of not enforcing exact incompressibility, $\nabla \cdot \boldsymbol{u} = 0$, and instead adding an evolution equation for $\nabla \cdot \boldsymbol{u}$ so that any divergence is rapidly damped. The grid in the $y$ direction is non-uniform so that $\Delta y^+_{min} \approx 0.31$ at the wall and $\Delta y^+_{max} \approx 3.12$ at the centre of the channel. The domain is divided into three blocks in $y$ with different $x$- and $z$-resolutions. This multi-block technique allows a high number of grid points to be set in the region near the walls, with a coarser resolution in the core of the channel. In this latter region the resolution is set to resolve all turbulent scales. In the blocks containing the walls it is also necessary to resolve the flow around the roughness elements as obstacles which requires a higher resolution for small $k^+$. The resolution in the central block is $\Delta^+_{x_c} \approx 6$ and $\Delta^+_{z_c} \approx 3$ in the $x$ and $z$ directions respectively, while in the other two blocks a finer resolution is used, $\Delta^+_{x_w} = \Delta^+_{z_w} \approx$ 1-2.

The simulations are run at constant flow rate, and the value of the viscosity is adjusted a priori for each simulation to keep the friction Reynolds number close to $\mathrm{Re}_\tau \approx 185$. $\mathrm{Re}_\tau$ is computed using $\delta$ and the friction velocity $u_\tau$ obtained by extrapolating the total shear to an effective channel height $\delta'$. This height is set so that $u_\tau^2 = -\delta' \partial p / \partial x$ (García-Mayoral and Jiménez, 2011*b*; Chung et al., 2015), which is the relationship that would hold for a smooth channel at $\delta$. For the geometry under consideration this implies that $\delta'$ is measured from a virtual origin $y_0 = 1/8\,k$ below the roughness tips.

The roughness elements are introduced through a direct-forcing immersed boundary method (Mohd-Yusof, 1997; Fadlun et al., 2000; Iaccarino and Verzicco, 2003). The roughness geometry is depicted in figure 2. The unit element is a cube of side $k$, repeated along $x$ and $z$ in both walls with a texture period $s = 2k$.

## 3. Results

We conduct six DNSs with values of $k^+$ spanning the transitional regime. The parameters for the simulations are summarised in table 1.

Roughness is usually classified into k- and d-roughness. In k-roughness, the effective roughness $k_s$ is proportional to $k$, while in d-roughness it is proportional to the boundary layer thickness (Jiménez, 2004). In the range of $k^+$ studied, our geometry displays conventional k-roughness behaviour, i.e. friction increases with $k^+$ and the mean velocity profile is shifted downwards, as shown in figure 3(a), increasing drag. In order to quantify this shift on the

**Table 1.** Parameters of the simulations. $s^+$ is the post spacing in both the streamwise and spanwise directions. The number of collocation points in the streamwise and spanwise directions are $N_{x_c}$ and $N_{z_c}$ in the middle block, and $N_{x_w}$ and $N_{z_w}$ in the top and bottom blocks. The resulting resolution is $\Delta^+_{x_w}$ and $\Delta^+_{z_w}$ in the middle block, and $\Delta^+_{x_c}$ and $\Delta^+_{z_c}$ in the top and bottom walls. $N_y$ is the number of points in the $y$ direction between roughness tips.

| Case | $k^+$ | $s^+$ | $\Delta U^+$ | $\mathrm{Re}_\tau$ | $N_{x_c}$ | $N_{z_c}$ | $N_{x_w}$ | $N_{z_w}$ | $N_y$ |
|---|---|---|---|---|---|---|---|---|---|
| 0C | – | – | – | 184.5 | 192 | 192 | 192 | 192 | 153 |
| 6C | 6.1 | 12.1 | 0.63 | 184.9 | 192 | 192 | 1152 | 576 | 153 |
| 9C | 9.0 | 18.0 | 1.07 | 183.1 | 192 | 192 | 768 | 384 | 153 |
| 12C | 12.0 | 24.0 | 1.90 | 183.6 | 192 | 192 | 1152 | 576 | 153 |
| 18C | 17.8 | 35.5 | 3.81 | 181.0 | 192 | 192 | 768 | 384 | 153 |
| 24C | 23.5 | 47.0 | 5.29 | 179.4 | 192 | 192 | 576 | 288 | 153 |
| 36C | 37.9 | 75.8 | 7.37 | 193.0 | 192 | 192 | 768 | 384 | 153 |

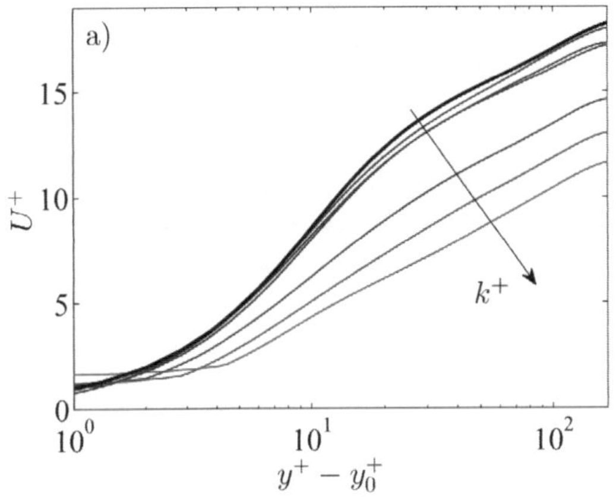


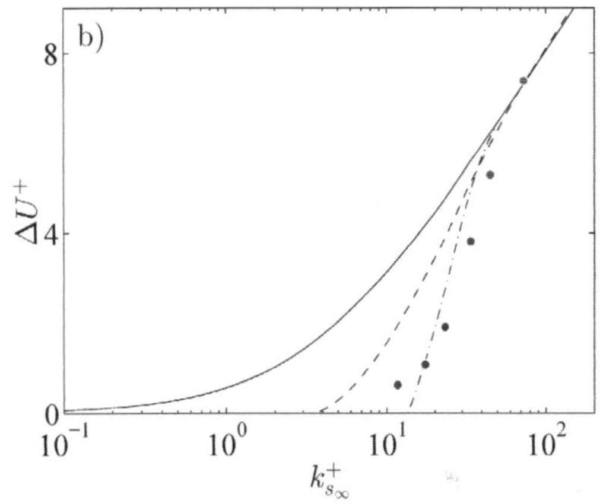


**Figure 3.** (a) Mean velocity profiles measured from $y_0^+ = 1/8\,k^+$. Thick black line, case 0C; blue to red, cases 6C through 36C of table 1. (b) Roughness function $\Delta U^+$; —, correlation for Colebrook results (Jiménez, 2004); ----, correlation of sand grain roughness (Nikuradse, 1933); –·–, correlation of sphere roughness (Ligrani and Moffat, 1986); o present DNSs using the same colors as in (a).

velocity profile and also be able to establish a comparison with other surfaces, we portray in figure 3(b) the roughness function, $\Delta U^+$, against $k^+_{s_\infty}$, which for the geometry considered satisfies $k_{s_\infty}/k \approx 0.5$. This representation collapses the fully rough regime to a single asymptote, so the variations in the transitional regime for different surfaces can be compared. The roughness function of our cases is also compared to the results of Colebrook and White (1937), Nikuradse (1933), and Ligrani and Moffat (1986). In Colebrook and White (1937), Colebrook (1939), and Nikuradse (1933) the transition is smooth and begins at low values of $k^+_{s_\infty}$. In our case, the transitional regime resembles more closely that of the packed spheres of Ligrani and Moffat (1986), evolving rapidly from low to high values of $\Delta U^+$. This is in agreement with the observation by Colebrook (1939) that regular roughness departs from the hydraulically smooth regime more gradually than regular roughness.

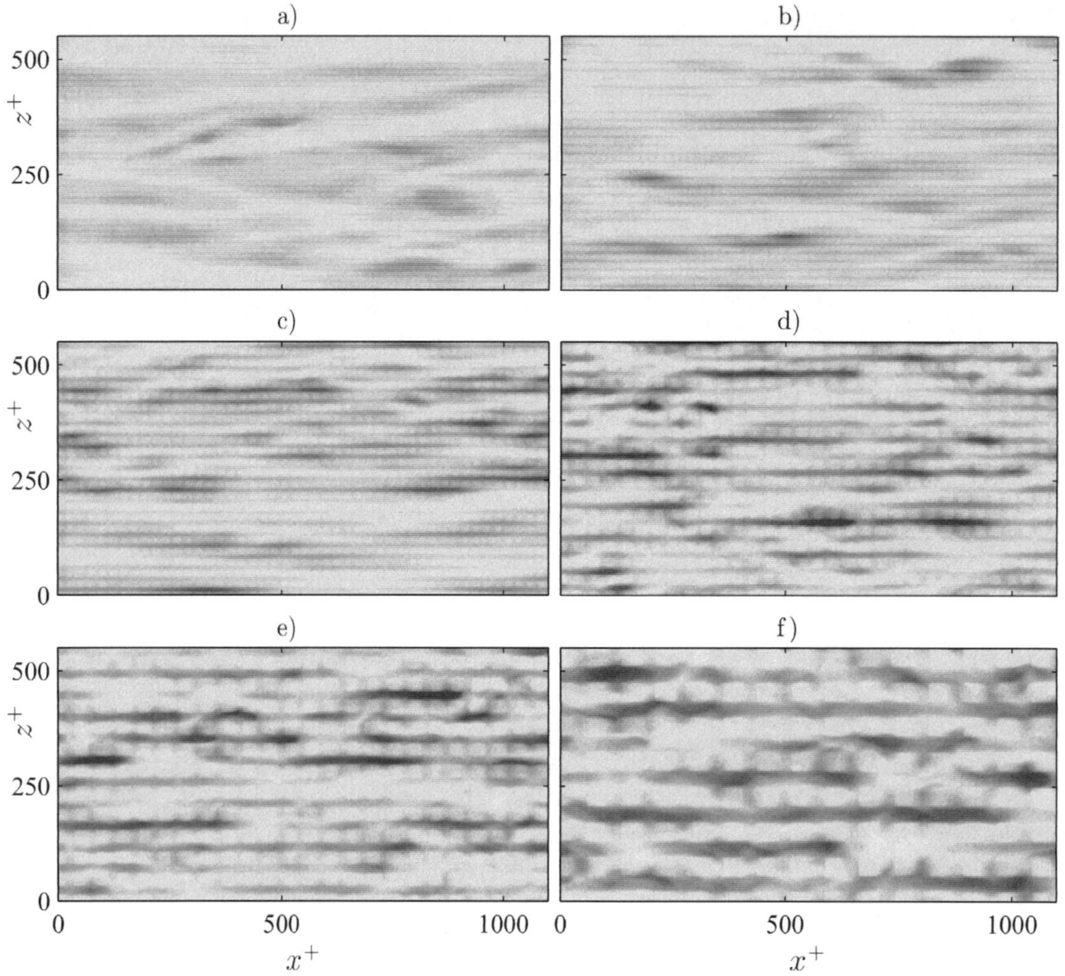


**Figure 4.** Instantaneous streamwise velocity at $y^+ \approx 1.3$ above roughness crest, for cases (a) 6C, (b) 9C, (c) 12C, (d) 18C, (e) 24C, and (f) 36C. The same scale of colour is used in all subfigures, with dark red and light yellow indicating high and low velocity, respectively.

### *3.1. Effect of roughness on the velocity fluctuations*

To understand how the flow is modified by the presence of roughness, producing the changes in the mean flow described above, we focus our analysis on the velocity fluctuations. For the present paper, we restrict ourselves to the fluctuations of streamwise velocity. The analysis of the other velocity components and of the Reynolds stresses are left for future work.

The effect of roughness is limited to the vicinity of the wall, and decays rapidly away from it. As illustrating examples, figure 4 portrays instantaneous realizations of the streamwise velocity for cases 6C through 36C at $y^+ \approx 1.3$ above the roughness peaks, close enough to the wall to observe the effect of the roughness elements. The figure shows that the fluctuating velocity signal has two distinct contributions. The first is the typical background signal characteristic of wall turbulence, which consists of streamwise elongated streaks of high and low $u$-velocity, of length $500 - 1000$ wall units and width $50 - 100$ wall units (Kim et al., 1971; Smith and Metzler, 1983). This contribution remains essentially unmodified, at least for $k^+ < 15$ (or $\Delta U^+ < 3$). The second contribution is directly caused by the presence of the roughness elements, and consists of

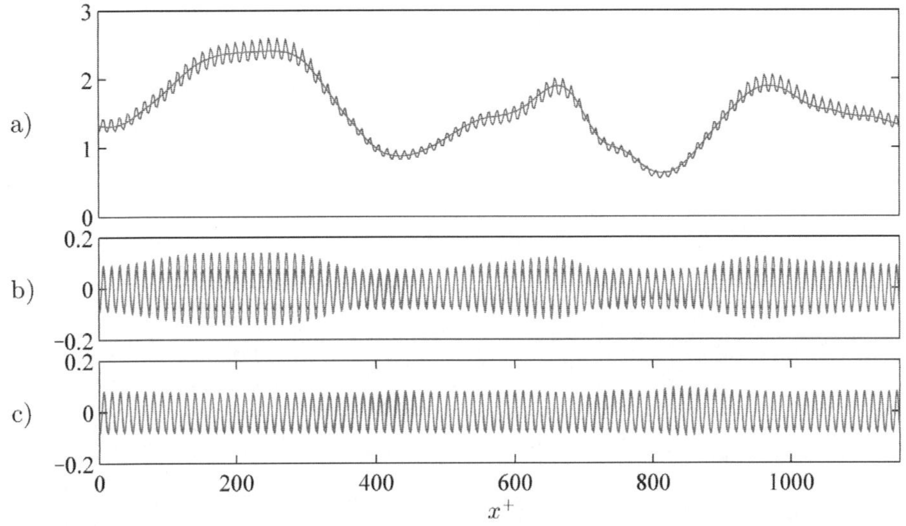


**Figure 5.** Instantaneous components of the streamwise velocity for case 12C at $y^+ \approx 0.7$ above roughness crest and a given $z$- section half-way between two rows of roughness elements. (a) In black, $u$; in red, $u_T$. (b) In black, $\overline{u}_R$; in red, $u_R$. (c) In black, $\overline{u}_R$ as obtained by averaging using equation 3; in red, $\overline{u}_R$ as obtained from equation 4.

alternating regions of relatively low velocity, right over the protruding elements, and relatively high velocities, over the valleys between elements. This signal is attached to the roughness geometry, and is therefore not advected over time as the turbulent contribution. It would instead be essentially repeated periodically in a lattice of streamwise and spanwise periodicity $s$, so that its value depends only on the relative position within each periodic unit, that is, the coordinates $\tilde{x}$ and $\tilde{z}$, with values between 0 and $s$, and the wall normal coordinate $y$. The lengthscales of this roughness contribution would scale with $k^+$, and its intensity would also increase with the roughness size.

To characterize these two contributions, we could try to decompose the flow as

$$u(x, y, z, t) = u_T(x, y, z, t) + \overline{u}_R(\tilde{x}, y, \tilde{z}) = U(y) + u_S(x, y, z, t) + \overline{u}_R(\tilde{x}, y, \tilde{z}), \tag{2}$$

where $\overline{u}_R$ would be the spatially fluctuating, time-independent component due to the roughness, and $u_T$ the background turbulent signal, which would include the mean velocity $U(y)$ and the fluctuating component $u_S$, analogous to that over a smooth surface. The geometry-coherent signal $\overline{u}_R$ can be obtained by averaging over time and over the roughness lattice,

$$\overline{u}_R(\tilde{x}, y, \tilde{z}) = |u(x, y, z, t) - U(y)|_{t,N_R}, \tag{3}$$

where $N_R$ is the number of roughness elements in the simulation domain.

The above decomposition is analogous to that of Reynolds and Hussain (1972) for coherent waves in turbulence, and is widely used in flows over complex surfaces (Choi et al., 1993; Jiménez et al., 2001; García-Mayoral and Jiménez, 2011*b*; Jelly et al., 2014; Seo et al., 2015). However, the actual roughness-induced contribution, $u_R$, turns out to be different from the time-averaged $\overline{u}_R$, so we drop the decomposition of equation (2). The problem is illustrated in figure 5, which portrays the different contributions at a given instant and a given pair of coordinates $z$ and $y$, that is, along an $x$-line, for case 12C. For this case $k^+$ is still small enough for the rough and turbulent components to have clearly separate wavelengths, so that $u_T$ can be obtained

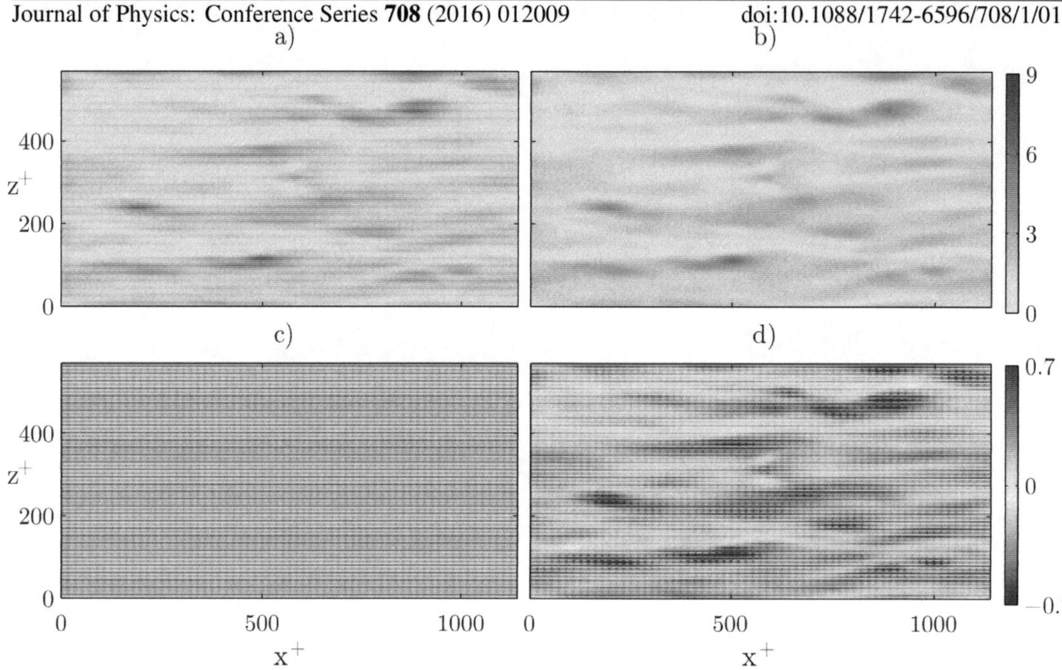


**Figure 6.** Instantaneous components of the streamwise velocity at $y^+ \approx 1.3$ above roughness crest for case 9C. (a) Full velocity signal $u$. (b) Background turbulent contribution $u_T$. (c) Coherent, time average signal induced by the presence of roughness, $\overline{u}_R$. (d) Instantaneous, $u_T$-modulated signal induced by the roughness, $u_R$.

simply by filtering out small wavelengths spectrally. Panel 5(a) shows the full velocity signal $u$ superimposed with the filtered $u_T$, and illustrates the starting hypothesis that the velocity signal is made up of a turbulent signal, analogous to that over smooth walls, plus a small-intensity, small-wavelength, roughness-coherent signal. However, when $u_T$ is subtracted from $u$, as portrayed in panel 5(b), the result is not exactly $\overline{u}_R$. The resulting $u_R$, is not exactly $\overline{u}_R$, but the modulation in amplitude of $\overline{u}_R$ by the turbulent $u_T$,

$$u_R = \frac{u_T}{U}\overline{u}_R. \tag{4}$$

Note that, on average, $u_T/U$ is 1. Panel 5(c) shows how, once the modulation is removed from $u_R$, the coherent $\overline{u}_R$ is recovered.

Taking the above discussion into account, the correct decomposition that replaces equation (2) is

$$u = u_T + u_R = u_T + \frac{u_T}{U}\overline{u}_R, \tag{5}$$

which is essentially equation (2) with $\overline{u}_R$ replaced by $u_R$.

When both $x$ and $z$ are considered, the decomposition is a bit more complicated, but the discussion above serves to illustrate it. Furthermore, the spectral filtering only works when the wavelengths of $u_T$ and $\overline{u}_R$ are clearly separated, which does not hold for the cases with larger $k^+$. However, it is always possible to obtain $\overline{u}_R$ by using equation (3) over the whole sampling history, and $u_T$ can then be obtained algebraically from equation (5). The $u_T$-modulated, full roughness contribution, $u_R$, can be obtained once $u_T$ and $\overline{u}_R$ are known using equation (4).

The result from decomposing the streamwise velocity using this procedure is shown for case 9C in figure 6. The comparison of panels (c) and (d), which portray respectively $\overline{u}_R$ and $u_R$, shows

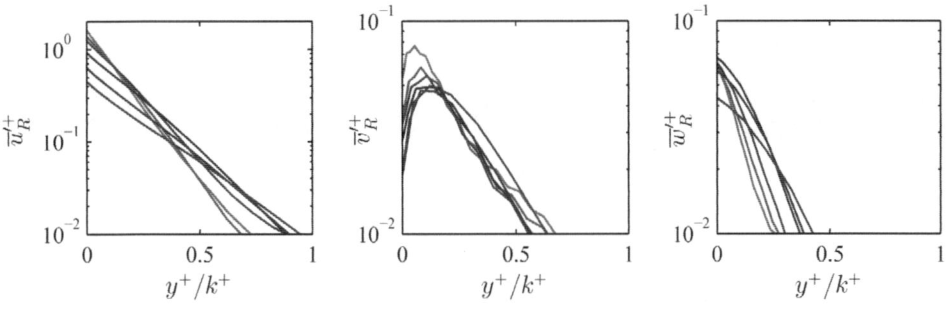


**Figure 7.** Root mean square of the coherent part of the velocity components. Blue to red, cases 6C through 36C.

how neglecting the amplitude modulation of $u_R$ can lead to underpredicting its contribution to the total intensity of the fluctuations. Nevertheless, the perturbation caused by the surface on the flow decays rapidly away from the wall. Figure 7 shows the r.m.s. values, denoted by a 'prime' superscript, of the coherent part of the three velocity components, $\overline{u}_R$, $\overline{v}_R$ and $\overline{w}_R$. The three of them exhibit an exponential behaviour with a logarithmic slope roughly proportional to $-y/k$, which is to be expected from perturbations induced at the wall (Seo et al., 2015).

The present modulation of the near-wall, roughness-induced flow by the overlying turbulent flow is similar to the modulation of buffer-layer turbulence by outer-layer large structures, as presented by Marusic et al. (2010), except for the difference in the scales involved. In the present study, the modulating signal is actually the buffer-layer turbulence. The low Reynolds number of our DNSs prevents the development of large-scale turbulence, but if this was present it could be expected to modulate the buffer-layer flow. In turn, the total turbulent velocity above each roughness element would modulate the local $u_R$. This modulation of $u_R$ by the overlying turbulence is also connected to the concept of protrusion height in riblets (Luchini et al., 1991; García-Mayoral and Jiménez, 2011*a*). For vanishing riblet spacing, the flow near each texture element is produced by the quasi-uniform shear induced by the turbulent eddy just above. In the scale of the texture, this eddy can be represented as quasi-infinite and quasi-steady. The overlying turbulent velocity therefore sets the scale for the local velocity within the riblet grooves, just like in our case the overlying $u_T$ sets the scale for the local $u_R$.

The above discussion is only valid for vanishingly small roughness, when $u_T$ modulates $u_R$ but is itself unaffected by the presence of roughness. However, as the roughness size increases and becomes comparable to the turbulent eddies, this assumption ceases to hold. For the geometry considered in this paper, we would expect this to happen for $s^+ \approx 15 - 20$, when the spacing between roughness elements becomes comparable to the typical diameter of the quasi-streamwise vortices that are characteristic of near-wall turbulence (Kim et al., 1987). The validity of the assumption that $u_T$ remains essentially unaffected by the presence of roughness can be checked in figure 8, which compiles the r.m.s. values of $u^+$, $u_T^+$ and $u_R^+$, denoted by a 'prime' superscript, as a function of $y^+$. Panel (c) shows how $u_R'^+$ increases gradually with $k^+$ from the smallest roughness, while significant changes in $u_T'^+$ only appear for $k^+ > 15$, or its equivalent roughness function $\Delta U^+ > 3$. This implies that the onset of roughness effects, that is, the beginning of the transitionally rough regime, can occur for roughness sizes at which the buffer-layer turbulence is essentially unaffected, so that it may be possible to estimate the changes in the flow independently of the turbulence.

This argument is further supported by the effect of the roughness under consideration on the r.m.s. of the streamwise vorticity $\omega_x'^+$, shown in figure 9. For the cases 6C, 9C and 12C,

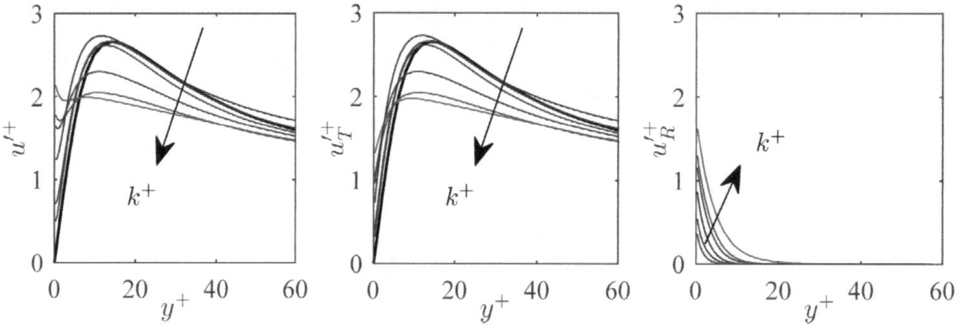


**Figure 8.** Root mean square of the different contributions to the streamwise velocity. From left to right, full $u$-signal, background-turbulence velocity $u_T$ and roughness-induced velocity $u_R$. Thick black line, smooth channel (Hoyas and Jiménez, 2008); blue to red, cases 6C through 36C. The arrows indicate increasing $k^+$.

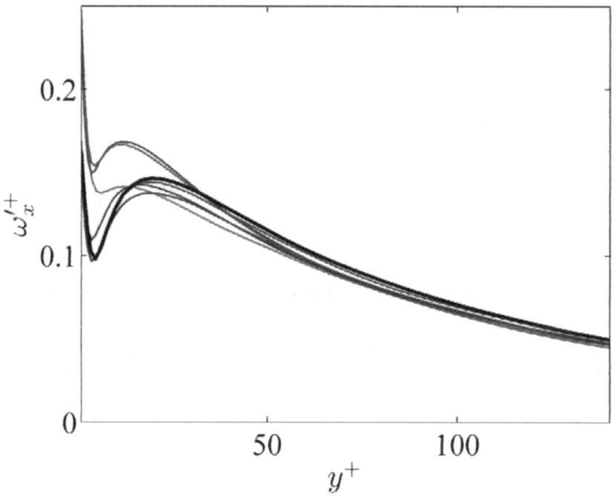


**Figure 9.** Root mean square of the streamwise vorticity with the origin of $y$ placed at roughness crest. Lines coloured as in figure 8.

with $\Delta U^+ < 3$, the curves display no significant differences with that over a smooth wall, with similar values and the same distinctive maximum at $y^+ \approx 15$, which marks the average height of quasi-streamwise vortices. For slightly larger values of the roughness function, $\Delta U^+ \approx$ 4-6, the local maximum is shifted towards the wall and increases in magnitude. This may indicate an intensification and partial entrainment into the roughness troughs of the near-wall vortices, or a more complex interaction. Verifying this would require further studies, but in any event the change in the maximum marks a modification of the smooth-wall-like turbulence dynamics. For even larger roughness function, $\Delta U^+ > 7$, which is already in the fully rough range, the trend changes and the magnitude of the maximum decreases further, which could indicate a weakening of the quasi-streamwise vortices, and a more complete obliteration of the near-wall cycle by the roughness, as is typical of fully rough surfaces.

## 4. Conclusions and future work

We have focussed the present work in the transitionally rough regime of turbulent flows over rough walls. As a starting benchmark case, we have selected a surface made up of evenly spaced, uniform cubes, and we have conducted a series of DNSs in which we have varied the size of the cubes, $k^+$, while maintaining the surface shape, so that the whole transitionally rough regime could be studied. We have hypothesised that the fluctuating flow could be decomposed into a smooth-wall-like turbulent contribution and a roughness-induced one. We have checked this assumption for the fluctuating streamwise velocity. For the small values of $k^+$ in the regime studied the decomposition is valid, except that the roughness-induced component is modulated in amplitude by the turbulent one. For somewhat larger $k^+$, the turbulent component differs from that over smooth walls, but in the onset of the transitionally rough regime the turbulent component remains essentially canonical, and all the changes in the flow can then be attributed to the geometry-induced component. This result suggests that it may be possible to predict the onset of roughness effects without considering the interaction of the roughness geometry with the turbulence, but further studies are required to verify this.

**Acknowledgments**
This work was supported by the European Research Council through the II Multiflow Summer Workshop, and by the British Engineering and Physical Sciences Research Council through grant number EP/M506485/1. We are grateful to Daniel Chung and Michael MacDonald for their careful critique of this manuscript.

**References**

Choi H, Moin P and Kim J 1993 Direct numerical simulation of turbulent flow over riblets *J. Fluid Mech.* **255**, 503–539.

Chung D, Chan L, MacDonald M, Hutchins N and Ooi A 2015 A fast direct numerical simulation method for characterising hydraulic roughness *J. Fluid Mech.* **773**, 418–431.

Colebrook C F 1939 Turbulent flow in pipes, with particular reference to the transitional region between smooth and rough wall laws *J. Inst. Civ. Eng.* **11**, 133–156.

Colebrook C F and White C M 1937 Experiments with fluid friction in roughened pipes *Proc. R. Soc. London Ser. A* **161**, 367–381.

Darcy H 1857 *Recherches expérimentales relatives au mouvement de l'eau*, Mallet-Bachelier Paris.

Fadlun E A, Verzicco R, Orlandi P and Mohd-Yusof J 2000 Combined immersed-boundary finite-difference methods for three-dimensional complex flow simulations *J. Comput. Phys.* **161**, 35–60.

Ferziger J H and Perić M 2002 *Computational methods for fluid dynamics* 3 edn Springer.

Flack K A and Schultz M P 2010 Review of hydraulic roughness scales in the fully rough regime *J. Fluids Eng.* **132**, 041203.

Flack K A and Schultz M P 2014 Roughness effects on wall-bounded turbulent flows *Phys. Fluids* **26**, 101305.

Flack K A, Schultz M P and Rose W B 2012 The onset of roughness effects in the transitionally rough regime *Int. J. Heat Fluid Flow* **35**, 160–167.

García-Mayoral R and Jiménez J 2011*a* Drag reduction by riblets. *Phil. Trans. R. Soc. A* **369**, 1412–1427.

García-Mayoral R and Jiménez J 2011*b* Hydrodynamic stability and breakdown of the viscous regime over riblets *J. Fluid Mech.* **678**, 317–347.

Hama F R 1954 Boundary-layer characteristics for smooth and rough surfaces *Trans. Soc. Naval Archit. Mar. Eng.* **62**, 333–358

Hoyas S and Jiménez J 2008 Reynolds number effects on the Reynolds-stress budgets in turbulent channels *Phys. Fluids* **20**, 101511.

Iaccarino G and Verzicco R 2003 Immersed boundary technique for turbulent flow simulations *Appl. Mech. Rev.* **56**, 331–347.

Jelly T O, Jung S Y and Zaki T A 2014 Turbulence and skin friction modification in channel flow with streamwise-aligned superhydrophobic surface texture *Phys. Fluids* **26**, 095102.

Jiménez J 2004 Turbulent flows over rough walls *Annu. Rev. Fluid Mech.* **36**, 173–196.

Jiménez J and Moin P 1991 The minimal flow unit in near-wall turbulence *J. Fluid Mech.* **225**, 213–240.

Jiménez J, Uhlmann M, Pinelli A and Kawahara G 2001 Turbulent shear flow over active and passive porous surfaces *J. Fluid Mech.* **442**, 89–117.

Kim H T, Kline S J and Reynolds W C 1971 The production of turbulence near a smooth wall in a turbulent boundary layer *J. Fluid Mech.* **50**, 133–160.

Kim J, Moin P and Moser R D 1987 Turbulence statistics in fully developed channel flow at low Reynolds number *J. Fluid Mech.* **177**, 133–166.

Le H and Moin P 1991 An improvement of fractional step methods for the incompressible Navier-Stokes equations *J. Comput. Phys.* **92**, 369–379.

Leonardi S and Castro I P 2010 Channel flow over large cube roughness: a direct numerical simulation study *J. Fluid Mech.* **651**, 519.

Leonardi S, Orlandi P and Antonia R A 2007 Properties of d- and k-type roughness in a turbulent channel flow *Phys. Fluids* **12**, 125101.

Ligrani P M and Moffat R J 1986 Structure of transitionally rough and fully rough turbulent boundary layers *J. Fluid Mech.* **162**, 69–98.

Luchini P, Manzo F and Pozzi A 1991 Resistance of a grooved surface to parallel flow and cross-flow *J. Fluid Mech.* **228**, 87–109.

Marusic I, Mathis R and Hutchins N 2010 Predictive model for wall-bounded turbulent flow. *Science* **329**, 193–196.

Mohd-Yusof J 1997 Combined immersed-boundary/B-spline methods for simulations of flow in complex geometries *Annu. Res. Briefs Center Turbul. Res.* pp. 317–327, Stanford Univ.

Nikuradse J 1933 Laws of flow in rough pipes *VDI-Forsch.* **361**

Nordström J, Mattsson K and Swanson C 2007 Boundary conditions for a divergence free velocity-pressure formulation of the Navier–Stokes equations *J. Comput. Phys.* **225**, 874–890.

Raupach M R, Antonia R A and Rajagopalan S S 1991 Rough-wall turbulent boundary layers *Appl. Mech. Rev.* **44**, 1–25.

Reynolds W C and Hussain A K M F 1972 The mechanics of an organized wave in turbulent shear flow. Part 3. Theoretical models and comparisons with experiments *J. Fluid Mech.* **54**, 263–288.

Schlichting H 1936 Experimental investigation of the problem of surface roughness *NACA Tech. Memo* 823.

Schultz M P and Flack K A 2007 The rough-wall turbulent boundary layer from the hydraulically smooth to the fully rough regime *J. Fluid Mech.* **580**, 381–405.

Seo J, García-Mayoral R and Mani A 2015 Pressure fluctuations in turbulent flows over superhydrophobic surfaces *J. Fluid Mech.* **783**, 448–473.

Smith C and Metzler S 1983 Characterization of low-speed streaks in the near-wall region of a turbulent boundary layer *J. Fluid Mech.* **129**, 27–54.

# The minimal channel: a fast and direct method for characterising roughness

**Michael MacDonald[1], Daniel Chung[1], Nicholas Hutchins[1], Leon Chan[1], Andrew Ooi[1], Ricardo García-Mayoral[2]**

[1] Department of Mechanical Engineering, University of Melbourne, Victoria 3010, Australia
[2] Department of Engineering, University of Cambridge, CB2 1PZ, UK

E-mail: michael.macdonald@unimelb.edu.au

**Abstract.** Roughness only alters the near-wall region of turbulent flow and leaves the outer-layer unaffected, making it a prime candidate for the minimal-span channel framework which only captures the near-wall flow. Recently, Chung *et al.* (*J. Fluid Mech.*, vol. 773, 2015, pp. 418–431) showed that the minimal-span channel can accurately characterise the hydraulic behaviour of roughness. Following on from this, we aim to further optimise the minimal-span channel framework by primarily noting that the outer layer it produces is inherently incorrect, and as such modifications to this region can be made to improve performance. Firstly, a half-height channel with slip wall is shown to reproduce the near-wall behaviour seen in a standard channel, but with half the number of grid points. Next, a forcing model is introduced into the outer layer of a half-height channel. This reduces the high streamwise velocity associated with the minimal channel and allows for a larger computational time step. The streamwise length of the channel is also investigated independent of the previous improvements, and suggests the minimum length should be at least 3 times the spanwise width and also 1000 viscous-units long, whichever is longer. Finally, an investigation is conducted to see if varying the roughness Reynolds number with time is a feasible method for obtaining the full hydraulic behaviour of a rough surface, instead of running multiple simulations at fixed roughness Reynolds numbers.

## 1. Introduction

Numerical simulations of wall-bounded turbulent flows represent a challenging computational problem, as both small and large scales need to be represented. The former requires a fine grid to resolve the small viscous scales, while the latter requires a large domain to capture the large outer-layer motions. However, pioneering work by [1, 2] into minimal-flow units showed that only a small computational domain can be used to exclusively capture the turbulent near-wall cycle, independent of the large outer scales. This is achieved by simply restricting the domain of the channel to a small size, where the spanwise and streamwise lengths are prescribed in terms of viscous units. Ref. [1] showed that turbulence could be maintained in the form of the near-wall cycle of the buffer layer when the spanwise domain width was only on the order of $100\nu/U_\tau$; here $\nu$ is the kinematic viscosity and $U_\tau = \sqrt{\tau_w/\rho}$ is the friction velocity, defined using the wall shear stress $\tau_w$ and the fluid density $\rho$ . This was further supported by future studies into minimal-flow units [3–5]. In particular the work of [3] showed that the minimal-span channel can also capture the logarithmic layer of turbulent flows.

An important aspect of minimal-span channels is that the near-wall flow is accurately captured up until a critical wall-normal location, $z_c$. Above this point, the streamwise velocity increases compared to a full-span channel. This occurs as the narrow spanwise domain width of the minimal-span channel acts as a filter which limits the largest spanwise scale of energy-containing eddies. Note that the flow retains turbulent scales smaller than the spanwise domain width, so it is not laminar either. The critical value where the minimal-span channel departs from the full-span channel has been shown to scale with the spanwise domain width, $z_c \approx 0.4L_y$ [5], although a constant of 0.3 is suggested in [3]. Following [3, 6], we will refer to the flow below $z_c$ as 'healthy' turbulence, as it is the same as in a full-span channel. The largest captured eddy present in the domain would have a spanwise size $\lambda_y$ that is equal to channel width, $\lambda_y = L_y$. In the logarithmic region of the flow, the streamwise length of quasi-streamwise vortices scale as roughly 2–3 times the spanwise width, $\lambda_x \approx 2\text{–}3\lambda_y$ [3, 7].

In the context of roughness, the central question is how to relate the geometry of a rough surface to its hydraulic behaviour; namely, what value the (Hama) roughness function, $\Delta U^+$, takes [8]. This quantity reflects the flow retardation, or velocity shift, that the roughness has on the flow when compared to a smooth wall, for matched friction Reynolds numbers, $Re_\tau = U_\tau h/\nu$ (here $h$ is the half-channel height, boundary layer thickness, or pipe radius). For a given surface, $\Delta U^+$ can be obtained from various models and approximations, or directly from full-scale experiments and numerical simulations. The former are of varying accuracy and depend on the model selected and the rough surface in question, while the latter can be prohibitively expensive. In particular, full scale numerical simulations suffer from the drawbacks mentioned above, in that both small- and large-scale motions need to be captured. However, roughness is thought to primarily alter only the near-wall flow, in a region called the roughness sublayer which extends 3–5$k$ from the wall, where $k$ is some characteristic height of the roughness [9]. Outside the roughness sublayer, in the outer layer, the flow is only changed insofar as it depends on the friction velocity $U_\tau$. This is the basis of Townsend's outer-layer similarity hypothesis [10] which has received significant attention and has been supported by several rough-wall studies [9, 11, 12] and also numerical simulations with modified boundary conditions [13–15]. If we assume Townsend's hypothesis holds, then it follows that the roughness only alters the near-wall region of the flow which therefore makes it a prime candidate for use in the minimal-flow framework.

Recently, we have applied the idea of minimal-span channels to the roughness problem [5]. It was shown that the roughness function could be accurately computed from minimal-span channels, when compared with full-span channels. However, the use of minimal-span channels has some drawbacks; namely the flow is 'bursty', resulting in large fluctuations in the flow statistics over time, which requires a long averaging time to obtain statistically stationary data. Another issue is that the streamwise velocity is extremely large in the altered outer layer above $z_c$, which places a restriction on the time step due to the Courant-Friedrichs-Lewy ($CFL$) condition.

In this paper, we aim to improve the minimal-span channel method for roughness. This is achieved by noting that the outer layer in minimal-flow units is meaningless as it does not represent a true turbulent flow, so alterations to this region are made while ensuring the near-wall behaviour remains unchanged. To this end, a series of optimisations to the minimal channel are performed. Firstly, only half of the channel is simulated, with a slip wall placed at the channel centreline. Second, a forcing model is applied to the outer layer of the half-height channel, to reduce the centreline velocity and improve the computational time step. Finally, an investigation of the streamwise domain length is performed, independent of the previous improvements. While the flow will still be bursty, which is an intrinsic aspect of minimal flow units [16, 17], these improvements reduce the impact that it has as the long time-scale burstiness can be simulated faster. Finally, an investigation independent of the previous improvements is conducted in which the bulk velocity (and hence roughness Reynolds number) is varied with time, to see if the full hydraulic behaviour of the rough surface can be obtained. This is of interest as currently multiple (steady) simulations are needed to obtain the $\Delta U^+$ vs $k^+$ curve.

## 2. Numerical Method

We conduct Direct Numerical Simulations (DNS) using the same finite difference code as in [5], which uses a fully conservative fourth-order staggered-grid scheme. Time integration is performed using the third-order low-storage Runge–Kutta scheme of [18], and the fractional-step method of [19] is used. The Navier–Stokes equations are

$$\nabla \cdot \mathbf{u} = 0, \qquad \frac{\partial \mathbf{u}}{\partial t} + \nabla \cdot (\mathbf{u}\mathbf{u}) = -\frac{1}{\rho}\nabla p + \nabla \cdot (\nu \nabla \mathbf{u}) + \mathbf{F} + \mathbf{G} + \mathbf{K}, \tag{1}$$

where $\mathbf{u} = (u, v, w)$ is the velocity in the streamwise ($x$), spanwise ($y$) and wall-normal ($z$) directions, $t$ is time, and $p$ is pressure. $\mathbf{G} = G_x(t)\mathbf{i}$ is the spatially invariant, time-varying streamwise forcing term which drives the flow at constant mass flux. The flow is solved in a moving reference frame so that the mass flux is zero at all times, although all equations in this paper are given in the stationary reference frame. Solving in a moving reference frame permits a larger computational time step [20], as well as reducing high-wavenumber convective disturbances produced by finite difference schemes [21]. $\mathbf{F} = -\alpha F(z, k)u|u|\mathbf{i}$ is a roughness model based on the work of [22] which applies a forcing in the streamwise direction which opposes the flow. The roughness factor $\alpha = 1/(40k)$ is kept constant for all simulations, with the roughness height $k = h/40$. The function $F(z, k)$ is a simple step function which applies the roughness forcing close to the top and bottom no-slip channel walls at $z = 0$ and $z = 2h$,

$$F(z,k) = \begin{cases} 1, & \text{if } z < k \text{ or } 2h - k < z. \\ 0, & \text{otherwise.} \end{cases} \tag{2}$$

This roughness model removes any spanwise or streamwise roughness length scales from the problem and allows for simpler computations as the same smooth-wall grid can be used. This is a similar approach to [23], although there the author used an $\mathbf{F} \propto u\mathbf{i}$ scaling. The final term in (1), $\mathbf{K}$, is a forcing function which damps the velocity in the outer layer of the minimal channel. It has the form

$$K_i = -\lambda \Gamma(z, z_d)\Big(u_i(x,y,z,t) - \langle u_i\rangle(z_d,t)\Big)\Big|u_i(x,y,z,t) - \langle u_i\rangle(z_d,t)\Big|, \tag{3}$$

where angled brackets denote the spatial average of the instantaneous velocity over a wall-parallel plane. The factor $\lambda$ has units of inverse time and in this study has a value of $\lambda h/U_\tau \approx 1$. The parameter $\Gamma(z, z_d)$ is similar to the function $F(z, k)$ of the roughness forcing model in that it prescribes where the damping is applied, namely in the outer layer of the channel,

$$\Gamma(z, z_d) = \begin{cases} 1, & \text{if } z > z_d \text{ and } z < 2h - z_d. \\ 0, & \text{otherwise.} \end{cases} \tag{4}$$

The value of $z_d$ should be greater than the critical wall-normal location, $z_c$, of the minimal-span channel, so that the forcing does not contaminate the healthy turbulence of the near-wall flow. Several different values of $z_d$ are tested and will be presented in the following section. In (3), the term $\langle u_i\rangle(z_d, t)$ is the wall-parallel spatially averaged velocity at $z_d$. This is present to reduce the magnitude of the streamwise velocity; the minimal channel has a very high streamwise velocity which presents a time step restriction from the *CFL* number. The forcing is in this form so that the velocity is forced to remain at the same level as at $z = z_d$. This outer-layer damping is similar to the masks employed in [6], which damp fluctuations in the outer-layer region of minimal channels. The current forcing model is fairly crude in that it simply sets the streamwise velocity to be approximately that at $z = z_d$, however it will be shown to work reasonably well for the

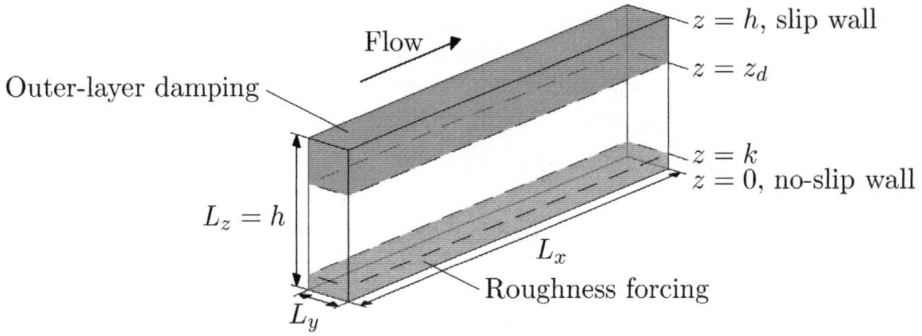


Figure 1: Half-height minimal-span channel, showing the roughness forcing zone (2) and the outer-layer damping zone (3). No-slip wall at $z = 0$ and slip wall at $z = h$.

purpose of obtaining $\Delta U^+$. Note that the average velocity in the spanwise and wall-normal directions is zero, so that the forcing in these directions can simply be $K_i = -\lambda\Gamma(z, z_d)u_i|u_i|$, for $i = 2, 3$. Figure 1 shows the two forcing regions that are employed in the current study.

The streamwise and spanwise grid is evenly spaced, while the wall-normal grid is stretched with a cosine mapping. Periodic boundary conditions are applied in the spanwise and streamwise directions. In the case of standard-height channels, no-slip walls are located at $z = 0$ and $z = 2h$. For clarity, the word 'full' will be used to refer to the span (full span), while the case with two no-slip walls will be referred to as 'standard' height. For half-height (open) channels, a no-slip wall is still positioned at $z = 0$, however now the top domain surface is a slip wall with $\partial u/\partial z = \partial v/\partial z = 0$, positioned at $z = h$. This slip wall maintains the impermeability constraint, $w = 0$. The spanwise width, $L_y$, of the minimal-span channel follows the guidelines of [5]. Simulation parameters are given in table 1. For each entry, both a smooth wall and modelled rough wall are simulated with the stated parameters.

### *2.1. Temporal sweep*

The equivalent sandgrain roughness, $k_s^+$, is a single dynamic parameter that is used to describe the hydraulic behaviour of roughness. It is defined so that the roughness function collapses for all data in the fully rough regime, when the friction factor no longer depends on the bulk-velocity Reynolds number [24]. However, each rough surface will have a unique behaviour in the transitionally rough regime. It is an expensive process to determine the full hydraulic behaviour, as a range of simulations need to be conducted for the different roughness Reynolds numbers, each requiring a different body-fitted grid. Moreover, for constant mass flux simulations as in this study, the mass flux is altered until the friction Reynolds number is close to the desired value. The log layer of the flow is not easily identifiable for low Reynolds number simulations, so the roughness function is calculated assuming the smooth- and rough-wall flows have matched $Re_\tau$. This means that differences in $Re_\tau$ introduce a level of uncertainty into the estimate of $\Delta U^+$. It is desirable to conduct a simulation in which only a single grid is used, and the bulk velocity is changed over time. This is a similar approach to how experimental studies are performed, and will allow a sweep through a range of roughness Reynolds numbers. In order to obtain statistics at a desired friction Reynolds number, the flow is averaged over a small window in which the instantaneous friction Reynolds number is close to the desired one. Within this window, the flow is assumed to be quasi steady. An important consideration which will be investigated here is whether acceleration effects become significant and distort the estimation of $\Delta U^+$, that is, how quickly can the bulk velocity be varied.

The sweep will start with the highest friction number to be tested, $Re_{\tau,start}$ and will then be decelerated via an adverse pressure gradient. This results in the *CFL* restriction gradually reducing as the simulation progresses and also ensures an adequate grid resolution at the start

Table 1: Description of the simulations performed. All simulations done as both smooth and rough walls. Entries: $Re_\tau$, nominal friction Reynolds numbers; $L$, domain length; $N$, number of grid points; $\Delta$, grid-spacing in viscous units, subscript $w$ and $h$ refers to the wall-normal grid spacing at the wall and channel centre; $z_d^+$, location where the outer-layer damping begins (3). $L_z/h = 2$ denotes standard-height (two no-slip walls), $L_z/h = 1$ denotes half-height channel (one no-slip wall and one slip wall). $\Delta t_S^+$ and $\Delta t_R^+$ is the smooth- and rough-wall time step. Roughness height fixed at $k = h/40$ so $k^+ \approx 15$ at $Re_\tau = 590$, and $k^+ \approx 50$ at $Re_\tau = 2000$.

| $Re_\tau$ | $\frac{L_x}{h}$ | $L_x^+$ | $L_y^+$ | $\frac{L_z}{h}$ | $N_x$ | $N_y$ | $N_z$ | $\Delta x^+$ | $\Delta y^+$ | $\Delta z_w^+$ | $\Delta z_h^+$ | $z_d^+$ | $\Delta t_S^+/\Delta t_R^+$ |
|---|---|---|---|---|---|---|---|---|---|---|---|---|---|
| $S_1$: Full-span channel | | | | | | | | | | | | | |
| 590 | $2\pi$ | 3707 | 1854 | 2 | 384 | 384 | 256 | 9.7 | 4.8 | 0.04 | 7.2 | - | 0.26/0.29 |
| 590 | $2\pi$ | 3707 | 1854 | 1 | 384 | 384 | 128 | 9.7 | 4.8 | 0.04 | 7.2 | - | 0.29/0.32 |
| $S_2$: Minimal-span channel | | | | | | | | | | | | | |
| 590 | $2\pi$ | 3707 | 118 | 2 | 384 | 24 | 256 | 9.7 | 4.9 | 0.04 | 7.2 | - | 0.35/0.40 |
| 590 | $2\pi$ | 3707 | 118 | 1 | 384 | 24 | 128 | 9.7 | 4.9 | 0.04 | 7.2 | - | 0.36/0.41 |
| 590 | $2\pi$ | 3707 | 236 | 2 | 384 | 48 | 256 | 9.7 | 4.9 | 0.04 | 7.2 | - | 0.37/0.39 |
| 2000 | $0.59\pi$ | 3707 | 300 | 1 | 384 | 60 | 384 | 9.7 | 5.0 | 0.02 | 8.2 | - | 0.24/0.31 |
| $S_{3a}$: Minimal-span, half-height channel with outer-layer damping | | | | | | | | | | | | | |
| 590 | $2\pi$ | 3707 | 118 | 1 | 384 | 24 | 128 | 9.7 | 4.9 | 0.04 | 7.2 | 100 | 0.48/0.59 |
| 590 | $2\pi$ | 3707 | 118 | 1 | 384 | 24 | 128 | 9.7 | 4.9 | 0.04 | 7.2 | 150 | 0.46/0.54 |
| 590 | $2\pi$ | 3707 | 118 | 1 | 384 | 24 | 128 | 9.7 | 4.9 | 0.04 | 7.2 | 200 | 0.44/0.51 |
| 590 | $2\pi$ | 3707 | 118 | 1 | 384 | 24 | 128 | 9.7 | 4.9 | 0.04 | 7.2 | 300 | 0.41/0.47 |
| $S_{3b}$: Minimal-span, half-height channel with outer-layer damping (wider span) | | | | | | | | | | | | | |
| 590 | $2\pi$ | 3707 | 236 | 1 | 384 | 48 | 128 | 9.7 | 4.9 | 0.04 | 7.2 | 200 | 0.43/0.48 |
| 590 | $2\pi$ | 3707 | 236 | 1 | 384 | 48 | 128 | 9.7 | 4.9 | 0.04 | 7.2 | 250 | 0.43/0.47 |
| 590 | $2\pi$ | 3707 | 236 | 1 | 384 | 48 | 128 | 9.7 | 4.9 | 0.04 | 7.2 | 300 | 0.42/0.46 |
| $S_{3c}$: Minimal-span, half-height channel with outer-layer damping (increased $Re_\tau$) | | | | | | | | | | | | | |
| 2000 | $0.59\pi$ | 3707 | 300 | 1 | 384 | 60 | 384 | 9.7 | 5.0 | 0.02 | 8.2 | 300 | 0.25/0.37 |
| 2000 | $0.59\pi$ | 3707 | 300 | 1 | 384 | 60 | 384 | 9.7 | 5.0 | 0.02 | 8.2 | 400 | 0.24/0.35 |
| $S_{4a}$: Minimal-span, standard-height channel, varying streamwise length | | | | | | | | | | | | | |
| 590 | $0.1\pi$ | 185 | 118 | 2 | 24 | 24 | 256 | 7.7 | 4.9 | 0.04 | 7.2 | - | 0.28/0.32 |
| 590 | $0.16\pi$ | 297 | 118 | 2 | 48 | 24 | 256 | 6.2 | 4.9 | 0.04 | 7.2 | - | 0.24/0.27 |
| 590 | $\pi/4$ | 463 | 118 | 2 | 48 | 24 | 256 | 9.7 | 4.9 | 0.04 | 7.2 | - | 0.36/0.42 |
| 590 | $\pi/2$ | 927 | 118 | 2 | 96 | 24 | 256 | 9.7 | 4.9 | 0.04 | 7.2 | - | 0.36/0.42 |
| 590 | $\pi$ | 1854 | 118 | 2 | 192 | 24 | 256 | 9.7 | 4.9 | 0.04 | 7.2 | - | 0.36/0.41 |
| $S_{4b}$: Minimal-span, standard-height channel, varying streamwise length (wider span) | | | | | | | | | | | | | |
| 590 | $0.1\pi$ | 185 | 354 | 2 | 24 | 72 | 256 | 7.7 | 4.9 | 0.04 | 7.2 | - | 0.37/0.45 |
| 590 | $0.16\pi$ | 297 | 354 | 2 | 48 | 72 | 256 | 6.2 | 4.9 | 0.04 | 7.2 | - | 0.31/0.40 |
| 590 | $\pi/4$ | 463 | 354 | 2 | 48 | 72 | 256 | 9.7 | 4.9 | 0.04 | 7.2 | - | 0.46/0.52 |
| 590 | $\pi/2$ | 927 | 354 | 2 | 96 | 72 | 256 | 9.7 | 4.9 | 0.04 | 7.2 | - | 0.43/0.46 |
| 590 | $\pi$ | 1854 | 354 | 2 | 192 | 72 | 256 | 9.7 | 4.9 | 0.04 | 7.2 | - | 0.38/0.40 |
| $S_5$: Temporal sweep (at initial conditions) | | | | | | | | | | | | | |
| 590 | $2\pi$ | 3707 | 354 | 2 | 384 | 80 | 256 | 9.6 | 4.4 | 0.04 | 7.2 | - | 0.38/0.43 |

of the simulation. This means that at the final friction Reynolds number, $Re_{\tau,end} < Re_{\tau,start}$, the grid resolution would be finer than necessary, which for the current simulations would have 4–5 times more cells than if a conventional steady simulation were conducted.

Models for how to vary the bulk velocity over time, $U_b(t)$, could be made to ensure that there occurs a certain number of turnover times of the largest captured eddy in the minimal-span

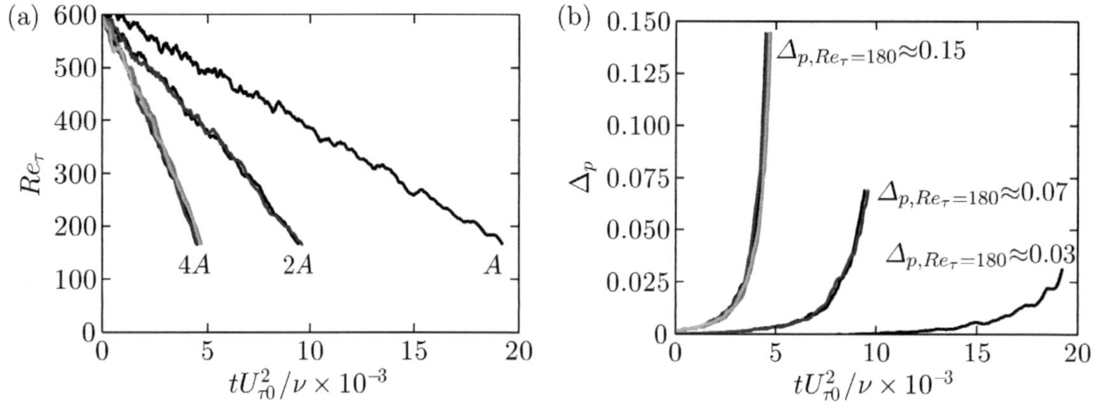


Figure 2: (a) Instantaneous Reynolds number and (b) pressure gradient parameter (6) of the smooth-wall temporal sweep, against viscous time. The cases with faster sweeps have multiple runs (shown by grey lines) and are ensemble averaged. $A$ is the linear rate of change of $U_b$ (5).

channel. This would require information on the characteristic time scale of the eddies, which would change with $Re_\tau$. However, as a preliminary study into the feasibility of this approach, we will vary it linearly with time,

$$\mathrm{d}U_b/\mathrm{d}t = A = (U_{b,end} - U_{b,start})/\Delta T, \tag{5}$$

where $U_{b,end}$ is the desired end point of the simulation (the bulk velocity corresponding to $Re_\tau = 180$), and $\Delta T$ is a time scale which should be sufficiently long enough that acceleration effects are not significant to the flow. The initial bulk velocity, $U_{b,start}$, is set corresponding to a friction Reynolds number of $Re_{\tau,start} = 590$ and different values of $A$ are tested. An initial value of $A$ is selected such that over the course of the sweep $tU_{\tau0}/h = 30$, with subsequent runs using $2A$ and $4A$. In these cases of a higher rate of change of $U_b$, multiple sweeps are run from different initial conditions with the results ensemble averaged, to ensure that statistics are obtained over the same amount of simulation time. For example, the case where the gradient is quadrupled, four sweeps are conducted with four different initial conditions. This is visualised in figure 2(a), which shows the change in the friction Reynolds number as a function of time. An important consideration with this technique is that the friction velocity and hence the normalised spanwise domain width $L_y^+$ will vary with time. The domain width must remain larger than 100 viscous units at all times to ensure the turbulent flow is sustained, which may necessitate multiple stages in the sweep if $L_y^+$ becomes too small. Each stage in the sweep should be set up such that $L_y^+ \gtrsim 100$.

A non-dimensional measure of the flow acceleration is the pressure gradient parameter,

$$\Delta_p = \frac{\nu}{\rho U_\tau^3}\frac{\mathrm{d}p}{\mathrm{d}x}. \tag{6}$$

Various experimental studies of decelerating boundary layers show that mild decelerations of $\Delta_p < 0.01$ have small ($< 3\%$) errors in calculating $U_\tau$ from methods based on assuming zero pressure gradients [25, 26]. While we are interested in channel flow as opposed to boundary layers, this value should still be useful in providing an indication as to whether acceleration effects will be significant. Ref. [27], meanwhile, compared a step acceleration with a slow ramp up in which the bulk velocity was linearly varied in a channel such that $-\Delta_p \approx 0.73$ (favourable pressure gradient). Acceleration effects were still seen in this slow ramp up. Note

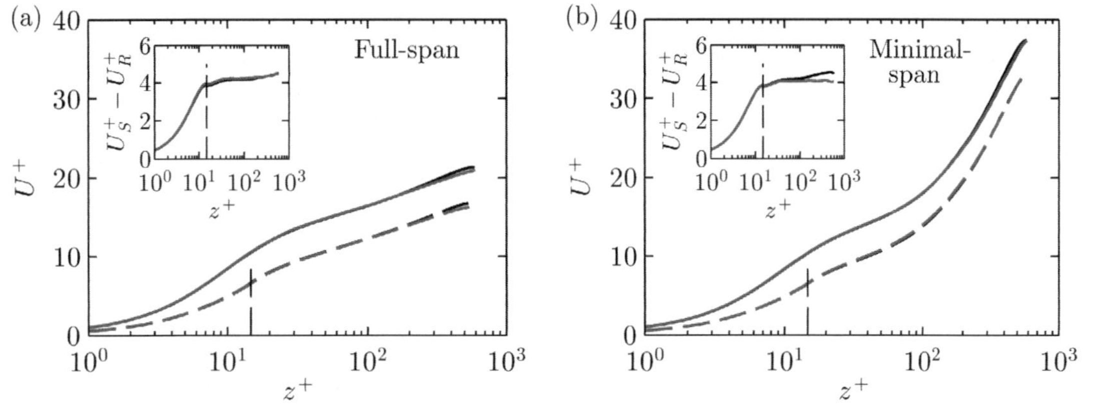


Figure 3: Mean streamwise velocity against wall-normal position for (a) full-span channel (set $S_1$) and (b) minimal-span channel (set $S_2$). Line styles are: black, standard-height channel; grey, half-height channel; solid, smooth wall; dashed, rough-wall model. Vertical dashed line shows roughness crest, $k = h/40$. Inset shows difference in smooth- and rough-wall velocity, $U_S^+ - U_R^+$.

that for steady channel flow, the pressure gradient parameter is $\Delta_p = -1/Re_\tau < 0$, which for the current simulations at $Re_\tau = 590$ takes a value of $-0.0017$. In the current study, three different sweep rates are studied. As shown in figure 2(b), the linear variation in $U_b$ results in the pressure gradient parameter being maximum at the end of the simulation, when $Re_\tau \approx 180$. The different sweeps will therefore be referred to by their maximum pressure gradient parameter, which are $\Delta_{p,Re_\tau=180} \approx (0.03, 0.07, 0.15)$. These values place the sweeps considered in this study into the range between negligible ($\Delta_p < 0.01$) and noticeable ($-\Delta_p > 0.73$) pressure gradient effects, and are much larger than the steady value ($-\Delta_p = 0.0017$).

## 3. Results

### 3.1. Half-height channel (sets $S_1$ and $S_2$)

The effect of using a half-height (open) channel is investigated first. This channel has a slip wall positioned at the channel centreline, so only half of the channel needs to be simulated. This reduces the size of the grid by a half when compared to a conventional standard-height channel with two no-slip walls. The simulations are detailed in table 1 as sets $S_1$ and $S_2$. Here, the streamwise length is fixed at $L_x^+ \approx 3700$ and there is no outer-layer damping, $K_i = 0$.

Figure 3 shows the effect of the half-height channel in terms of the mean velocity profile, when compared to the standard-height channel. This is done for both smooth and rough walls, as well as full-span and minimal-span channels. For clarity, the full-span and minimal-span channels are shown in different figures, however the near-wall behaviour is identical up until $z_c = 0.4L_y \Rightarrow z_c^+ \approx 50$, as observed in [5]. Both sets of figures show that the use of the half-height channel with slip wall has a negligible effect on the flow. The main difference is seen in the wake region, where the half-height channel restricts the outer-layer structures, resulting in a slight decrease in the mean velocity. Crucially, this difference is the same for both smooth and rough walls, meaning that the difference between them, $U_S^+ - U_R^+$, is the same. Moreover, the change in the half-height channel occurs above the critical wall-normal location, where the minimal-span flow is already altered compared to the full-span flow. The root-mean-squared velocity fluctuations (not shown) also show good agreement between half- and standard-height channels in the near-wall region, in agreement with previous open channels [28–30]. There is some variation between the two in the outer region although this is to be expected.

Essentially, the near-wall flow is preserved, despite the imposition of an outer-layer boundary

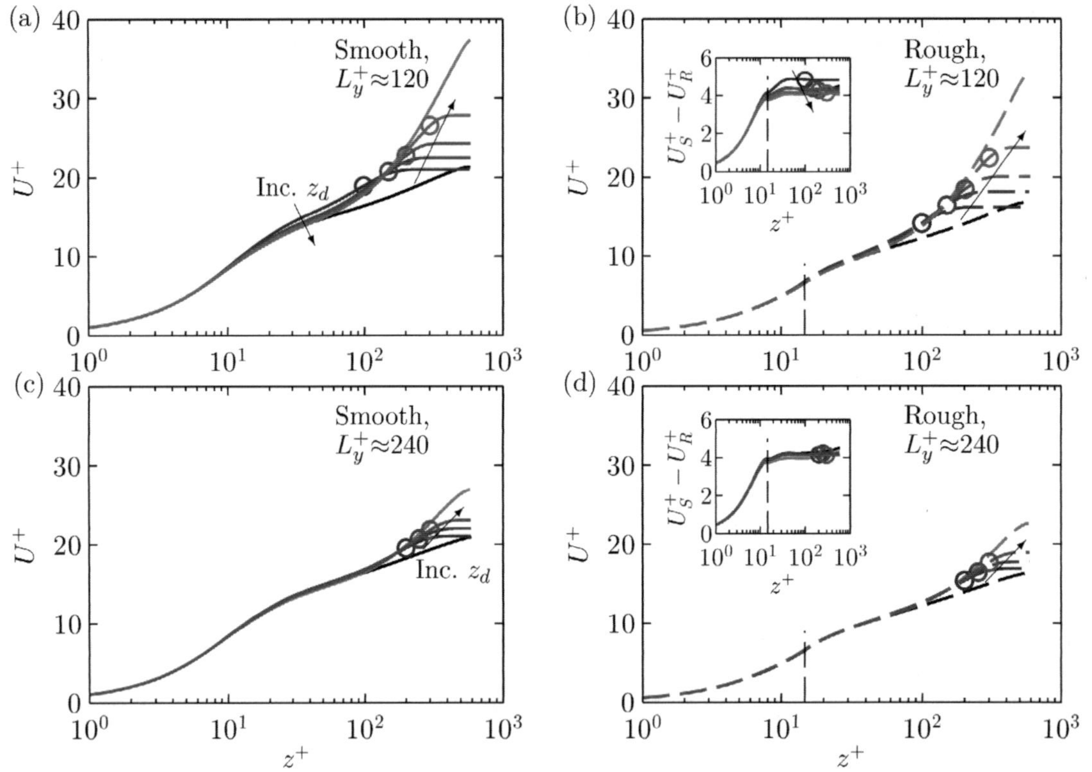


Figure 4: Streamwise velocity profile for outer-layer damping for (a,c) smooth- and (b,d) rough-wall half-height channels. Spanwise domain width is (a,b) $L_y^+ \approx 120$ (set $S_{3a}$) and (c,d) $L_y^+ \approx 240$ (set $S_{3b}$). Line styles are: black, full-span channel; grey, minimal-span channel with outer-layer damping (3) for varying damping heights of (a,b) $z_d^+ = [100, 150, 200, 300]$ and (c,d) $z_d^+ = [200, 250, 300]$, shown by circles; light-grey, minimal-span with no damping. Vertical dashed line indicates the roughness height $k = h/40$. Inset of (b,d) shows the difference in smooth- and rough-wall velocity. Arrows indicates trend of increasing $z_d$.

condition. This echoes the work of [6], who showed that the near-wall cycle is preserved despite damping of outer-layer fluctuations. The half-height channel is effectively a less restrictive boundary condition than this damping, as the flow retains velocity fluctuations in the outer-layer. Importantly, it shows that only half the domain needs to be simulated without significant near-wall detriment.

### *3.2. Outer-layer damping (sets $S_{3a}$, $S_{3b}$, and $S_{3c}$)*

The outer-layer flow is now damped through use of the $K_i$ forcing term (3) in a half-height channel. This is done to reduce the large centreline velocity (figure 3b) of the minimal-span channel, which places a restriction on the time step due to the *CFL* number. The streamwise domain length is fixed at $L_x^+ \approx 3700$, and other relevant parameters are detailed in table 1 under set $S_3$. The height where the damping begins, $z_d$, is investigated with several different positions in channels of two different spanwise widths; $L_y^+ \approx 120$ (set $S_{3a}$) and $L_y^+ \approx 240$ (set $S_{3b}$). These minimal channels will have a healthy turbulence region up to $z_c^+ \approx 47$ and $z_c^+ \approx 94$, respectively.

Figure 4 shows the mean streamwise velocity profile for smooth- and rough-wall minimal-span channels, for both of the spanwise widths. The smallest spanwise width, $L_y^+ \approx 120$ (figure 4a,b)

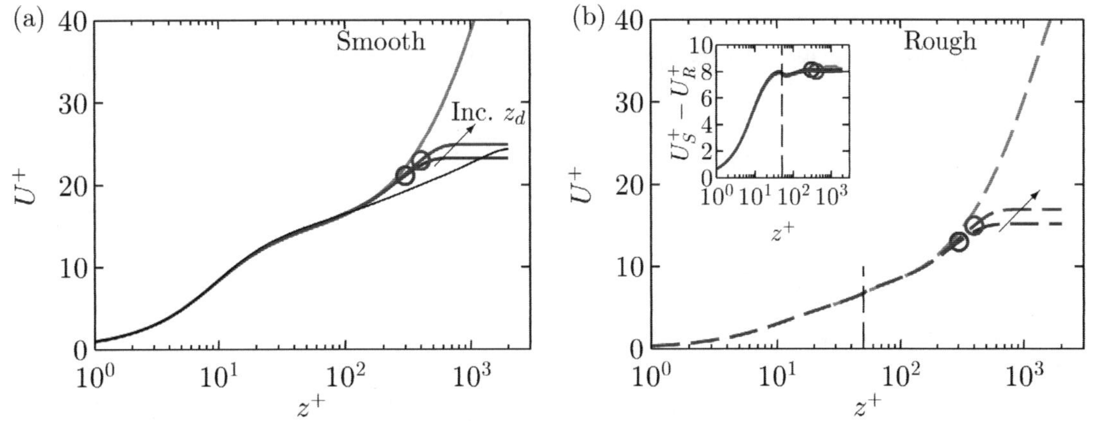


Figure 5: Streamwise velocity profile for (a) smooth- and (b) rough-wall half-height channels at $Re_\tau = 2000$ (set $S_{3c}$). Line styles are same as figure 4, however full-span smooth-wall channel data taken from [31] and damping heights are $z_d^+ = [200, 300]$, shown by circles.

has four different positions for $z_d$ examined, with values of $z_d^+ = (100, 150, 200, 250)$. In all cases, the damping tries to restricts the streamwise velocity above $z > z_d$ to the value of $U(z = z_d)$. It is clear in the smooth-wall flow (figure 4a) that the two smallest values of $z_d^+ = 100$ and 150 are too close to the wall, resulting in an increase in the streamwise velocity below $z_d$. In particular, $z_d^+ = 100$ results in an obvious overestimation of $\Delta U^+$, as seen in the inset of figure 4(b). A value of $z_d^+ = 150$ has a similar, although not as strong of an effect. This is clearly not desirable and suggests that a value of $z_d^+ \geq 200$ is recommended for this spanwise width.

To investigate how the location of $z_d$ is related to $z_c$ will require cases with a larger spanwise domain width, as $z_c \approx 0.4L_y$. This is done in figure 4(c,d), where the spanwise width is now $L_y^+ \approx 240$. Here, three different positions for $z_d$ are analysed, with values of $z_d^+ = (200, 250, 300)$. There is little difference in the near-wall profiles of the three different $z_d$ positions. This suggests that, even though $z_c$ has doubled, we still require the damping to begin at least 200 viscous-units away from the wall. Having the outer-layer damping begin closer to the wall (i.e. reducing $z_d$) seems to lead to an increase in viscous stress in the log-layer and even buffer layer (figure 4a), resulting in an overestimation of the mean streamwise velocity.

Next, the friction Reynolds number is increased to $Re_\tau = 2000$ (set $S_{3c}$). Since the roughness height is fixed as a ratio of the channel half-height, $k = h/40$, then the roughness Reynolds number increases to $k^+ = 50$. The spanwise width also needs to be increased to submerge the roughness sublayer in healthy turbulence [5], here it takes a value of $L_y^+ = 300$ so that the critical wall-normal location is $z_c^+ \approx 120 = 2.4k$. Figure 5 shows the mean velocity profile for these higher $Re_\tau$ simulations in which $z_d^+ = (200, 300)$. Only smooth-wall full-span channel data is available from [31] to compare with the minimal-span channel, and good agreement is seen up until $z^+ \approx 120$ (figure 5a). Here, both positions of $z_d$ produce velocity profiles which compare well with the minimal-span channel with no outer-layer damping.

Recall that the purpose of this outer-layer damping is to reduce the computational time step, which is limited by the $CFL$ number, $CFL = \max_{1\leq i\leq 3}(|u_i|\Delta t/\Delta x_i)$. Here $|u_i|$ is the maximum instantaneous velocity in the $i$th direction and $\Delta x_i$ is the corresponding grid spacing at that location. Note that the viscous $CFL$ number does not become significant in any of the simulations performed here. For minimal-span channels at $Re_\tau = 590$ with no outer-layer damping, the maximum $CFL$ number occurs in the outer layer due to the large streamwise velocity. We see the time step improves by roughly 20–24% with the use of the outer-layer damping (table 1, set

$S_{3a}$) than without (set $S_2$). This means the simulations can be completed quicker and hence use less computational resources. However, at higher friction Reynolds numbers of $Re_\tau = 2000$ (set $S_{3c}$), the time step for the smooth wall with outer-layer damping is unaltered compared to the case without (set $S_2$). This is due to the use of the cosine mapping to define the wall-normal grid, which produces a very fine grid near the wall especially for high Reynolds number cases. As a result of such a small $\Delta z$, the wall-normal *CFL* number is now maximum and this occurs in the near-wall region, which is independent of the outer-layer damping. If a more appropriate mapping was used for these high friction Reynolds numbers, we would expect a similar improvement in time step as at $Re_\tau = 590$.

The most appropriate position for $z_d$ is difficult to establish from the above data. A minimum of $z_d^+ \approx 200$ seems necessary to ensure that the damping does not interact with the near-wall flow. It also seems reasonable that $z_d$ should scale in some way on $z_c$, for larger domain widths. The three sets of simulations suggest that $z_d$ should be roughly two times $z_c$, so a tentative rule of thumb would be $z_d^+ \gtrsim \max(200,\ 2z_c^+)$. For an appropriate wall-normal mapping, this allows a 20–24% improvement in the computational time step.

### 3.3. Streamwise domain length (sets $S_{4a}$ and $S_{4b}$)

In the previous simulations, the streamwise length was fixed at $L_x/h = 2\pi$, which corresponds to a viscous length of about $3700\nu/U_\tau$ at $Re_\tau = 590$. This is regarded as being a suitable length for full-span simulations at this friction Reynolds number [32]. However, this length is only necessary due to the large outer-layer structures which are present in the full-span simulations, which have streamwise lengths on the order of the channel half-height, $h$. The minimal-span channel only captures the inner flow, so it is likely a shorter length can be used.

The near-wall cycle of the buffer layer produces streaky structures with streamwise lengths of 1000 viscous units [33]. However, these low- and high-speed streaks are nearly two-dimensional, which suggests they could be represented in the infinite ($k_x = 0$) mode and hence the domain length doesn't have to be this long. These streaks are accompanied by quasi-streamwise vortices [34], which extend into the logarithmic layer. These have streamwise lengths which scale as roughly 2–3 times their spanwise width [3, 7]. The minimum spanwise domain width that can sustain turbulence is around 100 viscous units [1], which means the smallest vortices that need to be captured would have a streamwise length of 200-300 viscous units. Simulations with larger spanwise domain widths would produce vortices with longer streamwise lengths, so that the largest captured eddy would have spanwise and streamwise lengths of $\lambda_y = L_y$ and $\lambda_x = 2$–$3L_y$.

Two different spanwise domain widths are investigated for varying the streamwise length (table 1, set $S_4$). Firstly, the smallest width of $L_y^+ \approx 120$ (set $S_{4a}$) would have the largest eddies having streamwise lengths of $\lambda_x^+ = 240$–$350$. Various streamwise domain lengths of $L_x^+ = (190, 300, 460, 930, 1850)$ are chosen to see how this affects the largest eddies. The second spanwise domain width of $L_y^+ \approx 350$ (set $S_{4b}$) would be able to capture larger eddies, the largest of which would have a streamwise length of roughly $\lambda_x^+ = 710$–$1100$. The same set of streamwise domain lengths are simulated for this wider domain. A standard-height channel is used, with no outer-layer damping, $K_i = 0$.

Figure 6 shows the mean velocity profile for the two spanwise widths, for all the streamwise lengths tested. For the smaller domain width (figure 6a,b), the shortest streamwise lengths of $L_x^+ \approx 190$ and 300 result in a reduction in the smooth-wall velocity above $z^+ \gtrsim 20$. There is also a slight increase in the centreline velocity. As a result, the difference in smooth- and rough-wall velocity (inset of figure 6b) decreases relative to the full-span and longer minimal-span cases. Note that while the grid resolution is different for these shorter channels (set $S_{4a}$, table 1) which may explain this effect, it will later be seen in the wider channels that have the same grid resolution differences (set $S_{4b}$) that this is not the case. Little discernible difference can be seen between the longer streamwise lengths of $L_x^+ \geq 460$, in agreement with the scalings

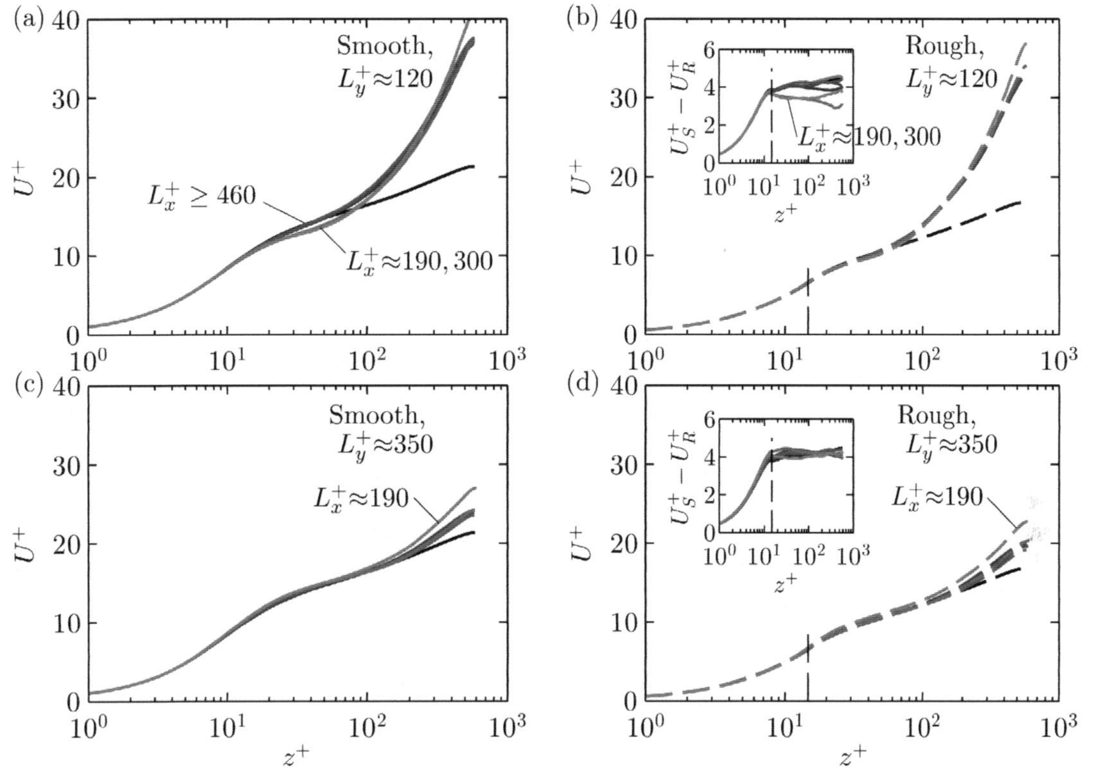


Figure 6: Streamwise velocity profile for varying streamwise length for (a,c) smooth- and (b,d) rough-wall standard-height channels. Spanwise domain width is (a,c) $L_y^+ \approx 120$ (set $S_{4a}$) and (b,d) $L_y^+ \approx 350$ (set $S_{4b}$). Line styles are: black, full-span channel; grey, minimal-span channel with varying streamwise length (dark grey is longer), refer table 1; Vertical dashed line indicates the roughness height $k = h/40$. Insets of (b,d) shows difference in smooth and rough wall flows.

discussed above. This also indicates that the low- and high-speed streaks, with lengths of 1000 viscous units, do not need to be explicitly captured. The case of $L_x^+ \approx 300 = 2.5L_y^+$ produces an incorrect profile and this suggests that a minimum streamwise length of $L_x^+ = 3L_y^+$ is required, especially for narrow channels in which $z_c$ is closer to the buffer layer than log layer of the flow.

The smallest streamwise length of $L_x^+ \approx 190$ was unable to sustain turbulence for a prolonged period. It would decay to a laminar state after $T^+ = tU_\tau^2/\nu \approx 14.8 \times 10^4$ in the smooth-wall channel, and after $T^+ \approx 9.7 \times 10^4$ in the rough-wall channel. The data shown in the previous figures for this streamwise length is averaged only when the flow is turbulent. This behaviour is similar to what was observed in [1], who showed that turbulence could not be maintained in channels with streamwise lengths of around 200 viscous units. As discussed above, this is because it is shorter than the quasi-streamwise vortices which have streamwise lengths of $\lambda_x^+ \approx 200$–$300$.

The larger spanwise domain (figure 6c,d) shows that the shortest streamwise length of $L_x^+ = 190$ restricts the size of the eddies so that the spanwise size of the largest eddy is smaller than the spanwise width of the domain. The resulting velocity profile more closely resembles that of the narrower spanwise width seen in figure 6(a,b). Interestingly, the streamwise length of $L_x^+ \approx 300$ appears to agree quite well with the cases with longer lengths. This is in contrast to the narrow domain (figure 6a), which shows a clear difference for $L_x^+ \approx 300$. Even the case with $L_x^+ \approx 460 = 1.3L_y^+$ is producing a velocity profile and roughness function comparable to that of channels with longer streamwise lengths, despite not having the requisite scaling of $L_x^+ \gtrsim 2$–$3L_y^+$ discussed above. This is similar to the results of [35] who looked at wide spanwise channels with very short streamwise lengths. The velocity profiles they obtained look similar to a full-span full-length channel, and there is no apparent increase in streamwise velocity in the outer layer that is characteristic of minimal-span channels. A possible explanation for the results seen here is that the wall-normal critical location $z_c^+ \approx 140 = 0.24h^+$ is outside the log-layer, which is generally believed to end at $z \approx 0.15h$ [36]. As such, the expected streamwise length scale of 2–$3L_y$ is no longer appropriate in the outer layer. In this case, the largest eddy at the edge of the logarithmic layer would have a streamwise length of roughly $\lambda_x^+ \approx$440–660, potentially explaining why the simulation with $L_x^+ \approx 460$ still produces a reasonable velocity profile.

Given the above results, it is possible to restrict the streamwise length to $L_x = 3L_y$. However, for very narrow channels this can result in a streamwise domain length less than 1000 viscous units. While it appears the buffer-layer streaks do not need to be captured in the domain, having such a small streamwise domain exasperates the bursty nature of the minimal-span channel [17]. Only one or maybe two quasi-streamwise vortices are present in the domain, which makes obtaining converged statistics difficult as the simulation needs to be run for a significantly long time. We believe that a minimum streamwise length of roughly 1000 viscous units seems a reasonable length in these cases, so that several quasi-streamwise vortices are present. As with the spanwise domain width recommendations in [5], the streamwise length should be larger than the characteristic streamwise length scale of the roughness, $\lambda_{r,x}$, if it were present. As such, we have $L_x \gtrsim \max(\,3L_y, 1000\nu/U_\tau, \lambda_{r,x})$.

### *3.4. Temporal Sweep (set $S_5$)*

In this section, the temporal sweep (§2.1) is performed in a standard-height channel, with no outer-layer damping ($K_i = 0$) and a streamwise length of $L_x/h = 2\pi$, corresponding to $L_x^+ \approx 3700$ at the start of the sweep (table 1). The roughness height is fixed at $k = h/40$, so the roughness Reynolds number varies as $15 \lesssim k^+ \lesssim 50$. Three different sweeps were conducted in which the pressure gradient parameter at the end of the sweep is $\Delta_{p,Re_\tau=180} = (0.03, 0.07, 0.15)$.

Figure 7 shows the mean velocity profiles for the fastest and slowest sweeps, comparing the sweep data with full-span steady-flow data at the highest and lowest friction Reynolds numbers tested. Figure 7(a,b) shows the mean velocity for the slowest sweep, when the pressure gradient parameter (6) at $Re_\tau \approx 180$ was $\Delta_{p,Re_\tau=180} = 0.03$. Figure 7(c,d) shows the mean velocity for the fastest sweep, with $\Delta_{p,Re_\tau=180} = 0.15$. To obtain statistics for a desired $Re_\tau$, time-averaging is performed over a window of $\pm 10\%$ of that value. At the initial friction Reynolds number of the simulation, $Re_\tau = 590$, both the sweeps (figure 7a,c) show good agreement with the full-span channel in the near-wall region. As the spanwise width is $L_y/h = 0.6$, then the critical wall-normal location is approximately $z_c^+ \approx 0.4L_y^+ = 142$ which agrees well with the figure. As $\Delta_p$ is not significant yet (figure 2b) then there is little difference between the two sweeps.

However, at the end of the sweep when the friction Reynolds number is close to $Re_\tau = 180$ and the $\Delta_p$ is maximum, then differences start to emerge. The slowest sweep (figure 7b) has reasonable agreement very close to the wall below $z^+ \lesssim 15$, but the pressure gradient has reduced the streamwise velocity compared to the full-span steady flow. Importantly though is that the difference between the smooth and rough-wall flows (inset of figure 7b) agrees quite well with the full-span data. Taking the difference between two flows with the same weak pressure gradient ostensibly still produces a correct estimate of $\Delta U^+$. However, when the pressure gradient is increased to to $\Delta_{p,Re_\tau=180} = 0.15$ (figure 7d) then larger differences are seen. The near-wall region has been noticeably changed, and the difference in smooth and rough wall velocity (inset) is seen to overestimate the full-span data. Note that the data have been ensemble averaged over four runs, so that the amount of time the flow is averaged over is the same as in the slowest sweep. This therefore indicates that the overestimation of $\Delta U^+$ is a direct result of the stronger

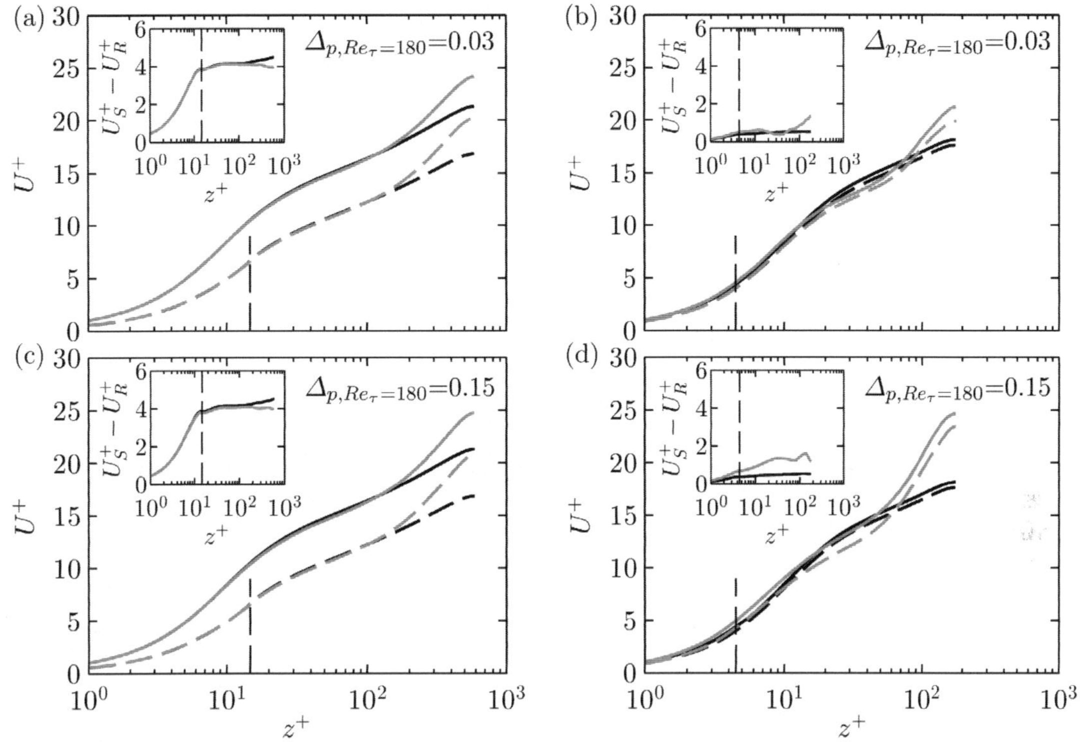


Figure 7: Mean streamwise velocity profile for (a,c) $Re_\tau \approx 590$ and (b,d) $Re_\tau \approx 180$, from full-span steady flow (black line) and temporal sweep (grey line). (a,b) slowest sweep ($\Delta_{p,Re_\tau=180} = 0.03$) and (c,d) fastest sweep ($\Delta_{p,Re_\tau=180} = 0.15$), §2.1. Line styles: Solid line, smooth wall; dashed line, rough wall; vertical dashed line, roughness crest, $k = h/40$. Inset shows difference in smooth- and rough-wall velocity.

pressure gradient.

The roughness function $\Delta U^+$ is now computed from the difference in smooth and rough-wall flows at matched $U_\tau$ values. As seen in the insets of figure 7, the smooth- and rough-wall velocity difference is not constant with $z^+$ as in the steady data. This indicates that the flow has not been averaged for long enough at the desired $Re_\tau$ value. As such, multiple sweeps from different initial conditions would need to be conducted to obtain a converged outer region. However, for the purposes of this investigatory study, the velocity difference is averaged from the crest of the roughness to the channel centre to obtain an estimate of $\Delta U^+$. This is done for a range of friction Reynolds numbers, with the resulting data plotted in figure 8 against the equivalent sandgrain roughness, $k_s^+$. Ideally, the temporal sweep data (lines) should agree with the steady flow data (black symbols). Indeed, the temporal sweep does show promise in this regard, especially for the weaker pressure gradient cases (slower sweeps). It is especially good considering the level of error that is normally associated with experimental studies (grey circles).

It appears as though the sweep approach to determining the hydraulic behaviour of roughness could work, although only for the slowest sweep tested. Hence, a pressure gradient parameter of $\Delta_p \leq 0.03$ is required, which is close to the recommendation in [25] that zero pressure gradient conditions can be assumed when $\Delta_p \leq 0.01$. A value of $\Delta_p \geq 0.07$ results in pressure gradient effects that lead to inaccuracies in $\Delta U^+$. Future studies could prescribe a constant value of $\Delta_p$ rather than linearly varying the bulk velocity, as this would ensure pressure gradient effects

Journal of Physics: Conference Series **708** (2016) 012010 doi:10.1088/1742-6596/708/1/012010

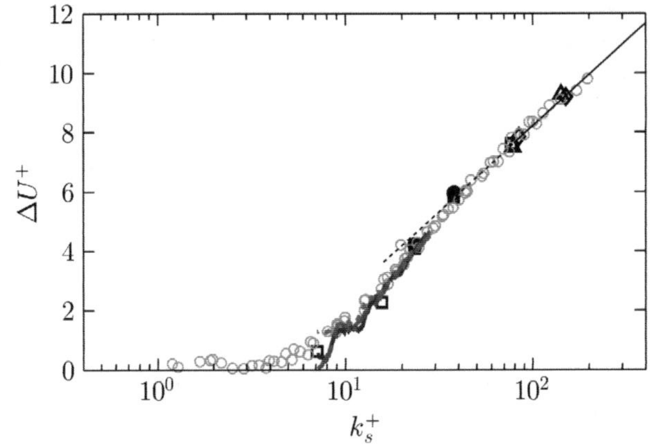


Figure 8: Hama roughness function for the temporal sweep with modelled roughness at fixed $k = h/40$, where $k_s^+ \approx 1.6k^+$. Symbols: grey circles, uniform sand grains of Nikuradse (1933) [37]; black symbols, steady flow data for the same modelled roughness [5]; blue solid line, sweep data for $\varDelta_{p,Re_\tau=180} = 0.03$ (slowest sweep); green dashed line, sweep data for $\varDelta_{p,Re_\tau=180} = 0.07$; red dotted line, sweep data for $\varDelta_{p,Re_\tau=180} = 0.15$ (fastest sweep).

remain negligible, while efficiently traversing the range of roughness Reynolds numbers.

## 4. Conclusions

In this paper, a series of improvements to the minimal-span channel were investigated. The goal is to obtain a direct measure of $\Delta U^+$ by only capturing the relevant near-wall scales. Three improvements are suggested. First, a half-height channel was used in which a slip-wall is positioned at the channel centre. This allows for the number of grid points to be halved, at the cost of running the simulation for twice as long to reach the same level of statistical convergence. It is up to the user to balance this trade off between memory and time. Second, a forcing was applied to the outer layer to reduce the centreline velocity. This allowed for an improvement in the time step of around 20–24%, although this did not extend to higher friction Reynolds numbers due to the grid using a cosine mapping in the wall-normal direction. The present data suggest that the location at which the forcing starts should be $z_d^+ \gtrsim \max(200, 2z_c^+)$. The first constraint is present for very narrow channels where $z_c^+$ is close to the wall, as the forcing will interfere with the near-wall flow otherwise. Third, the streamwise length was investigated, independent of the previous two improvements. It appears that the length should be long enough to capture the largest log-layer eddy which implies $L_x = 3L_y$, where $L_y \gtrsim \max(100\nu/U_\tau, k/0.4, \lambda_{r,y})$ is obtained from the guidelines in [5], $\lambda_{r,y}$ being some characteristic spanwise length scale. However, for very narrow channels the bursty nature of the minimal-span channel and small spatial domain makes obtaining converged statistics a very time-consuming process, so a minimum length of 1000 viscous units is necessary. As such, the recommendation for the streamwise domain length is $L_x \gtrsim \max(3L_y, 1000\nu/U_\tau, \lambda_{r,x})$, where $\lambda_{r,x}$ is the characteristic streamwise length scale of the roughness. Finally, a temporal sweep was conducted to see if the full roughness behaviour could be obtained by varying the roughness Reynolds number. A simple linear variation in the bulk velocity was tried, with three different rates of change investigated. This was independent of the previous improvements. It was shown that the slowest rate of change, with a final pressure gradient parameter value of $\varDelta_{p,Re_\tau=180} = 0.03$ could reasonably predict some of the $\Delta U^+$ vs $k_s^+$ curve. Larger values of $\varDelta_p$ resulted in substantial flow changes compared to the steady flow and so are not feasible.

It is important to consider the cost of the minimal-span channel relative to the full-span channel. Both the memory and computational time need to be analysed, where the former can be easily described by the grid size while the latter is more involved. In standard DNS full-span channel studies, at least 10 large-eddy turnover times are recommended to obtain converged first- and second-order statistics (see for example [31]). For a full-span channel this corresponds to $T = 10U_\tau/h$. However, if we only want to converge statistics up to the height $z = z_c < h$, then we only need 10 eddy turnovers of this $z_c$-sized eddy and so $T_{z_c} = 10U_\tau/z_c$. This argument assumes a fairly large spatial domain so that there are actually a large number of $z_c$-sized eddies present in the full-span channel at any instant. In a minimal-span channel, only a few such eddies could be present in the small domain at any instant. We would therefore need to run the simulation for long enough to observe the same number of $z_c$-sized eddies as would be seen in the full-span channel, presumably resulting in the same level of convergence. This costing argument together with user-decided trade-offs is to be detailed in a future paper.

**Acknowledgments**
Computational time was granted under the Victoria Life Sciences Computational Initiative, which is supported by the Victorian Government, Australia. The authors would also like to acknowledge the funding provided to the Multiflow program by the European Research Council.

## References

[1] Jiménez J and Moin P 1991 *J. Fluid Mech.* **225** 213–240
[2] Hamilton J M, Kim J and Waleffe F 1995 *J. Fluid Mech.* **287** 317–348
[3] Flores O and Jiménez J 2010 *Phys. Fluids* **22** 071704
[4] Hwang Y 2013 *J. Fluid Mech.* **723** 264–288
[5] Chung D, Chan L, MacDonald M, Hutchins N and Ooi A 2015 *J. Fluid Mech.* **773** 418–431
[6] Jiménez J and Pinelli A 1999 *J. Fluid Mech.* **389** 335–359
[7] Hwang Y 2015 *J. Fluid Mech.* **767** 254–289
[8] Hama F R 1954 *Trans. Soc. Naval Arch. Mar. Engrs* **62** 333–358
[9] Flack K A, Schultz M P and Shapiro T A 2005 *Phys. Fluids* **17** 035102
[10] Townsend A A 1976 *The Structure of Turbulent Shear Flow* 2nd ed (Cambridge University Press)
[11] Efros V and Krogstad P Å 2011 *Exp. Fluids* **51** 1563–1575
[12] Chan L, MacDonald M, Chung D, Hutchins N and Ooi A 2015 *J. Fluid Mech.* **771** 743–777
[13] Flores O and Jiménez J 2006 *J. Fluid Mech.* **566** 357–376
[14] Mizuno Y and Jiménez J 2013 *J. Fluid Mech.* **723** 429–455
[15] Chung D, Monty J P and Ooi A 2014 *J. Fluid Mech.* **742** R3
[16] Jiménez J, Kawahara G, Simens M P, Nagata M and Shiba M 2005 *Phys. Fluids* **17** 015105
[17] Jiménez J 2015 *Phys. Fluids* **27** 065102
[18] Spalart P R, Moser R D and Rogers M M 1991 *J. Comput. Phys.* **96** 297–324
[19] Kim J and Moin P 1985 *J. Comput. Phys.* **59** 308–323
[20] Lundbladh A, Berlin S, Skote M, Hildings C, Choi J, Kim J and Henningson D S 1999 An efficient spectral method for simulation of incompressible flow over a flat plate *TRITA-MEK* 1999/11 KTH, Stockholm
[21] Bernardini M, Pirozzoli S, Quadrio M and Orlandi P 2013 *J. Comput. Phys.* **232** 1–6
[22] Busse A and Sandham N D 2012 *J. Fluid Mech.* **712** 169–202
[23] Borrell G 2015 *Entrainment effects in turbulent boundary layers* Ph.D. thesis Technical University of Madrid
[24] Jiménez J 2004 *Annu. Rev. Fluid Mech.* **36** 173–196
[25] Patel V 1965 *J. Fluid Mech.* **23** 185–208
[26] Jones M, Marusic I and Perry A 2001 *J. Fluid Mech.* **428** 1–27
[27] Seddighi M, He S, Vardy A and Orlandi P 2014 *Flow Turbul. Combust.* **92** 473–502
[28] Lam K and Banerjee S 1992 *Phys. Fluids* **4** 306–320
[29] Handler R A, Saylor J R, Leighton R I and Rovelstad A L 1999 *Phys. Fluids* **11** 2607–2625
[30] Nagaosa R 1999 *Phys. Fluids* **11** 1581–1595
[31] Hoyas S and Jiménez J 2006 *Phys. Fluids* **18** 011702
[32] Chin C, Ooi A S H, Marusic I and Blackburn H 2010 *Phys. Fluids* **22** 115107
[33] Kline S J, Reynolds W C, Schraub F A and Runstadler P W 1967 *J. Fluid Mech.* **30** 741–773
[34] Jeong J, Hussain F, Schoppa W and Kim J 1997 *J. Fluid Mech.* **332** 185–214
[35] Toh S and Itano T 2005 *J. Fluid Mech.* **524** 249–262
[36] Marusic I, Monty J P, Hultmark M and Smits A J 2013 *J. Fluid Mech.* **716** R3
[37] Nikuradse J 1933 Laws of flow in rough pipes. English translation published 1950, NACA Tech. Mem. 1292

# Analysis and comparison between rough channel and pipe flows

**David Sassun[1], Oscar Flores[2] and Paolo Orlandi[1]**

[1] Dipartimento di Ingegneria Meccanica e Aerospaziale - Sapienza Universitá di Roma, Via Eudossiana 16, 00184, Roma, Italia

[2] Dep. Bioingenieria e Ing. Aeroespacial. Universidad Carlos III de Madrid. Avd Universidad 30, Leganes. 28911 Madrid, Spain

E-mail: david.sassun@uniroma1.it

**Abstract.** Direct numerical simulations of turbulent channel and pipe flows are presented to highlight the effect of roughness at low Reynolds number ($Re_\tau = 180 - 360$). Several surfaces are reproduced with the immersed boundaries method, allowing a one-to-one comparison of the two canonical flows. In general, all rough surfaces produce the same effect on the flow in pipes and channels, with small differences in the roughness function, RMS velocities and spectral energy density of pipes and channels. The only exception is for the rough surfaces made of longitudinal bars. In particular, the triangular bars (riblets) show drag reduction in the channel and drag increase in the pipe. This behaviour is linked to the development of spanwise rollers and wide $u$-structures near the plane of the crest of the pipe.

## 1. Introduction

According to the classical theory [1], the inner region of wall-bounded turbulent flows is essentially indistinguishable for the three canonical cases, namely boundary layers, pipes and channels. The differences between these three flows are expected to be limited to the outer region, where mean velocities and standard deviations are expected to scale with the friction velocity and the outer region length scale (boundary layer thickness, pipe radius, or the channel half-height).

However, experimental and numerical studies show differences between these flows in the inner region. For instance, Patel and Head [2] found that the mean velocity in the pipe flow does not follow the logarithmic profile, in contrast to the channel flow. Closer to the wall (i.e., in the viscous sublayer and in the buffer region), the mean velocities of channel and pipe flows collapse, consistent with the notion that sufficiently close to the wall (i.e., at a sufficiently large Reynolds number), the pipe surface appears to be flat. Other studies have confirmed these differences in the log-layer between channels and pipes, concluding that the von Kármán constant $\kappa$ might not be universal [3] and that the thickness of the logarithmic region might be different for internal and external flows [4].

In terms of the turbulent stresses, the literature also shows differences between the three canonical flows. One of the first studies comparing pipes and channels with Direct Numerical Simulation is Eggels et al. [5], who reported no significant differences of the velocity RMS but a meaningful difference in the skewness of the wall-normal velocity. They claimed that this difference is due to the impingement or splatting process, explained by Mansour et al. [6]. Monty et al. [7], on the other hand, stated that the wall normal and spanwise turbulence intensities differ in the core region of the flows, due to the outer structures that are larger in the pipe than in the channel. Other scholars have found differences in the turbulent kinetic energy between pipes, channels and boundary layers [8], and in the Reynolds stress budget of pipes and channels [9]. Several works [4, 10, 11] connect this differences to the fundamental differences in the large scale motions of the outer region of pipes, channels and boundary layers.

These differences in the outer region of pipes, channels and boundary layers are probably partly responsible for the disparity of results obtained for $k$-type and $d$-type roughness in pipes, channels and boundary layers. Based on the data summarized by Jiménez [12], it appears that the turbulent velocity fluctuations between channels and pipes are comparable to each other, while those obtained in turbulent boundary layers seem to scale differently. Borrell [13] argues that it is the entrainment on the turbulent/non-turbulent interface (outer region, large scales) which causes the differences between internal and external flows over rough walls: Kármán's integral equations links the boundary layer growth (i.e., the entrainment) and the friction at the wall (the roughness), resulting in a core region more dominated by entrainment in the case of rough walls. It is important to note that, beside this work, the number of studies analysing the differences between channels, pipes and boundary layers over rough-walls is scarce. To the best of our knowledge, there are no numerical studies comparing one-to-one turbulent flows in channels and pipes with rough walls.

It should be noted that, besides these differences between internal and external flows, there is a wide agreement in the community in that the main effect of roughness in pipes, channels and boundary layers is to shift downwards the mean velocity profile over a smooth wall by a constant amount in the log region, the so-called roughness function. In this paper, we will refer to the roughness function commonly found in literature as $\Delta U_c^+$, and to our definition of it as $\Delta U^+ = \Delta U_c^+ + U_R^+$ where $U_R^+$ is the slip velocity. The most pressing issue in this regard has been to connect the geometric parameters of the surface to the roughness function. A scaling in this sense has been found recently by Orlandi [14], who derived the relation $\Delta U^+ = B\kappa^{-1}\tilde{v}_R^+$, with $B$ and $\kappa$ being the log law constants for the smooth wall and $\tilde{v}_R^+$ the RMS of the vertical velocity fluctuations at the plane of the crests. The vertical stress $\tilde{v}_R^+$ is then the fundamental quantity to account for the effects of roughness: if it exceeds a threshold value, the transition to turbulence occurs (Orlandi [15]). A chart of the friction coefficient similar to that by Moody [16], can be calculated by replacing the equivalent sand grain roughness proposed by Nikuradse [17] with $\tilde{v}_R^+$.

The objective of the present work is to establish a one-to-one comparison between channels and pipes for different wall-roughness, in order to understand whether the differences between channels and pipes with smooth walls become more pronounced in presence of the wall corrugation. The paper is structured as follows. In Section 2 and 3, the details of the numerics and of the DNS database are briefly exposed. In Section 4 the results are shown and discussed. These primarily focus on the average velocity and turbulent intensities, the stress and the premultiplied spectra. Finally, in Section 5 the conclusions are presented.

Journal of Physics: Conference Series **708** (2016) 012011 doi:10.1088/1742-6596/708/1/012011

## 2. Physical and numerical model

The incompressible Navier Stokes equations written in non dimensional cartesian form are:

$$\frac{\partial u_i}{\partial t} + \frac{\partial u_i u_j}{\partial x_j} = -\frac{\partial p}{\partial x_i} + \frac{1}{Re}\frac{\partial^2 u_i}{\partial x_j^2} + \Pi\delta_{1i} \quad ; \quad \frac{\partial u_i}{\partial x_i} = 0 \tag{1}$$

where $\Pi$ is the pressure gradient required to maintain a constant mass flow rate, $u_i$ is the component of the velocity vector in the $i$ direction, and $p$ is the pressure. The reference values for the non dimensional quantities are either the channel half-height or the pipe radius, both termed $h$ in the hereinafter, and the laminar parabolic Poiseuille velocity at the centreline, $u_P$. An equivalent set can be written in cylindrical coordinates for the pipe flow, given in Chapter 10 of Orlandi [18]. The equations have been discretized in an orthogonal coordinate system, with a central staggered second-order finite difference method. The integration in time has been made with a third-order low storage Runge-Kutta algorithm, coupled with a second-order Crank-Nicolson scheme. To correct the non divergence-free field, the fractional step procedure is used. Further details are found in [18] and are not repeated here. No-slip boundary conditions are imposed in the non-homogeneous direction. For the cylindrical coordinates a particular treatment at the axis has been used. More details on the numerics for the pipe are found in Verzicco and Orlandi [19]. The periodicity in the homogeneous directions through a Fourier expansion allows to solve the elliptical equation for the pressure directly. The interaction between the flow and the roughness is reproduced by the immersed boundary technique, imposing zero velocity inside the body, and applying a correction to the viscous term at the first grid point near the solid surface. This method is described in detail in Orlandi and Leonardi [20].

## 3. DNS database

Several simulations of pipes and channels with different wall-roughness are performed at $Re = u_P h/\nu = 4900$. This $Re$ corresponds to a friction Reynolds number $Re_\tau = u_\tau h/\nu \approx 200$ for the smooth channel, where $u_\tau$ is defined based on the total stresses at the plane of the crests of the roughness. Note that since the bulk velocity is $u_b = 2/3u_P$ for the channel and $u_b = 1/2u_P$ for the pipe, the bulk Reynolds numbers are different for the two flows: $Re_b = 3266$ for the channel and $Re_b = 2450$ for the pipe. In all the simulations (see Table 1) the size of the computational domain is $L_x = 8h$ in the streamwise direction. The spanwise size of the computational domain of the channel is $L_z = 2\pi h$, in order to make the channel flow comparable to the pipe flow. The coordinates $y$ and $z$ will be used for both the channel and the pipe, where for the latter the wall-normal coordinate is $y = h - r$ and the spanwise coordinate is $z = \theta r$, where $r$ is the radial coordinate.

The database consists of eight surfaces including the smooth case, that will be referred to as SM. The geometry of the rough surfaces is shown in Figure 1 for the channel, where all rough cases have the same roughness height $k/h = 0.2$. For the channel cases, both the upper and lower walls are rough. Note that the roughness is carved into the smooth wall geometry, so that the distance from the plane of the crests to the centre of the channel/pipe is always $h$. The square and triangular bars placed transversely to the flow (Figures 1a and 1b respectively) will be referred to as ST and TT. For these cases, the cavity width (ST) and the tip spacing (TT) are equal to the height: $s = k$. In these cases, the disposition of the bars makes it necessary to have a large number of nodes in the streamwise direction (801), to accurately reproduce the flow inside

Journal of Physics: Conference Series **708** (2016) 012011 doi:10.1088/1742-6596/708/1/012011

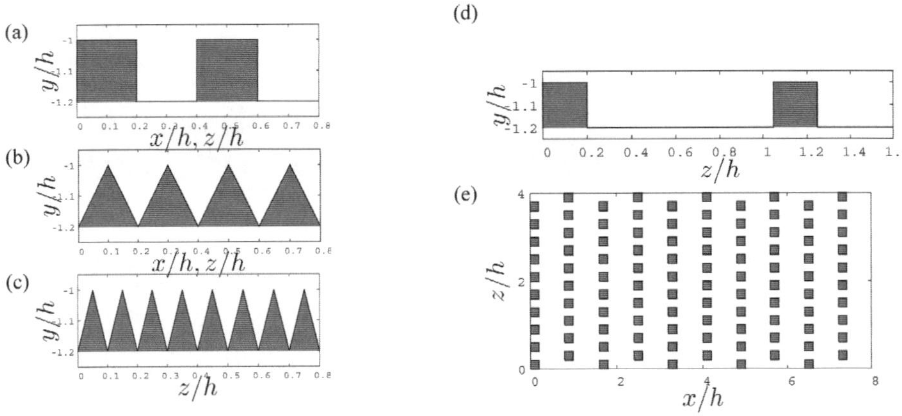


Figure 1: Sketch of the rough surfaces for the channel. (a) square bars longitudinal (SL) or transverse (ST), (b) triangular bars longitudinal (TL) or transverse (TT), (c) longitudinal triangular bars TLS, (d) longitudinal square bars SLL, (e) staggered cubes SC.

| | | | Channel | | | Pipe | |
|---|---|---|---|---|---|---|---|
| Case | $n_x \times n_z$ | $n_y$ | $Re_\tau$ | $U_R^+$ | $n_r$ | $Re_\tau$ | $U_R^+$ |
| SM | $801 \times 129$ | 257 | 204 | 0 | 129 | 169 | 0 |
| ST | $801 \times 129$ | 257 | 225 | 0.9 | 129 | 218 | 1.4 |
| TT | $801 \times 129$ | 257 | 290 | 1.8 | 129 | 294 | 2.2 |
| SL | $257 \times 513$ | 257 | 215 | 3.2 | 129 | 200 | 3.5 |
| SLL | $257 \times 513$ | 257 | 183 | 10.8 | 129 | 166 | 12.1 |
| TL | $257 \times 513$ | 257 | 192 | 6.8 | 129 | 178 | 9.5 |
| TLS | $257 \times 513$ | 257 | 157 | 6.6 | 129 | 162 | 6.5 |
| SC | $801 \times 513$ | 257 | 347 | 2 | 129 | 361 | 2.4 |

Table 1: Computational grid, values of the friction Reynolds numbers and of the slip velocity of the simulations

the grooves. The opposite holds for the longitudinal bars, where the number of points is reduced to 257 in the streamwise direction and increased to 513 in the spanwise/azimuthal direction.

Four cases with longitudinal bars have been simulated: two of square bars, with cavity widths $s = k$ (SL, same as Figure 1a) and $s = 4.236k$ (SLL, Figure 1d, chosen to have six bars in the domain), and two of triangular bars, with tip spacing $s = k$ (TL, same as Figure 1b) and $s = k/2$ (TLS, Figure 1c). A three dimensional geometry made of staggered rows of cubes (SC) is reproduced in Figure 1e. In this case, the number of grid nodes is increased in both directions (801 and 513).

## 4. Results

Table 1 gives the values of the friction Reynolds numbers for the different runs. Due to the difference in the bulk Reynolds number between the channel and the pipe, the values for SM differ significantly; however, the difference decreases in the rough cases. The transverse bars (ST, TT) and the three-dimensional roughness (SC) increase the value of the $Re_\tau$ with respect to the smooth wall, both for the channel and the pipe. However, the behaviour of the longitudinal

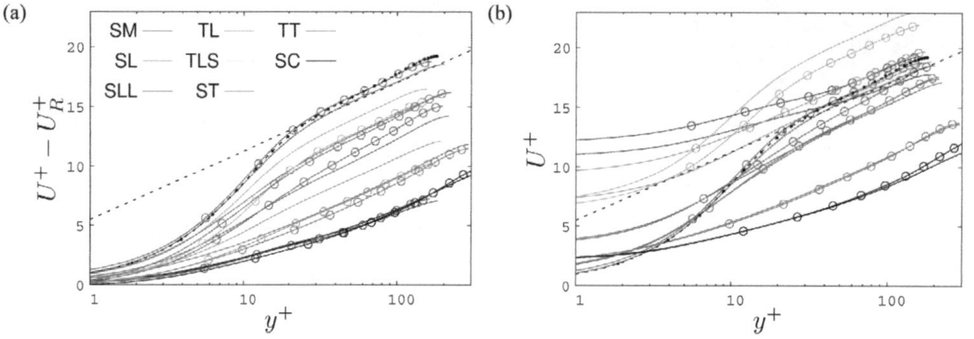


Figure 2: (a) Mean profile in wall units of the velocity minus the slip velocity at the wall $U_R^+$. The solid lines represent the channel flow, the dotted lines the pipe flow. The solid black symbols are the data by [21]. The roughness function $\Delta U^+$ is evaluated from this plot. (b) Same velocity profile without the subtraction of the slip velocity. The roughness function $\Delta U_c^+$ is evaluated from this plot.

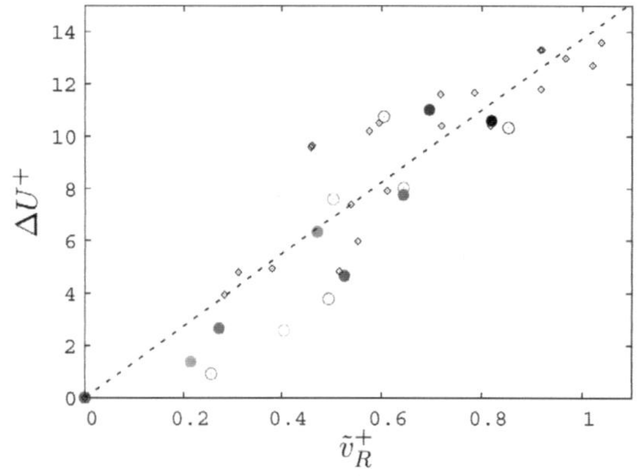


Figure 3: Roughness function $\Delta U^+$ versus the normal stress at the plane of the crests $\tilde{v}_R^+$. Solid circles are the channel, open circles are the pipe, open diamonds are the data by [14], the dashed line is $\Delta U^+ = B\kappa^{-1}\tilde{v}_R^+$, where $B = 5.5$ and $\kappa = 0.4$ are the constants in the log law for the smooth wall.

bars is more variable. The longitudinal square bars (SL and SLL) cause a moderate increase of $Re_\tau$ with respect to SM, similar to ST. On the other hand, TL shows an increase of $Re_\tau$ in the pipe and a decrease in the channel. For TLS, there is a small decrease in $Re_\tau$ in both configurations. We expect therefore that the the riblets will achieve drag reduction. The largest increase in $Re_\tau$ is observed for TT and SC, consistently with previous results [14].

The mean velocity profiles of the 16 simulations are presented in Figure 2, where the solid lines correspond to channels and the lines with circles to pipes. The wall normal coordinate is $y^+ = 0$ at the plane of the crests, which corresponds to the position of the wall in SM. The average value of the velocity at this plane is the slip velocity $U_R^+$ (reported in Table 1), which has been subtracted to the mean velocity profiles plotted in Figure 2a. It can be observed how the velocity profile of SM agrees with the log-law with $B = 5.5$, while the pipe has a slightly higher value of $B = 6$, consistent with Wu and Moin [21]. For transverse square bars ST, a small decrease of the average velocity appears, without an appreciable difference between the channel and the pipe. The transverse triangular bars TT and the staggered cubes

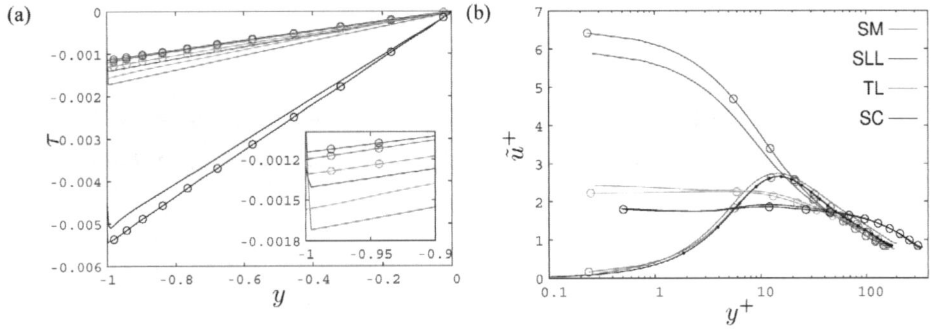


Figure 4: Profile of the the total stress $\tau$ (a), and of the normal stress $\tilde{u}^+$ (b), for the lower half of the channel (solid lines) and the pipe (dotted lines). The inset in (a) is an enlargement of the near wall region for the low drag cases.

SC also show the same velocity defect for the channel and the pipe, with the same increase of the roughness function $\Delta U^+$. The largest differences in $\Delta U^+$ between pipe and channel are found among the longitudinal bars cases, TL and SL. The latter also exhibits differences in the slope in the log region. The square bars with wide spacing SLL produce a large velocity defect, but similar for both pipe and channel. It is important to note that the roughness function computed here ($\Delta U^+$) is not the one found commonly in literature: the one evaluated from Figure 2b, that will be referred to as $\Delta U_c^+$ can be rather different. In fact, $\Delta U_c^+$ is the sum of two effects: the slip velocity at the wall $U_R^+$, representative of the viscous drag, and the roughness function $\Delta U^+$, linked to the form drag. Orlandi [14] gave the relationship between this roughness function and the vertical stress at the plane of the crests $\tilde{v}_R^+$ (the $\sim$ denotes an RMS value): $\Delta U^+ = \Delta U_c^+ + U_R^+ = B\kappa^{-1}\tilde{v}_R^+$, where $B = 5.5$ and $\kappa = 0.4$ are the constants in the expression of the log law for SM. Figure 3 demonstrates that this relationship holds for both channels and pipes.

To understand the difference in the behaviour of the friction, and the drag increase/decrease effect of the roughness on the flow, it is worth looking at the profiles of the total stress $\tau = \nu \partial U/\partial y - \langle uv \rangle$, shown in Figure 4a. In this plot, only the square bars SLL, the triangular bars TL and the staggered cubes SC are reported, since those are the cases with the greater differences between the channel and the pipe. As for the velocity profiles, both SC flows show large differences with respect to SM flows, leading to the conclusion that the wall curvature does not affect the total stress, and in particular the turbulent stress, which is much larger than the viscous stress. The square bars SLL, despite the large $\Delta U^+$ in Figure 2a, show drag reduction for both flows. In this case, the reduction is due to the large decrease of the viscous stress, due to the large $U_R^+$ at the plane of the crests. The turbulent stress is large, but does not compensate the reduction of $\nu dU/dy$. The large $\Delta U^+$ and the low friction coefficient are not contradictory. In fact, due to the large slip velocity, $\Delta U_c^+$ is small, and the relationship between the roughness function and the friction coefficient (Hama [22]) can be satisfied. The TL flows are the more surprising, showing drag reduction in the channel and drag increase in the pipe.

The profiles of the RMS of the streamwise velocity component are reported in Figure 4b, showing no substantial differences between channel and pipe runs for a given roughness. The

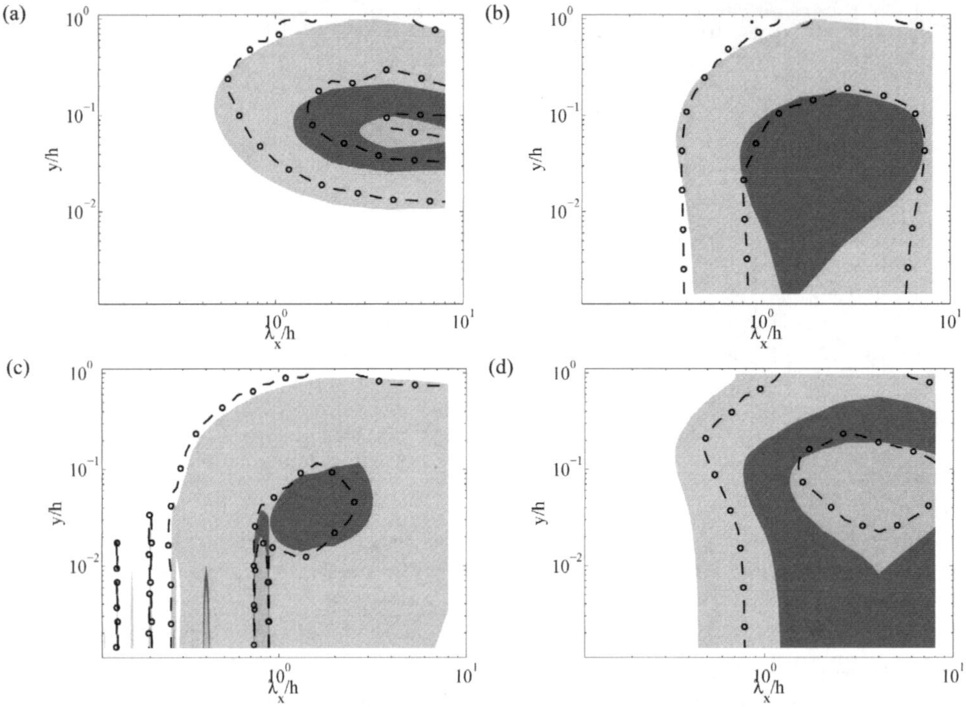


Figure 5: Premultiplied one dimensional spectra of $u$ in the streamwise direction $k_x^+ E_{uu}(\lambda_x^+)$, as a function of the distance from the wall. The filled contour represents the channel and the black contour lines represent the pipe. (a) SM, (b) TL, (c) SC, (d) SLL. The levels are the same in all figures, and are 10%, 50% and 90% of the maximum in SM.

SLL case exhibits the largest $\tilde{u}^+$ at the wall, consistent with a relatively large value of the slip velocity (see $U_R^+$ in Table 1). This behaviour is consistent with the recent results of Vanderwel and Ganapathisubramani [23], who investigated the effect of the spanwise spacing of longitudinal square bars on boundary layers. They found that a spacing of $s/\delta = 0.88$ results in the largest secondary flow, very similar to the spanwise spacing of case SLL, $s/h = 0.85$. They also reported that the profile of $\tilde{u}$ is maximum in the vicinity of the roughness, which is confirmed by our results. The vertical and spanwise stresses show similar trends as those for $\tilde{u}^+$ and therefore are not presented here.

The RMS profiles suggest that the distribution of velocity fluctuations with $y$ is very similar in pipes and channels with the same roughness. It is however interesting to check if the same can be said about the characteristic wavelengths (sizes) of the structures containing that energy. That information can be obtained from the spectral energy density of the velocity fluctuations, presented in Figures 5-7. Only planes above the plane of the crests of the roughness elements are plotted.

Figure 5 shows the one dimensional premultiplied spectrum of the streamwise velocity fluctuation, $k_x E_{uu}(\lambda_x/h, y/h)$, normalised with $u_\tau$ and $h$. Coloured contours are for the channel flows, and the chain lines are for the pipe. The differences between the smooth channel and pipe (Figure 5a) near the wall (below $y/h \approx 0.5$) are mostly due to the difference in $Re_\tau$. Both cases show a peak peak located at $y/h \approx 0.07$ ($y^+ \approx 15$), at a wavelength $\lambda_x/h \approx 8$, that corresponds to the position and length of the near wall streaks. However, the differences in the outer region also ($y/h = 0.7 - 1$, $\lambda_x \approx h$) do not seem to be a $Re_\tau$ effect, since similar differences have been reported between smooth-walled channels and pipes at higher Reynolds numbers [9, 24].

The spectra for the longitudinal bars (triangular in Figure 5b, square in Figure 5d) show the presence of long energetic structures near the plane of the crests, characterized by a broad peak around $\lambda_x/h \approx 2 - 3$. This peak seems to be shifted towards longer wavelengths for the pipes. It is interesting to note that, for case SLL (Figure 5d) the spectrum of the pipe flow is less intense than the channel's, despite the agreement of the profiles of the RMS of the streamwise velocity perturbation in Figure 4b. This apparent contradiction between the RMS profiles and the spectra is explained by the energy in the $k_x = 0$ modes, which is not represented in Figure 5: for SLL, the pipe flow has more energy in infinitely large structures than the channel flow, an effect that apparently persists through the whole flow thickness. Comparing TL and SLL for pipes and channels, there is no apparent correlation between the streamwise length of the $u$-structures near the plane of the crests and the drag reduction/increase behaviour of the roughness.

Comparing the longitudinal bars with the staggered cubes (Figure 5c), the latter show short isotropic structures near the plane of the crest. The sharp vertical peaks in the spectrum correspond to the streamwise periodicity of the cubes and its harmonics. The peak about $y/h \approx 0.05$ is damped in case SC compared to SM, consistent with the damping of the buffer region cycle by the roughness [25].

The information of the streamwise/wall-normal spectra is complemented with the spanwise/wall-normal spectra shown in Figure 6. Again, the one dimensional premultiplied spectrum for the streamwise velocity is shown, $k_z E_{uu}(\lambda_z/h, y/h)$. Note that for the pipes, the spanwise wavelength is defined in terms of the arc length in the circumferential direction. Hence, as $y$ tends to $h$ (i.e., as we move from the wall to the center of the pipe), the available spanwise wavelengths shrink. This is clearly observed in SM (Figure 6a), where the channel spectrum is roughly aligned with $\lambda_z \propto y$, while the pipe spectrum turns towards smaller wavelengths near the centre of the pipe. It is possible that this effect is responsible for the shorter streamwise wavelengths for $y/h \gtrsim 0.7$ discussed before (Figure 5a). Near the wall, both the pipe and channel over smooth walls show a peak at $\lambda_z/h \approx 0.5$, which roughly corresponds to the spanwise spacing of the near wall streaks, $\lambda_z^+ \approx 100$.

The spectrum for TL (Figure 6b) shows the direct effect of the surface roughness in a sharp energy peak at $\lambda_z \approx 0.2h$, which corresponds to the spanwise spacing of the bars. This peak extends up to $y/h \approx 0.07$ ($y^+ \approx 12$). Note that the peak for the pipe appears at a slightly smaller wavelength, which is probably a consequence of the curvature of the wall. For the channel, we can observe several harmonics at smaller wavelengths. For the pipe, these harmonics are less intense, and do not show in the spectrum. Moreover, the spectrum for the pipe is wider near the plane of the crest than that of the channel. For the SLL case (Figure 6d), we can also observe a dominant peak at $\lambda_z \approx h$ (close to the spacing of the bars), with stronger harmonics than in TL. However, in contrast to TL, for SLL the $u$-structures of the channel near the plane of the crest are wider in the channel than in the pipe. In general, the flow structures generated by SLL are stronger than those generated by TL, and have a stronger influence in the outer flow.

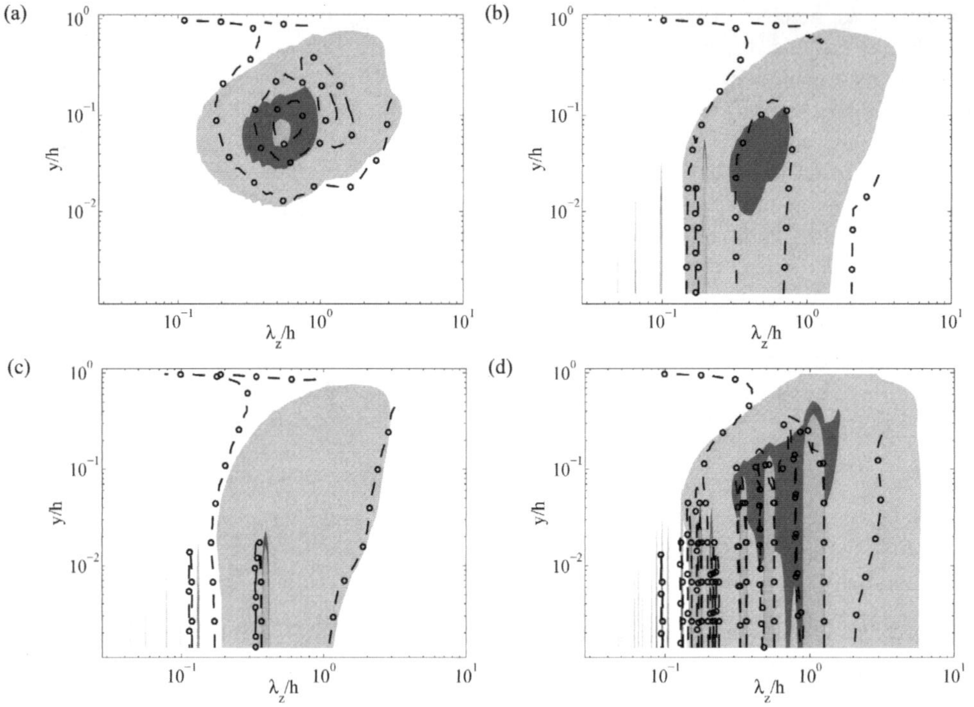


Figure 6: Premultiplied one dimensional spectra of $u$ in the spanwise direction $k_z^+ E_{uu}(\lambda_z^+)$, as a function of the distance from the wall. The filled contour represents the channel and the black contour lines represent the pipe. (a) SM, (b) TL, (c) SC, (d) SLL. The levels are the same in all figures, and are $10\%$, $50\%$ and $90\%$ of the maximum in SM.

For SC, (Figure 6c), it is possible to observe an intense peak near the plane of the crests at $\lambda_z = 0.4h$ ($\lambda_z^+ \approx 140$), a wavelength equal to that of the spanwise spacing of the staggered cubes. In this case, the peak is localized closer to the plane of the crests than in TL. The effect of SC on the near-wall peak is stronger than TL, consistent with the previous discussion of Figure 5. Although it is not visible in Figure 6c because of the levels chosen for the contour, the near-wall peak is still present around $y^+ \approx 15$ and $\lambda_z^+ \approx 100$, although with a lower energy content than in SM and TL.

Finally, it is interesting to note that the SC case shows a very similar behaviour for pipes and channels near the plane of the crests. However, the spectra of TL and SLL for pipes and channels show differences near the plane of the crests: TL in pipes develops wider $u$-structures at the plane of the crest than in channels, while the opposite is true for SLL (at all heights). It is interesting to try to link these differences to the drag increase/decrease behaviour of these cases. According to García-Mayoral and Jiménez [26], the break-up of the drag reducing regime of riblets can be tracked to the appearance of spanwise rollers on the crest of the riblets, due to an inflectional instability of the mean velocity profile. This instability is given by the permeability condition of the plane of the crests, which is why these rollers are not present over a smooth

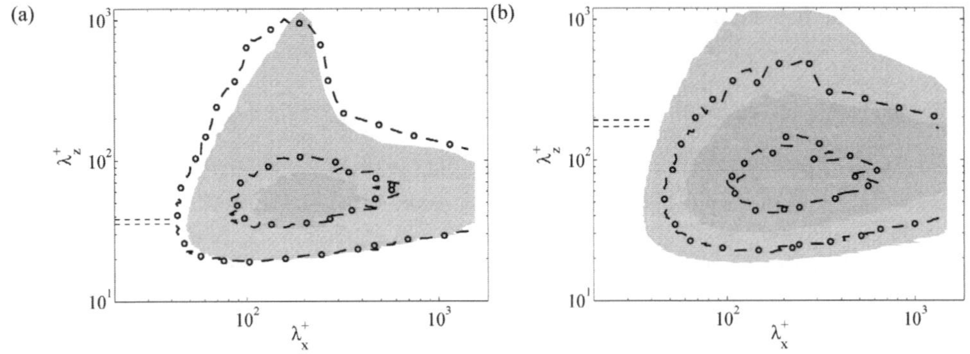


Figure 7: Premultiplied two dimensional spectra of $v$, as a function of the streamwise and spanwise wavelengths, normalised in wall-units. The levels are 10% and and 50% of the maximum of the spectra of case TL. The spectra are plotted at $y^+ = 5$. (a) Case TL. (b) Case SLL. The horizontal dashed lines mark the wavelength of the roughness. The filled contour represents the channel and the black contour lines represent the pipe.

wall. The analysis of García-Mayoral and Jiménez shows that these rollers are more apparent in the two-dimensional premultiplied spectra of the wall normal velocity, which is plotted for cases TL and SLL in figure 7. The plot shows that for SLL (Figure 7b), the spectra is essentially limited by the spanwise wavelength of the bars (indicated by the horizontal dashed lines in the figure). The most apparent difference between the pipe and the channel is the overall intensity of the spectra, consistent with the differences in the RMS observed in Figure 4b just above the plane of the crests. For case TL, the spectrum is mostly on the sub-harmonics of the spanwise spacing of the bars, and it shows the presence of rollers: very wide structures, with streamwise wavelengths of the order of $\lambda_x^+ \approx 150$, or $\lambda_x \approx 0.8h$. The rollers are more intense in the pipes than in the channel, which might explain why the former are drag increasing while the latter are drag reducing. Stronger rollers might be also responsible for wider $u$-structures at the plane of the crest for TL in the pipe as compared to the channel.

## 5. Conclusions

Direct numerical simulations of the flow in turbulent channels and pipes with several surface roughnesses have been performed, allowing a one-to-one comparison between pipes and channels for different wall-roughness at $Re_\tau = 180-360$. The comparison has been established in terms of global indicators of the flow ($Re_\tau$, $\tau$), mean velocity and RMS profiles, and the spectral energy density of the velocity fluctuations.

In terms of global indicators, our results show that the scaling of the roughness function with the vertical velocity fluctuations at the plane of the crest (proposed by Orlandi [14]) works also for pipes. Of the roughness geometries explored here, the largest roughness functions are obtained for staggered cubes and longitudinal square bars with large spacing ($s \approx 5k \approx h$). These roughness geometries produce the largest perturbations in the overlying flow. While for the cubes the perturbations seem not to be coherent at large distances from the plane of the crests, for the longitudinal bars with large spacing the footprint of the roughness is visible in the spectra up to $y \approx 0.7h$, well into the core region of the flow.

In general, the spectral energy density of the velocity fluctuations of pipes and channels are similar, at least for the wall roughness analysed here. The largest qualitative differences between channels and pipes are found for the longitudinal triangular bars (riblets), which are drag reducing for the channel and drag increasing for the pipe. The streamwise velocity spectrum shows that the $u$-structures at the plane of the crest are wider in the pipe than in the channel. The analysis of the two-dimensional spectrum for the wall-normal velocity shows the presence of spanwise rollers in both cases, but the rollers in the pipe (drag increasing) are stronger than in the channel (drag reducing), in agreement with García-Mayoral and Jiménez [26].

## Acknowledgments

The authors would like to acknowledge the support of the European Research Council, that funded the Multiflow program within which this work was developed.

## References

[1] Tennekes H and Lumley J L 1972 *A First Course in Turbulence* (MIT Press)
[2] Patel V C and Head M R 1969 *J. Fluid Mech.* **38** 181–201
[3] Marusic I, McKeon B J, Monkewitz P A, Nagib H M, Smits A J and Sreenivasan K R 2010 *Phys. Fluids* **22** 0651031–24
[4] Jiménez J and Hoyas S 2008 *J. Fluid Mech.* **611** 215–236
[5] Eggels G J M, Unger F, Weiss M H, Westerweel J, Adrian R J, Friedrich R and Nieuwstadt F T M 1994 *J. Fluid Mech.* **268** 175–209
[6] Mansour N N, Kim J and Moin P 1988 *J. Fluid Mech.* **194** 15–44
[7] Monty J P, Hutchins N, Ng H C H, Marusic I and Chong M S 2009 *J. Fluid Mech.* **632** 431–442
[8] Mathis R, Monty J P, Hutchins N, and Marusic I 2009 *Phys. Fluids* **21** 1117031–4
[9] El Khoury G K, Schlatter P, Brethhouwer G and Johansson A V 2014 *J. Phys.: Conf. Ser.* **506** 012010
[10] Monty J P, Stewart J A, Williams R C and Chong M S 2007 *J. Fluid Mech.* **589** 147–156
[11] Lee J, Ahn J and Sung H J 2015 *Phys. Fluids* **27** 025101–16
[12] Jiménez J 2004 *Annu. Rev. Fluid Mech.* **36** 173–196
[13] Borrell G 2015 *Entrainment Effects in Turbulent Boundary Layers* Ph.D. Thesis U. Politécnica de Madrid
[14] Orlandi P 2013 *Phys. Fluids* **25** 110813–12
[15] Orlandi P 2011 *J. Turbul.* **12** N29
[16] Moody L F 1944 *Trans. ASME* **66** 671–684
[17] Nikuradse J 1933 *VDI-Forsch.* **361**
[18] Orlandi P 2000 *Fluid Flow Phenomena - A Numerical Toolkit* (Kluwer Academic Publisher)
[19] Verzicco R and Orlandi P 1996 *J. Comput. Phys.* **123** 402–414
[20] Orlandi P and Leonardi S 2006 *J. Turbul.* **7** N531–22
[21] Wu X and Moin P 2008 *J. Fluid Mech.* **608** 81–112
[22] Hama F R 1954 *Trans. Soc. Naval Archit. Mar. Eng.* **62** 333–358
[23] Vanderwel C and Ganapathisubramani B 2015 *J. Fluid Mech.* **774 R2** 1–12
[24] Ng H, Monty J P, Hutchins N, Chong M S and Marusic I 2011 *Exp. Fluids* **51** 1261–1281
[25] Flores O and Jiménez J 2006 *J. Fluid Mech.* **566** 357–376
[26] García-Mayoral R and Jiménez J 2011 *J. Fluid Mech.* **678** 317–347

# Large-eddy simulations of adverse pressure gradient turbulent boundary layers

**Alexandra Bobke, Ricardo Vinuesa, Ramis Örlü, Philipp Schlatter**
Linné FLOW Centre, KTH Mechanics, Stockholm, Sweden
E-mail: bobke@mech.kth.se

**Abstract.** Adverse pressure-gradient (APG) turbulent boundary layers (TBL) are studied by performing well-resolved large-eddy simulations. The pressure gradient is imposed by defining the free-stream velocity distribution with the description of a power law. Different inflow conditions, box sizes and upper boundary conditions are tested in order to determine the final set-up. The statistics of turbulent boundary layers with two different power-law coefficients and thus magnitudes of adverse pressure gradients are then compared to zero pressure-gradient (ZPG) data. The effect of the APG on TBLs is manifested in the mean flow through a much more prominent wake region and in the Reynolds stresses through the existence of an outer peak. The pre-multiplied energy budgets show, that more energy is transported from the near-wall region to farther away from the wall.

## 1. Introduction

Due to the difficulty of finding analytical solutions to the Navier–Stokes equations in turbulent flows, numerical simulations are performed from which relevant physics can be extracted and understood. Complex flows can then be reduced to canonical cases, which give the possibility to general statements and laws. A number of different flow types has been widely investigated, of which some are closer to industrial applications such as of pipe flow, while others are more idealized configurations which provide deeper understanding of some particular flow features. This is the case, for instance, of the zero pressure-gradient (ZPG) turbulent boundary layer (TBL). A less investigated case is the TBL with pressure gradient (PG). This type of TBL is observed in both external (such as in the flow around an airfoil) and internal (as is the case of a diffuser) configurations. With respect to such flows, many investigations are of experimental nature, as in Refs. [1, 2]. These studies have investigated the one-point statistics of PG flows in different ranges of Reynolds numbers ($Re$) and different strengths of PG, even comparably strong positive PGs leading to separation [3]. In more recent works the change of one-point statistics with increasing PG strengths (*e.g.* Ref. [4]) has been studied, in which a second peak was noticed in the wall-normal profiles of the root-mean-square (rms) of streamwise velocity fluctuations with strong adverse pressure-gradient (APG). Harun *et al.* [5] investigated the modification of the large-scale features of TBLs for different types of PGs. On the other hand, there are several relevant numerical studies of PG TBLs: cases close to separation were investigated [6, 7], while also direct numerical simulations (DNS) were performed with a wide range of PGs [8, 9, 10]. The Reynolds number was, however, low in all studies, except the more recent ones by Gungor *et al.* [10] and Kitsios *et al.* [11].

A number of parameters have been used in the literature to characterize PG TBLs. This complicates comparisons between the various studies, even if the particular PG parameters are matched between cases. History effects play an important role in the development of TBLs as well, and are not taken into account in these parameters. Therefore, one of our initial goals is to consider the impact of history effects on the state of the boundary-layer flows, in order to properly assess the effect of the pressure gradient and facilitate the determination of adequate scaling laws. The present paper does however not directly deal with this goal, but can be seen as a foundation for such a follow-up study. The set-up for PG flows in equilibrium (to be defined in section 2) is discussed, analyzed, and the influence of PGs in turbulent flows is shown.

The paper is organized as follows: Section 2 gives a description of the flow cases and introduces the equilibrium boundary layer. Section 3 explains the numerical method and boundary conditions. The preliminary simulations performed to assess the best set-up are presented in Section 4, together with a discussion of the results of the main simulations. Finally, Section 5 summarizes the paper and gives an outlook to forthcoming studies. The description of accompanying RANS computations together with the determination of the virtual origin can be found in the Appendix.

## 2. Definitions

Earlier studies of TBLs subjected to pressure gradients date back to the 1930s [12, 13]. Laminar boundary layers with pressure gradients are well known, and under some specific assumptions can theoretically be described by the Falkner–Skan similarity solution, which was developed in 1931 [14]. These solutions are valid when the free-stream velocity $U_\infty$ follows a power law of the form $U_\infty = ax^m$ (where $x$ is the streamwise coordinate and $m$ the power-law parameter), so that the boundary-layer approximation is valid [15]. The dimensionless velocity is $U = U_\infty(\mathrm{d}f/\mathrm{d}\eta)$, with an unknown dimensionless function $f$, and the dimensionless similarity parameter $\eta = y/\delta(x)$, with $\delta(x) = \sqrt{\nu x/U_\infty}$. The Hartree parameter $\beta_H$ from the Falkner–Skan equation [16] can be written in terms of $m$ as $m = \beta_H/(2-\beta_H)$. Note that, for instance, the range $0 < \beta_H < 1$ corresponds to the flow around a wedge. Therefore, the relation between $m$ and $\beta_H$ can be determined from the potential solution of such a flow. Clauser [17] found that near-equilibrium conditions are obtained in laminar case when the term $\delta^*(\mathrm{d}P/\mathrm{d}x)/\tau_w$ (where $\delta^*$ is the displacement thickness, $\tau_w$ the wall shear stress and $\mathrm{d}P/\mathrm{d}x$ the streamwise pressure gradient) is constant. This was then later referred to as the Clauser parameter $\beta$ and has been widely used. Several other studies [9, 18, 19] claim that a near-equilibrium state is reached in turbulent boundary layers when the Clauser parameter $\beta$ remains constant, as in the laminar case.

An additional condition for near-equilibrium, according to Ref. [9], is that the form parameter $G = (H_{12}-1)/(H_{12}\sqrt{c_f/2})$ (a measure of the velocity defect) has to be constant, where $H_{12} = \delta^*/\theta$ denotes the shape factor. Mellor and Gibson [19], as well as Townsend [18], found that a power-law distribution for the free-stream velocity leads to a near-equilibrium state and assessed the downstream evolution of the boundary-layer thickness using the Rotta–Clauser length scale $\Delta = (U_\infty/u_\tau)\delta^*$, where $u_\tau$ denotes the friction velocity. Skåre and Krogstad [4] produced a region of near-equilibrium conditions in their experimental study, where the Clauser parameter $\beta$ was constant over a significant part of the test section. Although they obtained their experimental free-stream velocity distribution by adjusting the wind tunnel ceiling, they could fit it to a power law defined as $U_\infty \approx (x - x_{0,e})^m$, where $x_{0,e}$ is the virtual origin.

A relation between the exponent $m$ and the Clauser parameter $\beta$ in turbulent flows was obtained in Ref. [20] by assuming $u_\tau/U_\infty \to 0$: $m = -\beta/(1+3\beta)$. To account for the empirical fact that $u_\tau/U_\infty$ approaches a constant rather than zero, Skote *et al.* [8] performed a non-linear analysis instead and obtained $m = -\beta/(H(1+\beta)+2\beta)$.

The non-dimensional acceleration parameter

$$K = \frac{\nu}{U^2}\frac{\mathrm{d}U_\infty}{\mathrm{d}x} \tag{1}$$

is also sometimes used [21] to describe the distribution of the free-stream velocity. However, it was shown in the study by Monty *et al.* [21] that characterizing the effect of the pressure gradient on the TBL was difficult with this parameter. The authors showed that the turbulence intensity was further increasing when $K$ was matched but $\beta$ increased, while for constant $\beta$ the profiles of the turbulence intensity showed a good collapse within the logarithmic region and a slight offset in the outer region. The parameter $K$ is also used to set the limit for relaminarization in favorable pressure-gradient TBLs as in Ref. [22].

An alternative definition of equilibrium was introduced in Refs. [23, 24] where instead of assuming the validity of the scaling laws, scaling parameters were derived from the conditions for similarity. The pressure gradient was defined as

$$\Lambda = \frac{\delta}{\rho U_\infty^2 \mathrm{d}\delta/\mathrm{dx}}\frac{\mathrm{d}P}{\mathrm{d}x} \approx \text{const.}, \tag{2}$$

with the boundary-layer thickness $\delta$. According to these authors the velocity scale for the velocity deficit law is $U_\infty$ and for the inner region it is $u_\tau$. The Clauser parameter $\beta$ is an asymptotic limit of $\Lambda$ for infinitely high $Re$, since $Re \sim U_\infty^2 \mathrm{d}\delta/\mathrm{d}x \sim u_\tau^2$ such as for ZPG TBLs. A more detailed description can be found in Ref. [24]. In the present work we will consider the region where $\beta$ remains constant to be in near-equilibrium conditions, as in Townsend [18] and Mellor and Gibson [19]. As mentioned in the Introduction, a wide range of studies are focused on APG TBLs, and in many of them the exponent $m$ was used to describe the pressure gradient. However, the results from the various studies cannot be easily compared, even if the value of $\beta$ is provided, due to history effects and the different conditions for equilibrium state. In the present study the pressure gradient will be imposed by considering a free-stream velocity following the power law with constant exponent $m$

$$U_\infty(x) = U_{\infty,0}\left(1 - \frac{x}{x_{0,l}}\right)^m \tag{3}$$

with $x_{0,l}$ being the virtual leading edge, where the displacement thickness is $\delta^* = 0$. The derivation of the free-stream velocity equation can be found in Ref. [8]. This case is referred to as the constant $m$ case. The corresponding free-stream distribution is shown in Figure 1(a). In a future follow-up study the history effects will be taken into account and the set up will be modified in order to impose a constant value of $\beta$. There, the PG will be described by a free-stream velocity distribution resulting in a constant pressure gradient coefficient $\beta$ over a long streamwise distance (as shown in Figure 1(c)) in order to eliminate upstream effects and to reach a near-equilibrium. During the entire study the integrated parameters are evaluated up to the boundary-layer thickness $\delta_{99}$. As shown by Vinuesa *et al.* [25], the diagnostic plot concept [26] is a robust method to determine the 99% boundary-layer thickness $\delta_{99}$ in pressure gradient TBLs. The boundary layer edge is found as the wall-normal location where $u_{rms}/(U\sqrt{H_{12}}) = 0.02$ corresponds to $U/U_\infty = 0.99$.

## 3. Numerical method

Numerical simulations were performed to investigate turbulent boundary layers with APGs. The spectral code SIMSON [27] developed at KTH was used to conduct large-eddy simulations (LES) in order to reduce the computational cost of the numerical campaign. The approximate

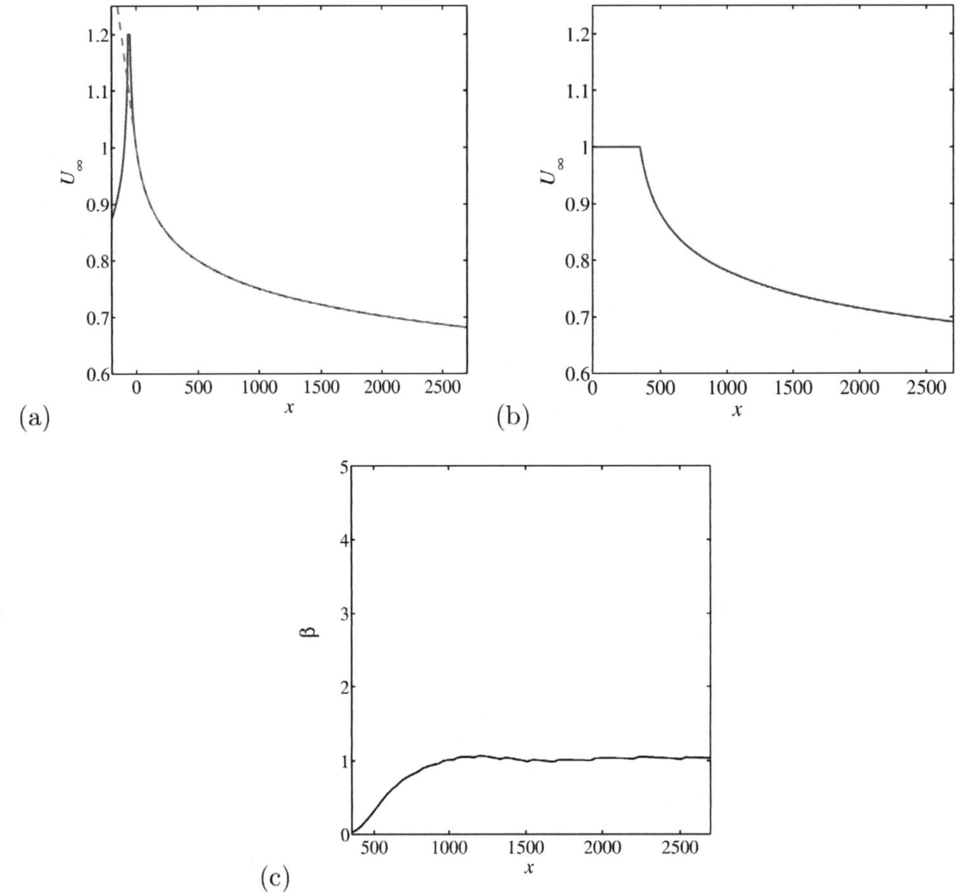


**Figure 1.** Free-stream velocity distribution (a) according to the power law (equation (3)) (blue) and with extended slope for $x < 0$ (red) for the LES with $m = -0.1$, (b) of the modified set-up with $m = -0.1$ including a ZPG region. (c) $\beta$ distribution of auxiliary RANS simulation (see Appendix).

deconvolution relaxation–term model (ADM-RT [27]) was used as the sub-grid scale (SGS) model to describe the small unresolved scales. In contrast to other SGS models such as the Smagorinsky model, here the SGS force acts directly on the resolved velocity components $\overline{u}_i$,

$$\frac{\partial \tau_{ij}}{\partial x_j} = \chi H_N * \overline{u}_i. \tag{4}$$

The symbol $*$ denotes the convolution and the overbar the implicit grid filter due to the lower resolution in the LES. The model coefficient $\chi = 0.2$ is chosen as constant similar to the time step of integration, $1/\Delta t$. The high-order filter $H_N$ with a cut-off frequency of $\omega_c \approx 0.86\pi$ affects only the small scales, whereas the energetic large scales are not taken into account by the model. As shown in earlier studies [28, 29], the flow predicted by this SGS model is in good agreement with DNS results in terms of mean flow, Reynolds-stress tensor and turbulence budgets. Even the budget terms are described well by LES data based on the current SGS model. The numerical method is based on Fourier discretization with dealiasing in the streamwise and spanwise directions as well as an expansion in Chebychev polynomials along the wall-normal

**Table 1.** Summary of the simulations of turbulent boundary layers with APG. Reynolds number $Re_{\delta_0^*}$, different domain sizes (in terms of the initial laminar displacement thickness $\delta_0^*$), grid points (before dealiasing), boundary conditions (BC) at the top boundary: 110 for asymptotic BC and 101 for Neumann BC, power-law exponent $m$ (equation (3)), and range of pressure-gradient parameter $\beta$.

| Case | $Re_{\delta_0^*}$ | Domain size $L_x \times L_y \times L_z$ | Grid points in $x \times y \times z$ | Top BC | m | $\beta$ |
|---|---|---|---|---|---|---|
| Skote [32] | 450 | 450×24×24 | 480×161×96 | 101 | -0.15 | [0.42; 0.68] |
| Fringe | 400 | 800×48×48 | 480×161×96 | 101 | -0.15 | [0.52; 1.40] |
| Blasius inlet | 450 | 800×48×48 | 480×161×96 | 101 | -0.15 | [0.58; 1.15] |
| Boxsize 1 | 450 | 3000×80×160 | 1024×121×128 | 101 | -0.1 | – |
| Boxsize 2 | 450 | 3000×120×160 | 1024×161×128 | 101 | -0.1 | – |
| Boxsize 3 | 450 | 3000×160×160 | 1024×201×128 | 101 | -0.1 | – |
| BC | 450 | 3000×120×160 | 1024×161×128 | 110 | -0.1 | – |
| m1 | 450 | 3000×120×160 | 2048×301×256 | 101 | -0.13 | [0.86; 1.49] |
| m2 | 450 | 3000×180×220 | 2048×361×384 | 101 | -0.16 | [1.55; 2.55] |

direction. A fringe region extends from $x = 2700$ up to the end of the box, where the flow is forced back to the Blasius inflow profile. The exact forcing is described in Ref. [27]. In this way, periodic boundary conditions can be imposed in the streamwise direction while simulating a spatially developing flow. At the wall the no-slip condition $(u, v, w)_{\text{lower}} = (0, 0, 0)$ was imposed, while at the upper boundary different boundary conditions were tested as described in Section 4.1. The inflow at $x = 0$ is a laminar Blasius profile with the Reynolds number based on the displacement thickness $Re_{\delta_0^*} = 450$. For the remainder of the paper, all length scales are given in terms of this displacement thickness $\delta_0^*$ and all velocities in terms of $U_{\infty,0}$ at $t = 0$ and $x = 0$, unless stated otherwise. At $x = 10$ the flow is tripped [30] and transitions thereafter into a turbulent flow. The length of the box is chosen as $L_x = 3000$ for all LES. During the study different box sizes and resolutions were tested before the final cases were defined. The chosen parameters of the auxiliary simulations as well as of the final cases are shown in Table 1. A coarse resolution was considered in the auxiliary cases in order to obtain general trends for the design of the final cases. For the main simulations the grid spacing was chosen to be of the same order as the zero pressure-gradient case by Schlatter *et al.* [31]: *i.e.* $\Delta x^+ = 21.5$, $\Delta y^+_{max} = 13.9$, $\Delta z^+ = 9.2$, and a total of 12 points below $y^+ = 10$.

## 4. Results

### 4.1. Auxiliary simulations

The final simulation parameters were determined by performing several auxiliary simulations, in which different set-ups and boundary conditions together with different box sizes were tested. The investigation started with the set-up as considered by Skote [32]. The box size and resolution (shown in Table 1) were chosen as in Case 2 from the DNS by Skote. At the inflow a Falkner–Skan profile with $m = -0.077$ was imposed, which was tripped at $x = 10$. The pressure gradient is imposed by setting the free-stream velocity distribution at the top of the box as in equation (3) over the entire box, where the exponent is chosen as $m = -0.15$ and $x_{0,l} = -60$. The exact condition at the upper boundary is described below. These results showed a strong dependence on the inflow conditions, which motivated further tests to investigate the influence of those. Hereby the test case *Fringe* was simulated once with the same fringe forcing as in Skote and another time with a stronger fringe forcing. The box was extended in all three directions to investigate the effect on the boundary-layer development at higher $Re$, while the number of grid points was kept constant. In Figure 2 the two cases (moderate and strong forcing) are compared with each other. In addition, the data of the case *Blasius inlet* (see Table 1) is shown, which differs from the moderate fringe forcing case by using a Blasius inflow instead of a Falkner–

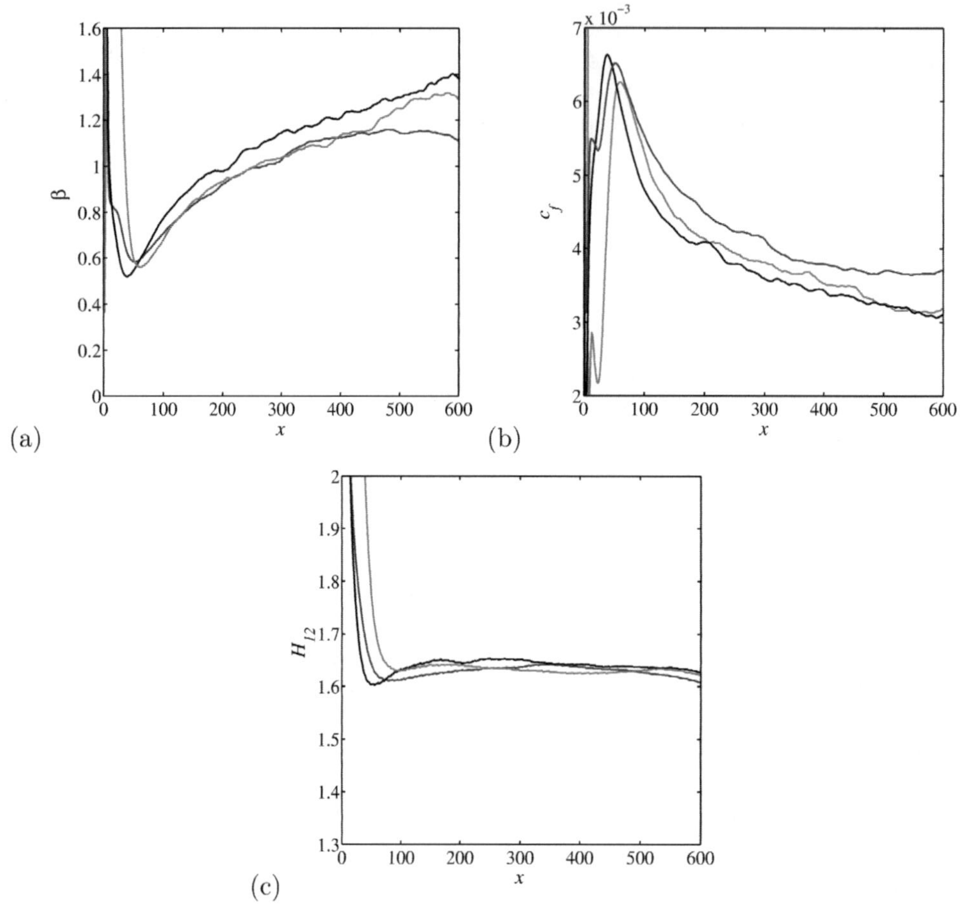


**Figure 2.** Comparison between the *Fringe* data set of moderate fringe forcing (black), the one with strong fringe forcing (red) and the data set with Blasius inlet profile (blue) of (a) the pressure gradient parameter $\beta$, (b) the skin friction coefficient $c_f$ and (c) the shape factor $H_{12}$.

Skan profile. The case with the strong fringe forcing shows an unphysical behavior in all three parameters $\beta$, $c_f$ and $H_{12}$ for $x < 100$. The effect of the fringe diminishes at higher Reynolds numbers. The resulting $\beta$ range is higher than in the *Skote* case, since higher *Re* are observed as well as resolution effects have to be considered. The test cases are not as finely resolved as the *Skote* case, but as mentioned an SGS model was employed instead. In all three cases $H_{12}$ converges towards a constant value of about 1.6, but shows different developments right after the inlet. The three parameters show a strong dependency on the inflow conditions. One should note that the statistics are not fully converged as it can be seen in the figures.

To ensure that the PG TBL is independent of the inflow conditions, the set-up had to be modified. Instead of a Falkner–Skan profile with pressure gradient at the inlet, the inflow was set to a Blasius profile. This approach emulates the procedure carried out in wind tunnel experiments. The fringe and tripping parameters remained unchanged. A ZPG region extends up to $x \approx 350$, henceforward the pressure gradient will be applied following the power–law, equation (3), with $x = [350, L_x]$. The exponent $m$ is initially chosen as zero at $t = 0$. Once the flow is fully turbulent over the whole box, the exponent is slowly adapted towards the final value. In comparison to the case of *Skote*, the computational box was extended in the streamwise direction to investigate higher Reynolds numbers. The free-stream velocity of the new set-up is shown in Figure 1(b). This modified set-up enables the study of turbulent boundary layers with various strengths of pressure gradient independent of the choice of the inflow profile and the fringe parameters. All the following simulations were performed with this new set-up.

Upon deciding the method to impose the PG, we investigate the effect of the box size on the statistics of the simulations. The streamwise box length $L_x$ was chosen in order to achieve the required Reynolds number before the fringe region. The spanwise box width $L_z$ was chosen according to earlier investigations, see *e.g.* Ref. [29], where $L_z > 2\delta_{99}$ (taken at the end of the domain) to ensure correct development of large-scale structures in the outer region.

The height of the box $L_y$ was tested in three different simulations (see Table 1), where the streamwise and spanwise extend was kept constant while the height of the box was varied. In Figure 3 the contour lines of the rms for all velocity components are shown for the three different box sizes. This figure shows that there is not a strong effect of the box height on the inner layer for all the components. The outer layer, on the other hand, is influenced strongly by the height of the box. In all three rms-components in *Boxsize 1* it can be seen that the large structures

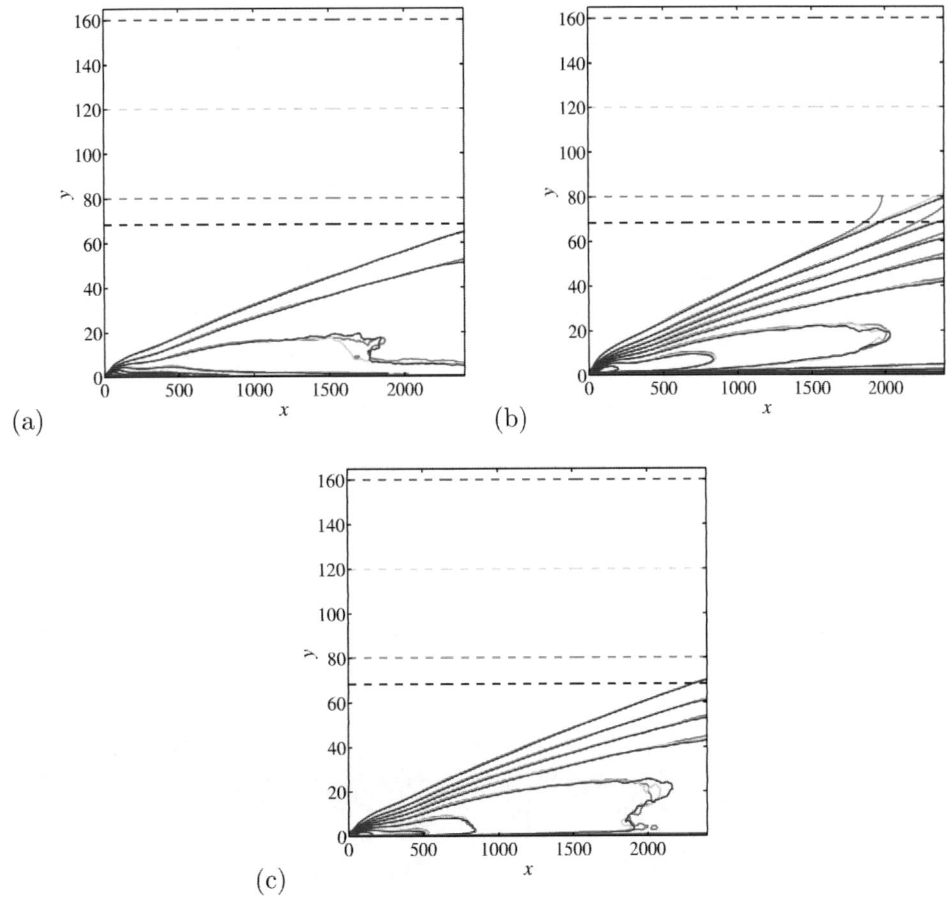


**Figure 3.** Contour lines (equidistantly spaced between 0 and the maximum rms values on 10 contour lines) of rms-profiles for cases *Boxsize 1* (red), *Boxsize 2* (green), *Boxsize 3* (blue) Table 1: (a) $u_{rms}$, (b) $v_{rms}$, (c) $w_{rms}$ scaled by $U_{\infty,0}$. The boundary-layer thickness $\delta_{99}$ (black) and the respective box heights are indicated through dashed lines.

are restricted by the height of the box, while no influence of the upper boundary is observed in *Boxsize 2* and *Boxsize 3*. The height of *Boxsize 2* is consequently a good choice and we conclude that a box height of $L_y = 2\delta_{99}$ will be deemed sufficient for further investigations.

As already mentioned, two different boundary conditions were tested for the upper boundary. One condition is the Neumann condition (denoted as 101 in Table 1), where the wall-normal derivatives of the streamwise and spanwise components are fixed to

$$\left.\frac{\partial u}{\partial y}\right|_{y=L_y} = \left.\frac{\partial w}{\partial y}\right|_{y=L_y} = 0. \tag{5}$$

at $y = L_y$. The wall-normal boundary condition is obtained from the continuity equation and expressed as:

$$\left.\frac{\partial v}{\partial y}\right|_{y=L_y} = -\frac{\partial U_\infty}{\partial x}. \tag{6}$$

The second investigated boundary condition is the so called asymptotic condition (denoted as 110 in Table 1) based on Fasel [33] and Malik *et al.* [34]. The condition is set by the asymptotic decay of the velocity perturbation in wall-normal direction. The base flow $\hat{\boldsymbol{U}}$ is calculated from the Blasius solution, which was chosen as the initial condition and is scaled for each streamwise position according to the outer velocity as imposed at the upper boundary. The boundary condition is written as

$$\left.\left(\frac{\partial \hat{\mathbf{u}}}{\partial y} + |k|\hat{\mathbf{u}}\right)\right|_{y=L_y} = \left.\left(\frac{\partial \hat{\boldsymbol{U}}}{\partial y} + |k|\,\hat{\boldsymbol{U}}\right)\right|_{y=L_y}, \tag{7}$$

where $(\hat{\cdot})$ denotes the horizontal Fourier transform with respect to the horizontal coordinates and $k^2 = \alpha^2 + \beta^2$ with the horizontal wave-numbers $\alpha$ and $\beta$ [27].

In Figures 4 and 5, first-order statistics are compared between simulations using the two boundary conditions. The free-stream distribution is defined exactly when using the Neumann condition. Conversely, in the case of the asymptotic condition $U_\infty$ does not follow the exact power law as specified. Overshoots in the velocity are observed in the transition from ZPG

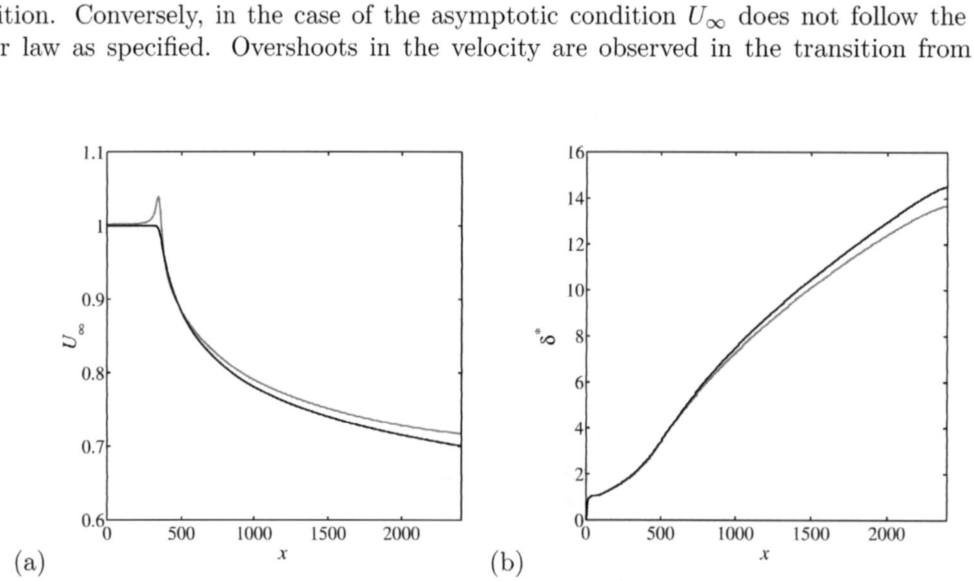


**Figure 4.** Comparison between the Neumann (black) and asymptotic (red) as in Ref. [33, 34] boundary condition. (a) shows the free stream velocity distribution $U_\infty$ at the top of the box and (b) the displacement thickness $\delta^*$ as a function of the streamwise position $x$.

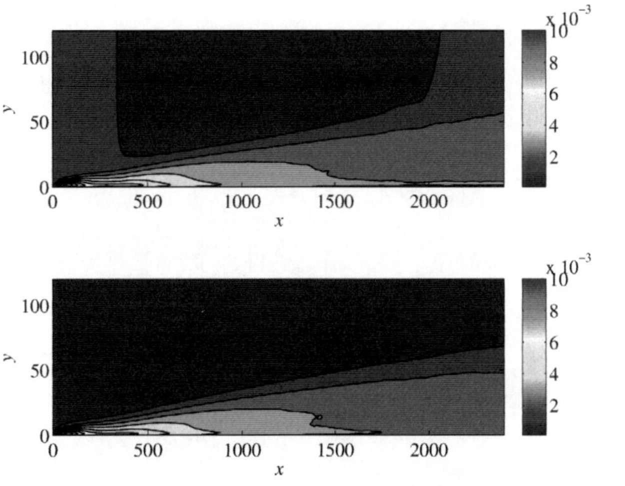


**Figure 5.** Comparison of Neumann (bottom) and asymptotic (top) as in Ref. [33, 34] boundary condition: rms of the pressure in terms of the reference quantities $U_\infty$ and $\delta_0^*$.

to APG and at the entrance of the fringe region. While it is possible to define a power law describing the distribution without the overshoots, it will not exactly match the prescribed one, and therefore the direct connection between the predetermined $m$ and the produced $\beta$ is not possible. The $U_\infty$ distribution also leads to differences in the displacement thickness $\delta^*$ as depicted in Figure 4(b). Since $U_\infty$ decreases less for the asymptotic condition downstream, $\delta^*$ increases slower and reaches lower values. The position of the maxima is at the same location but less pronounced when using the asymptotic condition. In Figure 5 the rms of the pressure fluctuations are compared in the $xy$-plane. An unexpected behavior can be noticed for the asymptotic condition in the free-stream, which cannot be further explained. This occurrence and the inexact definition of the free-stream velocity excludes the asymptotic condition as a boundary condition. While the asymptotic condition can be placed relatively close to the wall, the use of the Neumann condition is restricted by the size of the disturbance velocity: sufficiently high boxes have to be chosen, where the disturbance velocity is small. An alternative procedure to impose the pressure gradient on the TBL was developed by Spalart and Watmuff [35], who basically decomposed the velocity field $U$ into the sum of several fields as $U_0 + U_1 + U_2 + U_3$, $U_1$ being the unknown. Whereas $U_2$ is related to the fringe region and $U_3$ is a correction that depends on $U_1$, the pressure gradient is imposed indirectly through the velocity $U_0$. The idea is that $U_0$ sets a slip velocity throughout the boundary layer that basically is an extrapolation of the velocity in the rotational region across the boundary layer and all the way down to the wall. Nevertheless, in the present study the Neumann condition will be considered as the upper boundary condition.

### *4.2. Main simulations*

The statistics of the final cases with the constant power–law exponent $m = -0.13$ and $m = -0.16$ (*cf.* Table 1) are compared with each other and the influence of the strength of the PG on the flow statistics is investigated. The statistics were obtained after averaging over 11.5 eddy turnover times (ETT) in the case of $m = -0.13$ and over 2.3 ETT in case of $m = -0.16$. Here ETT is defined as $ETT = \delta_{99}/u_\tau$ at $x = 2300$, *i.e.* close to the end of the physical domain. According to Lozano-Durán and Jiménez [37], turbulent channel-flow simulations at similar $Re$ require an averaging time of around 11–12 ETT in order to provide fully converged statistics. In

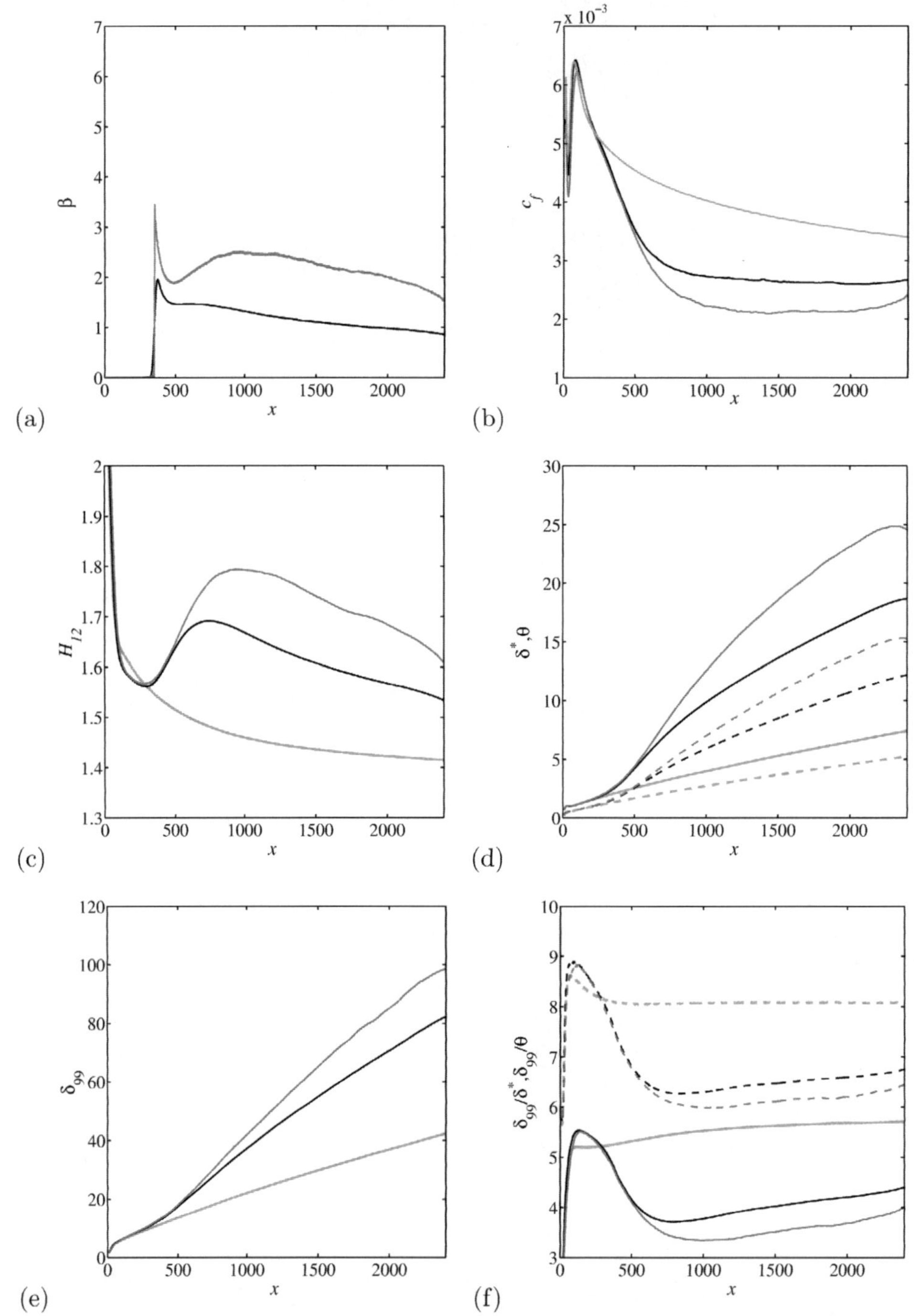


**Figure 6.** (a) Clauser parameter $\beta$, (b) skin-friction coefficient $c_f$, (c) shape factor $H_{12}$, (d) displacement thickness $\delta^*$ (solid) and momentum thickness $\theta$ (dashed), (e) boundary-layer thickness $\delta_{99}$, (f) ratio between $\delta_{99}$ and $\delta^*$ (solid) and $\delta_{99}$ and $\theta$ (dashed) for the two $m$-cases $m = -0.13$ (black), $m = -0.16$ (red) and ZPG [36] (grey) over the streamwise position $x$.

channel-flow simulations the streamwise and spanwise direction are homogeneous, which allows to increase the effective number of statistical samples by averaging in those directions. In the case of the spatially developing TBL only the spanwise direction is homogeneous, which means that converged statistics will require longer averaging times than those for a channel flow.

Therefore, the averaging time is not sufficient in the present simulation with $m = -0.16$, consequently some of the statistics presented here are not fully converged. Streamwise one-point statistics are shown in Figure 6 comparing the PG cases to a ZPG case [36]. In Figure 6(a) the streamwise distribution of the Clauser parameter $\beta$ is shown. A constant $m$ leads to a downstream decrease in $\beta$, while a larger absolute value of $m$ results in a larger value of $\beta$, since the velocity gradient over the streamwise direction is stronger. The skin-friction coefficient $c_f$, shown in Figure 6(b), first decreases downstream, but converges to a seemingly constant value, while for the ZPG $c_f$ continues to decrease over the whole box length as expected. Since the APG thickens the boundary layer, the velocity gradient at the wall is reduced and therefore APG TBLs exhibit a lower $c_f$ value than the corresponding ZPG case. The shape factor $H_{12}$, depicted in Figure 6(c), is higher for flows with APG as the ratio between the displacement thickness $\delta^*$ and momentum thickness $\theta$ increases as seen in Figure 6(d). In $\delta^*$ and $\theta$ upstream influences of the APG and effects of the different tripping [30] can be noticed at the streamwise positions $x > 110$. The boundary-layer thicknesses $\delta^*$ and $\theta$, as well as $\delta_{99}$ (Figure 6(e)), increase, since the APG decelerates the flow and it basically lifts up the boundary layer. The APG TBL is twice as thick as the ZPG case. For instance, comparisons with ZPG datasets (Ref. [30]) show that the boundary-layer thickness $\delta_{99}$ is five to six times $\delta^*$ and eight times $\theta$. In the present APG TBL the ratios are four to five times $\delta^*$ and seven times $\theta$. This clearly shows the effect of the APG in the development of the TBL.

In Figure 7 the mean velocity is shown and compared to a ZPG case at matched friction Reynolds number $Re_\tau = u_\tau \delta_{99}/\nu$, which fixes the inner-scaled thickness of the TBL, allowing the characterization of PG effects on the profiles. The profiles of two different APG strengths are compared to the ZPG case at $Re_\tau = 380$ and $Re_\tau = 600$. In Figure 7(a) the profiles of the mean velocity $U^+$ scale in the viscous sublayer as expected. In the log layer the APG profiles are consistently below the ZPG ones. In the outer region the influence of the APG is manifested in a stronger wake parameter $\Pi$, which is the consequence of more energetic

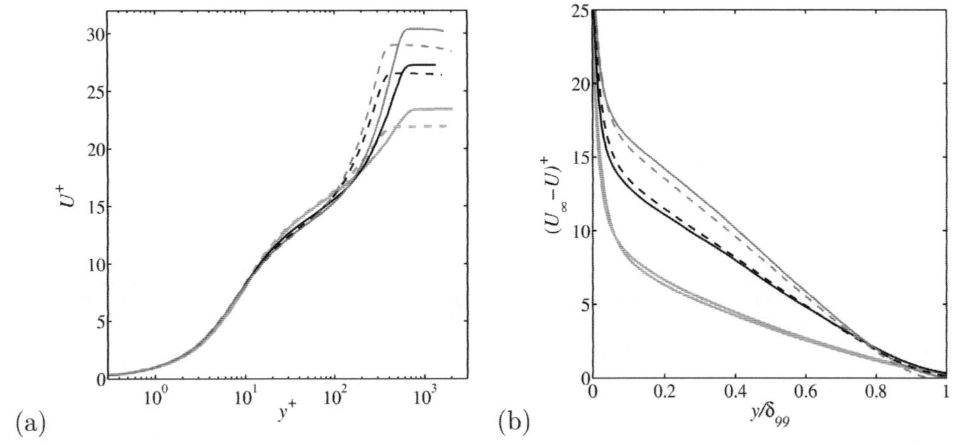


**Figure 7.** (a) Inner-scaled mean velocity profile and (b) velocity defect for the two $m$-cases; $m = -0.13$ (black), $m = -0.16$ (red) at $Re_\tau = 380$ (dashed) and $Re_\tau = 600$ (solid) and ZPG [36] (grey).

structures in the outer layer. Note that the velocity gradient is not necessarily zero for $y > \delta_{99}$, *i.e.* $U(y > \delta_{99}) \neq const.$ [25] as it can be observed in Figure 7(a). Further insight in the outer region can be gained by examining the defect mean velocity profile $(U^+_\infty - U^+)$ as in Figure 7(b). This figure shows clearly the large influence of the APG in the outer region.

The inner-scaled root mean square values $u^+_{rms}$, $v^+_{rms}$ and $\langle uv\rangle^+$ are depicted in Figure 8. The near-wall peak is higher for stronger PG when compared to ZPG TBLs, while the position of the $u^+_{rms}$ peak is invariant around $y^+ = 15$. In the outer region a second peak can be noticed as in Ref. [21]. It is interesting to observe that an outer peak emerges in all the components of the Reynolds-stress tensor, the magnitude of which increases with PG, as well as with $Re$. The strength of the outer peak in case $m = -0.16$ is of a comparable magnitude as the inner peak. In the case of Skote [32] with $m = -0.15$, a second hump can be noticed with values lower than 2. Even though the exponents $m$ are close to each other, it is difficult to compare these cases since both history and Reynolds number are different. In the study of Vinuesa *et al.* [38] the TBL around a wing was analyzed. Large PG with $\beta = 4.54$ lead to a strong outer peak in $u^+_{rms}$, however this peak amplitude is smaller than the inner peak. Low PG with $\beta = 0.53$ (comparable to Skote [32]), leads to a less pronounced outer peak and the shape of the $u^+_{rms}$ can be compared to the shape as presented in Skote. The possible connection between the outer peak in the turbulent intensity found in high $\beta$ and high $Re$ ZPG boundary layers will be further explored in future studies.

The decomposition of total shear stress $\tau$ into viscous stress and Reynolds shear stress is shown in Figure 8(b). In ZPG flows the maximum total shear stress is unity. In APG cases the decomposition of $\tau$ is the same, while the value of unity can be exceeded due to a much larger Reynolds shear–stress term in the outer region. The Reynolds shear stress $\langle uv\rangle^+$ is influenced strongly as indicated in Figure 8(b) with a strong outer peak, while the viscous stress term appears unaffected by PG effects. Since the free-stream velocity is defined with a power law, the TBL exhibits near-equilibrium features (Ref. [19]). This could be connected to the aforementioned fact that the viscous shear stress is independent of the PG.

The skewness and flatness factors for the streamwise velocity component are compared in

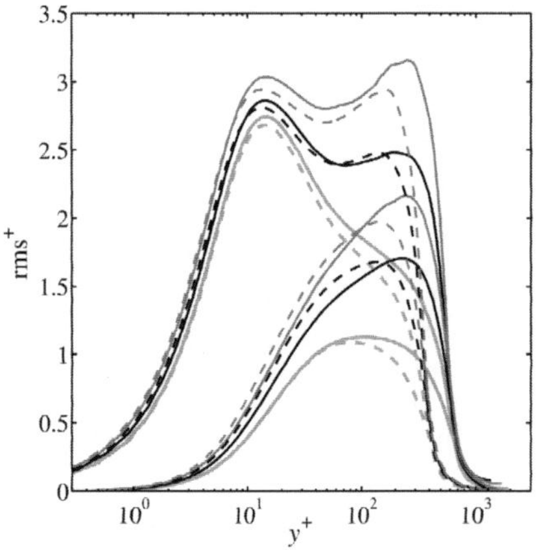

(a)

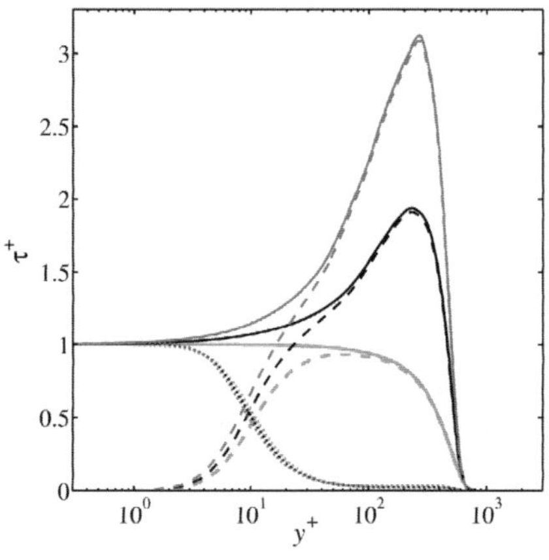

(b)

**Figure 8.** (a) Inner–scaled root mean square values ($u^+_{rms}$, $v^+_{rms}$) for the two $m$-cases: $m = -0.13$ (black), $m = -0.16$ (red) at $Re_\tau = 380$ (dashed) and $Re_\tau = 600$ (solid) and ZPG [36] (grey). (b) Decomposition of total shear stress $\tau^+$ into viscous (dotted) and Reynolds shear stress (dashed) for the two $m$-cases: $m = -0.13$ (black), $m = -0.16$ (red) at $Re_\tau = 600$ (solid) and ZPG [36] (grey).

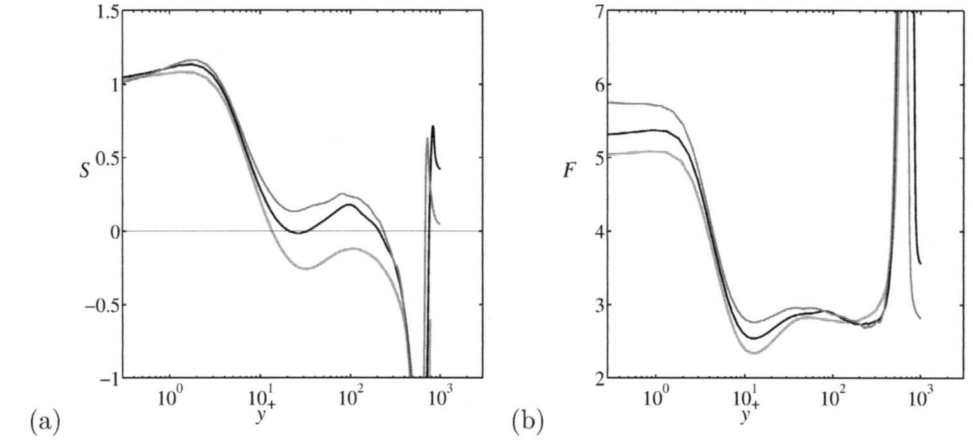


**Figure 9.** (a) Skewness and (b) flatness factor for the two $m$-cases: $m = -0.13$ (black), $m = -0.16$ (red) and ZPG [36] (grey) at $Re_\tau = 600$.

Figure 9. In general the values at the wall of skewness and flatness are higher in PG flows and are increasing with the strength of the APG. This is also seen in ZPGs with increasing $Re$. The increase of the values with the PG is also apparent throughout the buffer layer. As described in Refs. [28, 39] for large $Re$ in ZPGs, the large-scale motions become progressivly stronger at higher $Re$ in ZPG TBLs. It can therefore be assumed that large-scale motions also become stronger for increasing APGs, and consequently their influence on the small scales. In the flatness factor of the APG cases a second peak can be noticed farther away from the wall and the skewness does not become negative within the buffer layer. As stated in Ref. [40] the zero-crossing of the skewness factor and the minimum value of the flatness factor are at the same wall-normal position in ZPG TBLs (see also Eitel-Amor *et al.* [28]). In the present APG cases the first minimum of the flatness is at the same position as in the ZPG, while the second minimum coincides with the zero-crossing of the skewness. It is interesting to note that the

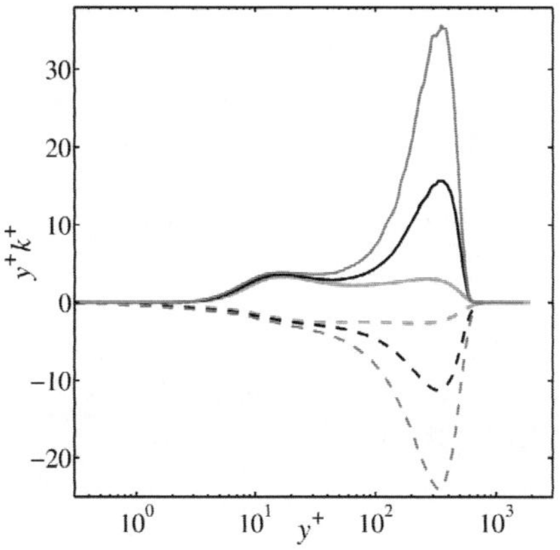


**Figure 10.** Inner-scaled pre-multipliedturbulent kinetic energy budget: production (solid) and dissipation (dashed) for the two $m$-cases: $m = -0.13$ (black), $m = -0.16$ (red) and ZPG [36] (grey) at $Re_\tau = 600$.

latter position is close to the one of the second peak occurring in the turbulence intensities. This behavior is in agreement with large *Re* ZPGs as reported in Vallikivi *et al.* [41].

Finally, the production and dissipation terms of the pre-multiplied kinetic energy budget are shown in Figure 10. The production exhibits a low inner peak and production and dissipation show a pronounced peak farther away from the wall at a position $y/\delta_{99} \approx 0.45$ in the APG cases, while the peak is stronger for stronger APG. The outer peak of the production was also noticed in Skåre and Krogstad [4]. They found a diffusion mechanism from the outer region of the boundary layer towards the wall. This is another manifestation of the impact of the APG on the large-scale structures of the flow. The dominant outer peak in the production, as seen in Figure 10, is a second energy source and dissipation is not restricted to the wall, but energy is transferred from large to small scales throughout the whole layer.

## 5. Conclusion and Outlook

In the present work we developed a suitable set-up to numerically study adverse pressure gradient turbulent boundary layers, with the future aim of assessing history and development effects on the flow. The set-up suggested by Skote *et al.* [8], where a power law was used to describe the velocity distribution at the free-stream, was modified in order to ensure the proper tripping and development of the TBL. The new approach mimics the procedure carried out in wind tunnel experiments. Here the APG is applied once the flow has reached a fully turbulent state. The computational box size has been assessed by testing different heights of the box and it could be concluded that the flow develops independently from the box height when $L_y > 2\delta_{99}$ as in ZPG TBLs. The spanwise length of the box was decided based on earlier investigations on ZPG flows as in Ref. [30], where a value of $L_z = 2 - 3\delta_{99}$ was appropriate. The correct definition of the boundary condition on the top boundary was systematically tested. Although the asymptotic condition suggested by Fasel [33] would in principle allow to use smaller box heights, we found that this BC does not lead to correct prescribed pressure distributions. The Neumann condition was found to describe the required velocity distribution exactly and was chosen as the upper boundary condition. After deciding on the final set-up, we performed the main simulations with following power–law exponents: $m = -0.13$ and $m = -0.16$, in the $U_\infty$ distribution given by equation (3). The effect of the APG on TBLs was manifested in the mean flow through a much more prominent wake region and in the Reynolds stresses through the existence of an outer peak. The pre-multiplied energy budget shows in the production term a pronounced second peak in the outer region, which underlines influence on the distribution of the turbulent kinetic energy transfer mechanism across the boundary layer. This might also indicate that the large-scale structures are stronger in APG TBLs than in ZPG TBLs. Note that the structures are not necessarily stronger when scaling with the local mean shear is considered. Another observation is that the displacement thickness and momentum thickness are around three times larger than the respective thicknesses in the ZPG case. The boundary layer decelerates by the influence of the APG, the skin friction is reduced and the boundary layer thickness increases. In this study we defined the set-up to investigate APG TBLs with constant $m$ and we also performed first tests to set-up a case with constant $\beta$ (see Appendix). In future work we will compare the two cases of constant $m$ and constant $\beta$ with each other to identify history effects and to further explain the characteristics of APG flows.

## Acknowledgments

Funded in part by the Multiflow program of the European Research Council and the Knut and Alice Wallenberg Foundation, which are both acknowledged. The computations were performed on resources provided by the Swedish National Infrastructure for Computing (SNIC) at PDC Center for High Performance Computing and High Performance Computer Center North (HPC2N).

## References

[1] Bradshaw P 1967 *J. Fluid Mech.* **29** 625–645
[2] Head M R 1976 *J. Fluid Mech.* **73** 1–8
[3] Simpson R L 1989 *Annu. Rev. Fluid Mech.* **21** 205–34
[4] Skåre P E and Krogstad P Å 1994 *J. Fluid Mech.* **272** 319–384
[5] Harun Z, Monty J P, Mathis R and Marusic I 2013 *J. Fluid Mech* **715** 477–498
[6] Spalart P R and Coleman G N 1997 *Eur. J. Mech. B-Fluid* **16** 169–189
[7] Na Y and Moin P 1998 *J. Fluid Mech* **374** 379–405
[8] Skote M, Henningson D S and Henkes R A W M 1998 *Flow Turbul. Combust.* **60** 47–85
[9] Lee J H and Sung H J 2008 *Int. J. Heat. Fluid Flow.* **29** 568–578
[10] Gungor A G, Maciel Y, Simens M P and Soria J 2014 *J. Phys.: Conf. Ser.* **506** 012007
[11] Kitsios V, Atkinson C, Sillero J A, Borrell G, Gungor A, Jiménez J and Soria J 2016 *Advances in Computation, Modeling and Control of Transitional and Turbulent Flows* 269–278
[12] Gruschwitz E 1931 *Ing.-Arch.* **2** 321–346
[13] Nikuradse J 1929 *Untersuchungen über die Strömungen des Wassers in konvergenten und divergenten Kanälen* (VDI-Verlag)
[14] Falkner V and Skan S W 1931 *Lond. Edinb. Dubl. Phil. Mag.* **12** 865–896
[15] Kundu P K, Cohen I M and Dowling D R 2012 *Fluid Mechanics* (Elsevier)
[16] Hartree D R 1937 *Proc. Camb. Phil. Soc.* **33** 223–239
[17] Clauser F H 1954 *AIAA J.* **21** 91–108
[18] Townsend A A 1961 *Equilibrium Layers and Wall Turbulence* vol 11 (Cambridge Univ Press)
[19] Mellor G L and Gibson D M 1966 *J. Fluid Mech.* **24** 225–253
[20] Tennekes H and Lumley J L 1972 *A First Course in Turbulence* (The MIT Press)
[21] Monty J P, Harun Z and Marusic I 2011 *Int. J. Heat. Fluid Flow.* **32** 575–585
[22] Piomelli U and Yuan J 2013 *Phys. Fluids* **25** 101304
[23] George W K and Castillo L 1993 *Near Wall Turbulent Flows* 901–910
[24] Castillo L 1997 *Similarity Analysis of Turbulent Boundary Layers* (State University of New York)
[25] Vinuesa R, Örlü R and Schlatter P 2016 *to be published J. Phys. Conf. Ser.*
[26] Alfredsson P H, Segalini A and Örlü R 2011 *Phys. Fluids* **23** 041702
[27] Chevalier M, Schlatter P, Lundbladh A and Henningson D S 2007 *SIMSON–A Pseudo-Spectral Solver for Incompressible Boundary Layer Flow* (KTH Mechanics)
[28] Eitel-Amor G, Örlü R and Schlatter P 2014 *Int. J. Heat. Fluid Flow.* **47** 57–69
[29] Bobke A, Örlü R and Schlatter P 2015 *J. Turbul.* **17** 155–178
[30] Schlatter P and Örlü R 2012 *J. Fluid Mech.* **710** 5–34
[31] Schlatter P, Li Q, Brethouwer G, Johansson A V and Henningson D S 2010 *Int. J. Heat. Fluid Fl.* **31** 251–261
[32] Skote M 2001 *Studies of Turbulent Boundary Layer Flow Through Direct Numerical Simulations* (KTH Royal Institute of Technology Stockholm)
[33] Fasel H 1976 *J. Fluid Mech.* **78** 355–383
[34] Malik M R, Zang T A and Hussaini M Y 1985 *J. Comput. Phys.* **61** 64–88
[35] Spalart P and Watmuff J H 1993 *J. Fluid Mech.* **249** 337–371
[36] Schlatter P, Örlü R, Li Q, Brethouwer G, Fransson J H M, Johansson A V, Alfredsson P H and Henningson D S 2009 *Phys. Fluids* **21** 051702
[37] Lozano-Durán A and Jiménez J 2014 *Phys. Fluids* **26** 011702
[38] Vinuesa R, Hosseini S M, Hanifi A, Henningson D S and Schlatter P 2015 *Proc. 9th Int. Symp. Turbulence & Shear Flow Phenomena, Melbourne, Australia*
[39] Mathis R, Marusic I, Hutchins N and Sreenivasan K 2011 *Phys. Fluids* **23** 121702
[40] Eckelmann H 1974 *J. Fluid Mech.* **65** 439–459
[41] Vallikivi M, Hultmark M and Smits A J 2015 *J. Fluid Mech.* **779** 371–389
[42] Spurk J H and Aksel N 2008 *Fluid Mechanics* (Springer-Verlag Berlin Heidelberg)
[43] Menter F R 1994 *AIAA J.* **32** 1598–1605

## Appendix: Design of a case with constant $\beta$

As discussed above, the pressure gradient is imposed on the TBL through the streamwise evolution of $U_\infty$. The $U_\infty(x)$ curve can be determined by the geometry, as it is the case in the flow around wings, by modifying the ceiling configuration in wind-tunnel experiments, or by setting its evolution as a power law with exponent $m$. All those flows have different histories, which are not taken into account in the analysis of scaling laws. With $\beta$ being constant over a long streamwise distance the same pressure gradient can be observed from low to high $Re$. It is therefore necessary to determine the $U_\infty(x)$ distribution leading to a constant $\beta$ over a long streamwise distance in the domain. Our initial idea was to mimic a wind-tunnel experiment by performing Reynolds-Averaged Navier–Stokes (RANS) computations, and to define the shape of the ceiling (initially chosen as in Ref. [4]) through an iterative process until a constant $\beta$ distribution was obtained in the $Re$ range of interest. Once a constant $\beta$ distribution was reached, the corresponding $U_\infty(x)$ could be extracted and then imposed in the LES.

The RANS computations were carried out by considering the two equation SST (shear stress transport) model [43], implemented in the commercially available CFD code Fluent (v.6.3). The inflow condition in the RANS is a uniform profile, which together with 0.5% turbulence intensity, leads to a turbulent flow few grid points downstream the inflow. However, in the LES the inflow is set as a laminar boundary layer defined by a Blasius profile with $Re_{\delta_0^*} = 450$, which is then tripped to turbulence. The tripping parameters were chosen as follows: time-dependent forcing amplitude $tampt = 0.4$, streamwise length scale of tripping $txsc = 4.0$, streamwise origin of tripping $tx0 = 10$, wall-normal length scale of tripping $tysc = 1.0$, time scale of tripping $tdt = 4.0$. It is important to ensure that the development of the boundary layer computed in the RANS is equivalent to the one in the LES. Since the RANS essentially starts from a turbulent profile, it is necessary to determine the value of $Re_{\delta^*}^{\text{turb}}$ that a turbulent boundary layer would have at the location where in the LES the laminar $Re_{\delta_0^*}^{\text{lam}} = 450$, *i.e.* at $x/\delta_0^*$. To this end, we considered the ZPG DNS data [36], which also has $Re_{\delta_0^*}^{\text{lam}} = 450$, and computed the virtual origin $x_0$ based on the data with $x/\delta_0^* > 400$. Note that although the flow is tripped at $x/\delta_0^* = 10$, we did not consider the part of the domain downstream the trip to ensure that a fully-developed turbulent region is used in the analysis. We initially considered the following

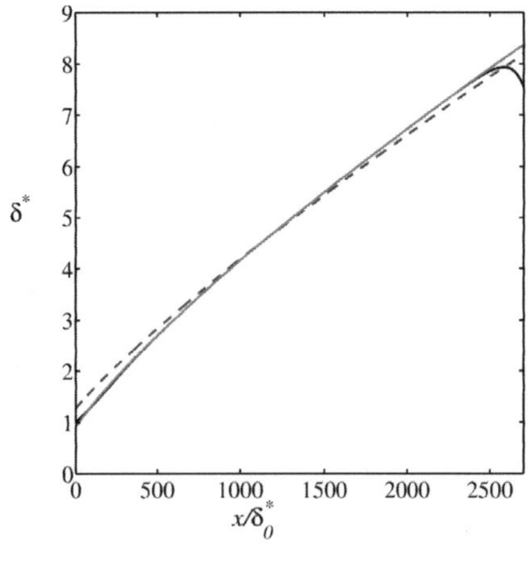


**Figure 11.** Displacement thickness $\delta^*$ of the ZPG DNS data (black solid), equation from Ref. [42] (blue dashed), fitted data (red solid).

empirical correlation by Spurk and Aksel [42] for the displacement thickness $\delta^*$,

$$\delta^* = 0.046 Re_x^{-1/5}(x - x_0) \qquad \text{with} \qquad Re_x = \frac{U_\infty}{\nu}(x - x_0), \tag{8}$$

and calculated, for each $x$ position, the corresponding virtual origin $x_0$. Interestingly, this method did not lead to a constant $x_0$ over $x$, which is an indication of the fact that equation (8) does not properly represent the DNS results in this low $Re$ range. Additional confirmation of this fact is given in Figure 11, where the displacement thickness from the DNS is shown as a function of $x/\delta_0^*$ together with the result of equation (8) with an average value of $x_0$. The insufficient agreement between both curves motivated the use of another method to determine $x_0$, and therefore we considered the following more general form of the $\delta^*$ curve,

$$\delta^* = A Re_x^B (x - x_0), \tag{9}$$

which has three fitting parameters: $A$, $B$ and $x_0$. Fitting equation (9) to the region of the DNS with $x/\delta_0^* > 400$, we obtained the values $A = 0.06527$, $B = -0.2209$ and the virtual origin of a TBL with $Re_{\delta_0^*}^{\text{lam}} = 450$ as inflow condition, *i.e.* $x_0 = -170.1$. In this case, equation (9) represents the DNS data very well, and interestingly, small deviations are observed between the fit and the DNS close to $x/\delta_0^* \simeq 0$. This is of course due to the fact that close to the inflow the simulated boundary layer is still laminar, and the fit proposed here is aimed at representing the turbulent region. We then used equation (9) to determine the value of $\delta^{*\text{turb}}$ at the origin of the LES domain ($\delta^{*\text{turb}} = 0.9259$), and consequently the value that will be considered at the inflow in the RANS simulation: $Re_{\delta_0^*}^{\text{turb}} = 417$. It is interesting to note that the displacement thickness of the turbulent boundary layer is in fact lower than the laminar one, which may come as a surprise, but this is a result of the different developments of both boundary layers: farther downstream where the turbulent boundary layer is the one with larger $\delta^*$ values.

After ensuring that both numerical boundary layers would theoretically exhibit the same streamwise development, we performed a RANS simulation of a ZPG TBL with $Re_{\delta_0^*}^{\text{turb}} = 417$. Since the length scale of the simulations is $\delta_0^*$, and they have different Reynolds numbers, the results from the RANS have to be scaled (taking into account that the ratio $U_\infty/\nu$ is still the same in both simulations) with the length $\ell = (Re_{\delta_0^*}^{\text{lam}}/Re_{\delta_0^*}^{\text{turb}})\delta_0^{*\text{turb}} = (450/417)\delta_0^{*\text{turb}}$. Figure 12

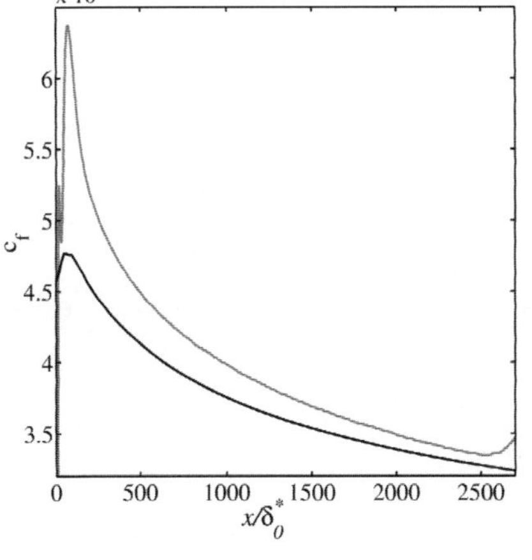


a)

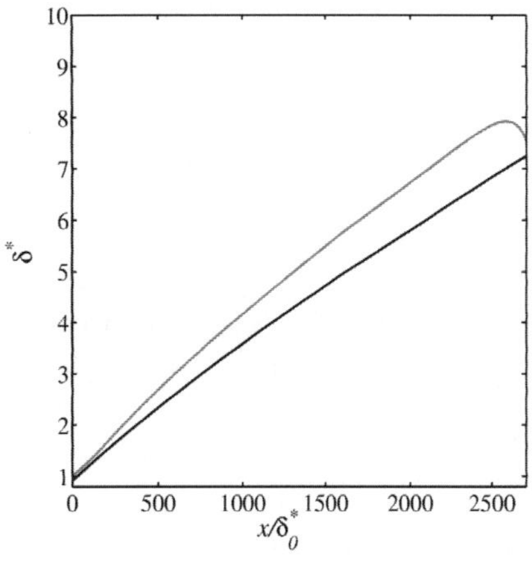


b)

**Figure 12.** Testing the correct scaling between RANS data (black) and DNS data [36] for ZPG (red): a) skin friction coefficient $c_f$ and b) displacement thickness $\delta^*$.

shows a comparison of the downstream evolution of skin friction coefficient $c_f$ and $\delta^*$ from the DNS and from the RANS, which are appropriately scaled as discussed above. Unfortunately, there are significant deviations between both boundary layers, and interestingly the RANS provides better predictions of skin friction at higher Reynolds numbers, whereas the boundary-layer growth deviates more from the DNS as one moves downstream. The first observation is explained by the fact that RANS models are calibrated with high-$Re$ datasets, and are focused on skin friction predictions, therefore the deviations decrease with $Re$. In the case of the displacement thickness, the two boundary layers start with similar initial conditions (note that $\delta^*$ is 0.9259 in the RANS), but the poor modeling of the boundary-layer development in the RANS leads to progressively larger deviations between the two simulations. As a consequence, the route of using the RANS to define the $U_\infty(x)$ distribution of the LES was not further pursued, and an alternative methodology where the free-stream velocity distribution is adapted dynamically during the simulation is currently being developed for future extensions of the present study.

# Coherent structures in a zero-pressure-gradient and a strongly decelerated boundary layer

**Mark P Simens**[1], **Ayse G Gungor**[2], **Yvan Maciel**[3]

[1] School of Aeronautics, Universidad Politécnica de Madrid, 28040 Madrid, Spain
[2] Faculty of Aeronautics and Astronautics, Istanbul Technical University, 34469 Maslak, Istanbul, Turkey
[3] Department of Mechanical Engineering, Laval University, Quebec City, QC, G1V 0A6 Canada

E-mail: ayse.gungor@itu.edu.tr

**Abstract.** Coherent structures in a strongly decelerated large-velocity-defect turbulent boundary layer (TBL) and a zero pressure gradient (ZPG) boundary layer are analysed by direct numerical simulation (DNS). The characteristics of the one-point velocity stastistics are also considered. The adverse pressure gradient (APG) TBL simulation is a new one carried out by the present authors. The APG TBL begins as a zero pressure gradient boundary layer, decelerates under a strong adverse pressure gradient, and separates near the end of the domain in the form of a very thin separation bubble. The one-point velocity statistics in the outer region of this large-defect boundary layer are compared to those of two other large-velocity-defect APG TBLs (one in dynamic equilibrium, the other in disequilibrium) and a mixing layer. In the upper half of the large-defect boundary layers, the velocity statistics are similar to those of the mixing layer. The dominant peaks of turbulence production and Reynolds stresses are located in the middle of the boundary layers. Three-dimensional spatial correlations of $(u, u)$ and $(u, v)$ show that coherence is lost in the streamwise and spanwise directions as the velocity defect increases. Near-wall streaks tend to disappear in the large-defect zone of the flow to be replaced by more disorganized $u$ motions. Near-wall sweeps and ejections are also less numerous. In the outer region, the $u$ structures tend to be shorter, less streaky, and more inclined with respect to the wall than in the ZPG TBL. The sweeps and ejections are generally bigger with respect to the boundary layer thickness in the large-defect boundary layer, even if the biggest structures are found in the ZPG TBL. Large sweeps and ejections that reach the wall region (wall-attached) are less streamwise elongated and they occupy less space than in the ZPG boundary layer. The distinction between wall-attached and wall-detached structures is not as pronounced in the large-defect TBL.

## 1. Introduction

The transposition of our knowledge acquired through the study of canonical turbulent wall flows to more complex flows such as adverse-pressure-gradient (APG) turbulent boundary layers (TBLs) is not straightforward. Consequently, a recognized and clear vision of the scaling, similarity laws and turbulence regeneration mechanisms in APG TBLs does not yet exist, which greatly impedes progress in the understanding and modeling of complex flows of this type.

For TBLs subjected to a prolonged or intense adverse pressure gradient, the layer structure and scaling parameters are known to differ from those of canonical wall flows because of their large momentum deficit and different mean strain field. Certain researchers consider that two layers still exist, but that the nature of these layers has changed, leading to different scales (see [1]). Other researchers have proposed a three-layer structure [2, 3]. Moreover in large-defect TBLs, turbulence activity and production is small near the wall and important in the outer region of the flow [4–6]. Marquillie et al. [7] have proposed that the outer peaks of Reynolds stresses and production could be related to a streak bursting process, whereas Elsberry et al. [6] suggest that it is due to the inflectional instability of the mean velocity profile, as in mixing layers. By analysing three different large-velocity-defect TBLs, Gungor et al. [8] have concluded that these boundary layers are globally less efficient in extracting turbulent energy from the mean flow than the zero-pressure gradient (ZPG) TBL. Furthermore, it is known that the outer-region turbulent statistics of TBLs close to detachment resemble those of mixing layers [5, 8–10]. These various observations suggest that the physical mechanisms and coherent structures responsible for the production and transport of turbulence might indeed be different in APG TBLs.

Information on the coherent structures found in APG TBLs is however rare. By analyzing the direct numerical simulation (DNS) data of a turbulent separation bubble of [5], Chong et al. [11] suggested that in the APG zone prior to detachment more of the eddies which contribute to the Reynolds shear stress are eddies which are not connected to the wall. In the case of an equilibrium APG TBL, Krogstad and Skåre [12] investigated the turbulent structures with two-point space-time correlations and with quadrant analysis. The latter is based on a quadrant decomposition in the $(u, v)$ plane, where $u$ and $v$ are respectively the streamwise and wall-normal fluctuating components of velocity. Intense second quadrant (Q2) motions $(u < 0, v > 0)$ are usually termed ejections, while intense fourth quadrant (Q4) motions $(u > 0, v < 0)$ are sweeps. These types of motions contribute significantly to the Reynolds shear stress. Krogstad and Skåre found that the lower part of the equilibrium APG boundary layer is strongly dominated by Q4 motions, while in a ZPG TBL second and fourth quadrant events are equally important. The streamwise correlation length of $u$ was also found to be considerably shorter in the APG case throughout the boundary layer, a result also obtained later in different large-velocity-defect TBLs by [13] and [8]. Rahgozar and Maciel [14] observed that the predominance of streaky $u$-structures in the outer region of a large-velocity-defect TBL is less than in the ZPG case. This predominance even disappears near detachment. By analysing the same flow, Rahgozar and Maciel [13] found that large-scale $u$-structures are less elongated than those of ZPG TBLs, especially in the lower part of the boundary layer.

In the present work, we investigate how a strong adverse pressure gradient affects the $u$ and $uv$ structures through comparisons with a ZPG TBL. To achieve this goal, we analyze the three-dimensional spatial correlations of $(u, u)$ and $(u, v)$ and the geometric and kinematic characteristics of three-dimensional sweeps and ejections. The present study provides new information on the three-dimensional properties of sweeps and ejections found in ZPG TBLs. The properties of the one-point velocity statistics are also investigated.

## 2. Numerical methodology

The two direct numerical simulations used to investigate the coherent structures have been performed with the same code. The ZPG TBL was simulated by Sillero et al. [15] while the APG TBL simulation is a new one carried out by the present authors. The DNS numerical scheme is described in detail in Refs. [16] and [15]. Both flows are simulated in a parallelepiped domain over a smooth no-slip wall, with spanwise periodicity and streamwise non-periodic inflow and outflow. The Navier-Stokes equations are integrated using a fractional step method on a staggered grid, with third-order Runge-Kutta time-integration, fourth order compact spatial discretization for the convective and viscous terms, and second order discretization for the pressure in the directions perpendicular to the span, which is spectral.

The DNS database of the ZPG TBL covers the Reynolds number range $Re_\theta = 2780 - 6680$. The DNS computational setup for the present APG TBL simulation is sketched in figure 1. It

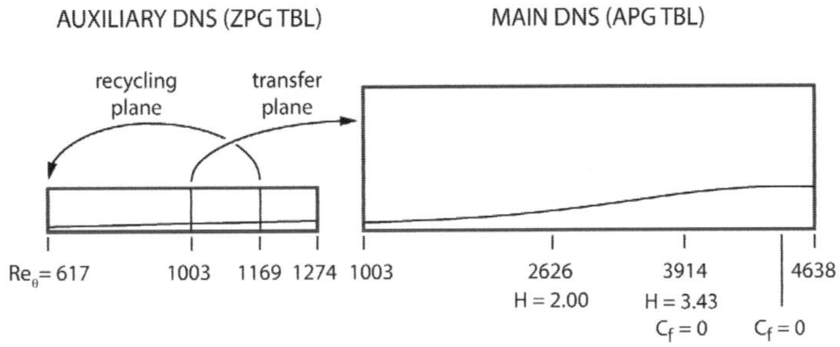


**Figure 1.** Schematic in the $xy$ plane of the numerical simulation setup showing the boundary layer thickness. The computational box dimensions are to scale

consists of two simulations running concurrently [15]. The first auxiliary simulation is that of a ZPG TBL and it is intended to provide a realistic turbulent inflow for the APG layer. The inflow of the auxiliary DNS is obtained by rescaling the velocity fluctuations of a downstream plane, while fixing the inflow mean velocity to a prescribed profile [16]. The recycling plane is located at $x \approx 398\theta_0 \approx 45\delta_0$, where $\theta_0$ and $\delta_0$ are respectively the momentum and boundary layer thicknesses at the beginning of the auxiliary ZPG DNS. A plane located at $x \approx 268\theta_0 \approx 30\delta_0$ of the first domain is transferred at each time step into the inlet of the second main domain. The velocities at the outflow of the two computational boxes are estimated by a convective boundary condition. The outflow streamwise velocities are corrected to compensate the minimal mass flux variations due to the time-dependent inflows [16]. Table 1 summarizes the simulation parameters for both layers. For the APG DNS, the box dimensions with respect to the boundary layer thickness at the outlet are $(L_x, L_y, L_z)/\delta_{exit} = (11.0, 3.4, 2.6)$. The resolutions in terms of the Kolmogorov length $\eta$ are $(\Delta x_g, \Delta y_g, \Delta z_g) < 4\eta$ except near the inlet very close to the wall where $\Delta x_g < 8\eta$. After an initial transient phase, statistics and flow fields were sampled over 53,600 time steps corresponding to a total time of $5250\theta_0/U_0$ or 10 flow throughs. A total of 134 instantaneous fields were kept, with a time interval between fields of $39.2\theta_0/U_0$.

The desired pressure gradient is controlled by imposing a streamwise dependent wall-normal velocity distribution at the upper boundary of the computational domain. The streamwise and spanwise velocities along the top boundary satisfy free-slip conditions. The imposed wall-normal velocity at the top boundary $V_{top}(x)/U_{e0}$ and the resulting streamwise velocity at the same boundary $U_{top}(x)/U_{e0}$ are illustrated in Figs. 2(a) and (b) respectively. In the design of the DNS setup, $U_{top}$ was chosen to give an almost linear increase of the shape factor $H$ in a central portion of the domain and to reaccelerate the flow at the end of the domain. It was obtained with boundary layer computations performed in inverse mode ($H$ is the input, $U_e$ is the output).

**Table 1.** Parameters of the APG TBL simulation. $L_x, L_y$, and $L_z$ are the box dimensions along the three axes. $N_x, N_y$, and $N_z$ are the collocation grid sizes. The momentum thickness $\theta$ is measured at the middle of each box.

| Case | $Re_\theta$ | $L_x, L_y, L_z$ | $(L_x, L_y, L_z)/\theta$ | $N_x, N_y, N_z$ |
|---|---|---|---|---|
| Auxiliary DNS (ZPG TBL) | 617-1274 | $88 \times 14 \times 34$ | $320 \times 49 \times 126$ | $1201 \times 191 \times 768$ |
| Main DNS (APG TBL) | 1003-4638 | $144 \times 45 \times 34$ | $118 \times 37 \times 28$ | $1921 \times 380 \times 768$ |

## 3. Results and discussion

### *3.1. Flow description*

The streamwise evolutions of $U_e$ and of the wall-friction velocity $u_\tau$ are depicted in Fig. 2(b), where $U_e$ is the maximum streamwise component of velocity in the wall-normal direction. The flow at the edge of the boundary layer decelerates over most of the domain but then reaccelerates at the end. As can be deduced from the evolution of $u_\tau$, the wall shear stress decreases and eventually becomes negative. The flow separates near the exit of the domain in the form of a very thin separation bubble. The height of the zone of negative mean streamwise velocity never exceeds $0.02\delta$. The reacceleration of the flow at the end reattaches the boundary layer and thereby avoids encountering problems with the outflow boundary condition.

Figure 2(c) shows the streamwise evolution of three different outer pressure gradient parameters $\beta_m$, $\beta_{zs}$ and $\beta_\tau$, as well as the inner pressure gradient parameter $\beta_i = \nu/(\rho u_\tau^3)(dp/dx)$, often denoted $p^+$ in the literature. The traditional outer pressure gradient parameter, Rotta-Clauser's $\beta_\tau = -(\Delta/u_\tau)(dU_e/dx)$, assumes the outer region velocity scale to be $u_\tau$, which is not the case for large-velocity defect TBLs such as the one considered here. $\beta_{zs}$ and $\beta_m$ are more appropriate outer pressure gradient parameters since they are based on velocity scales that are valid for both small and large defect TBLs. $\beta_m = -(\delta/U_m)(dU_e/dx)$ is expressed with the mixing-layer-type outer-velocity scale $U_m = 2(U_e - U(y = 0.5\delta))$ introduced by Gungor et al. [8]. $\beta_{zs} = -(\delta/U_{zs})(dU_e/dx)$ is based on the Zagarola-Smits velocity scale $U_{zs} = U_e\delta^*/\delta$. Both $U_m$ and $U_{zs}$ are proportional to the mean streamwise momentum deficit in the boundary layer. $\beta_{zs}$ and $\beta_m$ are not equivalent but their streamwise evolutions are qualitatively similar, as can be seen in Fig. 2(c).

Figure 2(c) shows that each pressure gradient parameter increases significantly in a different upstream portion of the flow. The positive gradient of these pressure gradient parameters is responsible for the increase in the streamwise mean momentum defect as shown in Fig. 2(d). $\beta_i$ and $\beta_\tau$ tend to infinity at separation since $u_\tau = 0$. The outer pressure gradient parameters $\beta_m$ and $\beta_{zs}$ start decreasing in the first half of the domain. The impact of the pressure force on the outer region is therefore diminishing but this change is not strong enough to reverse the situation in terms of mean momentum defect, which keeps increasing until flow separation. The increase in momentum defect can be appreciated with the increase in the shape factor $H$ shown in Fig. 2(d). At separation, $H = 3.43$.

### *3.2. One-point velocity statistics*

In this section, comparisons of the one-point velocity statistics of the ZPG and large-defect TBL with other flows of interest are made. The data for the present APG flow is taken at four streamwise positions where $H = 2, 2.5, 3$ and 3.43. The fourth one is the last mesh position before separation ($C_f \approx 10^{-6}$). Besides the ZPG TBLs of [15] and [16], the other turbulent boundary layers considered are the equilibrium APG TBL of Skåre and Krogstad [4] (SK) that has a relatively large velocity defect ($H \cong 2$) and is at high Reynolds numbers ($Re_\theta = 39000$–51000) and the non-equilibrium APG TBL of Maciel et al. [17] (MRL) with increasing velocity defect that includes data at detachment ($H = 1.72$–3.85). Finally, the single-stream mixing layer of Wygnanski and Fiedler [18] (WF) is also considered because the the velocity statistics of the outer region of large-defect boundary layers closely resemble those of the high speed side of a mixing layer [5, 8–10]. The key characteristics and parameter range of the boundary layer databases are given in table 2.

Streamwise mean velocity profiles of the various flows are shown in fig. 3(a). The mean velocity defect progressively increases in the present flow. The non-equilibrium APG TBL of MRL behaves in a similar manner but only the profile at separation is shown in figure 3(a) for clarity. Excluding the wall region, the velocity profiles of the large-defect boundary layers clearly resemble the single-stream mixing layer one, with large velocity gradients, similar curvatures

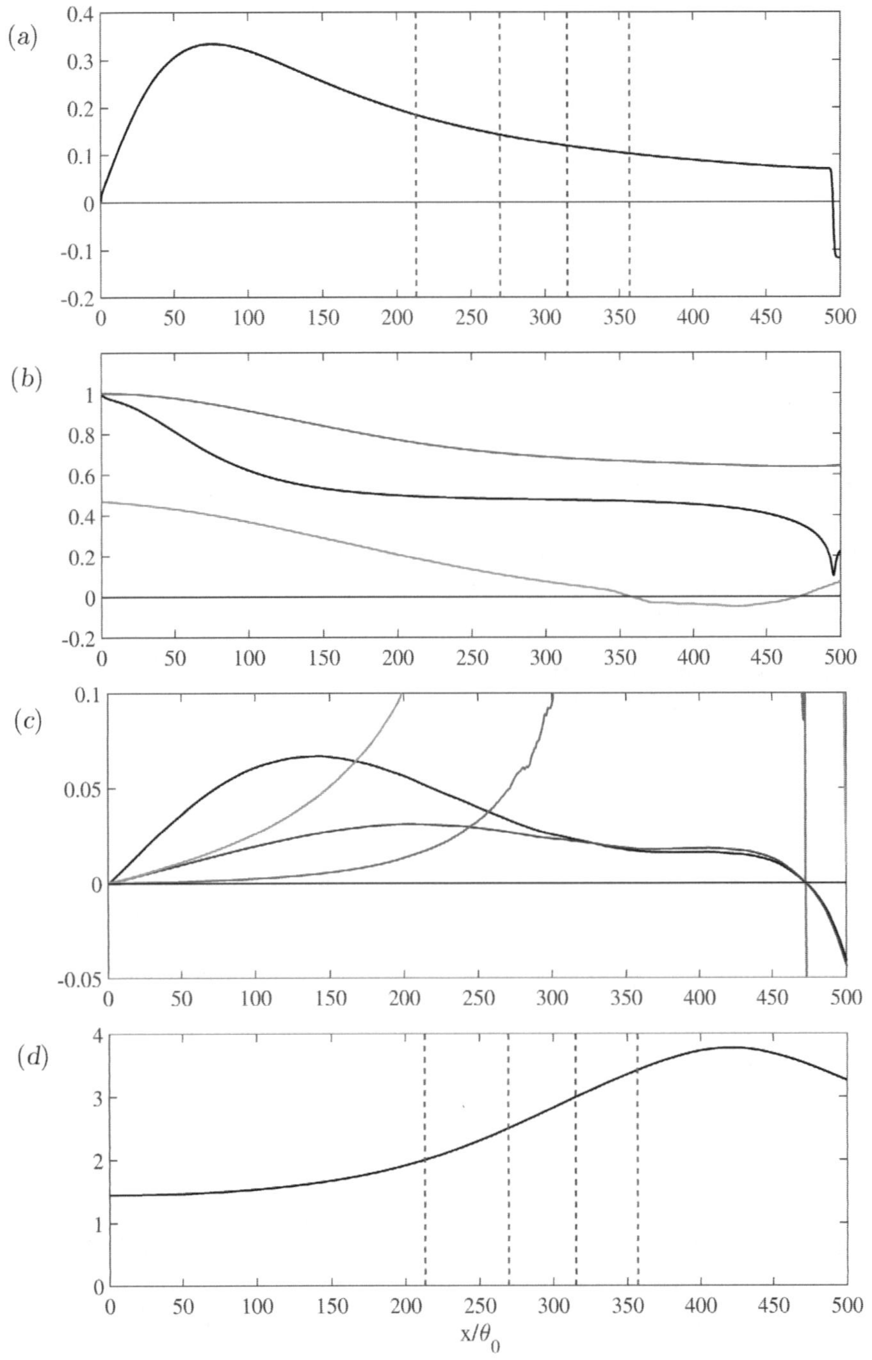


**Figure 2.** Mean flow parameters in the APG DNS. (a) Imposed wall-normal velocity along the top boundary. (b) Streamwise velocity at the top boundary, black; $U_e/U_{e0}$, red; $10u_\tau/U_{e0}$, green. (c) Pressure gradient parameters: $\beta_m$, black; $\beta_{zs}$, blue; $\beta_\tau \times 10^{-3}$, red; $\beta_i$, green. (d) Shape factor $H$. Vertical dashed lines in (a) and (d) denote the four streamwise positions where $H = 2$ (green), 2.5 (magenta), 3 (blue), and 3.43 (red)

**Table 2.** Boundary layer databases and the corresponding ranges of the parameters

| Flow | $Re_\theta$ | $H$ | $\beta_m$ | $\beta_\tau$ |
|---|---|---|---|---|
| equilibrium APG TBL (SK) | 39000–51000 | 1.99–2.01 | 0.092–0.094 | 19.6–21.4 |
| non-equilibrium APG TBL (MRL) | 3350–12691 | 1.72–3.85 | 0.099–0.043 | 3.2–$\infty$ |
| non-equilibrium APG TBL (present) | 1003–4655 | 1.45–3.77 | 0–0.067 | 0–$\infty$ |
| ZPG TBL [16] | 617–2139 | 1.41–1.54 | 0 | 0 |
| ZPG TBL [15] | 2780–6680 | 1.36–1.38 | 0 | 0 |

and the presence of an inflection point near the middle of the layer. Two of the boundary layer velocity profiles are at the same shape factor of 2, namely one for the present flow and the equilibrium one of SK, but the shapes of these profiles are very different. Since the present flow is a disequilibrium boundary layer, such a shape difference is expected. It is probably more related to the different streamwise evolution of the flows than to the dissimilarity in Reynolds number. Figure 3(a) also shows two profiles at separation, one for the present flow and one for the MRL flow. In this case, the different behaviour near the wall may be due to differences in both streamwise evolution of the TBLs and Reynolds number.

Figure 3(b) shows the mean velocity profiles normalized with the friction-viscous scales ($u_\tau$ and $\nu/u_\tau$) of the present flow, the equilibrium TBL of SK and the ZPG TBL. Profiles at separation cannot be normalized with the friction-viscous scales since in this case $u^+ \to \infty$ and $y^+ \to 0$. Even if at low Reynolds number, the ZPG TBL profile follows fairly closely the law of the wall throughout the inner region. As already shown by SK, the equilibrium APG TBL follows the logarithmic law in the overlap layer even if $C_f$ is relatively low. This is coherent with the fact that the inner pressure gradient parameter $\beta_i$ is low in this flow, $\beta_i = 0.013$, which is a

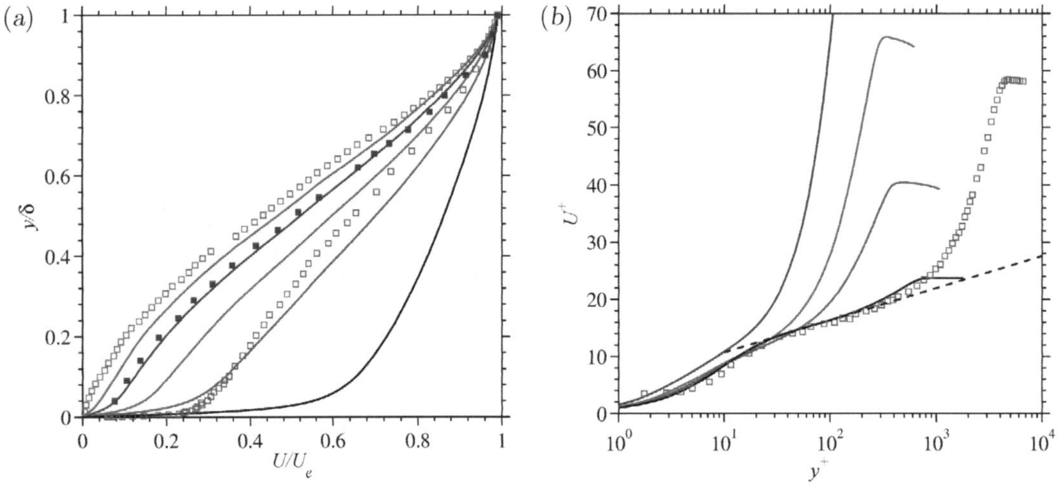


**Figure 3.** Mean velocity profiles normalized with (a) $U_e$ and $\delta$, and (b) friction-viscous scales. Present non-equilibrium APG TBL: $H = 2.0$, green line; $H = 2.5$, magenta line; $H = 3.0$, blue line; $H = 3.43, C_f = 0$, red line. ZPG TBL of [16] at $Re_\theta = 2000$, black line; equilibrium APG TBL of SK, green open square; profile at separation of the non-equilibrium APG TBL of MRL, red open square; and mixing layer of WF, blue full square. Dashed line, log law with $\kappa = 0.41$ and $B = 5.1$.

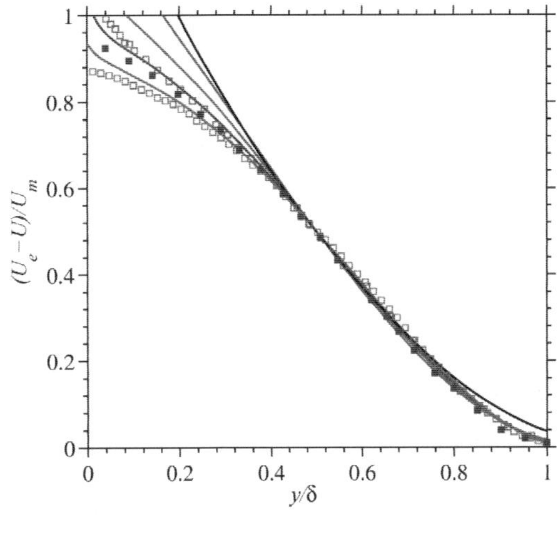


**Figure 4.** Velocity defect profiles normalized with $U_m$. Lines and symbols as in figure 3.

consequence of the high Reynolds number. The equilibrium state reached by the inner region in the SK flow, with $\beta_i$ nearly constant, is therefore very similar to that of canonical wall-bounded flows. In contrast, the non-equilibrium TBL deviates from the law of the wall throughout the inner region and the departure increases as $\beta_i$ increases. This confirms that the pressure force can play an important role even in the viscous sublayer in strong APG flows and that it has to be taken into account in the law of the wall [1, 19, 20]. However, as it was just mentioned, the Reynolds number also plays a role here. If the Reynolds number of the flow was higher, $\beta_i$ would be lower and the departure from the law of the wall would be probably less important [21].

The velocity defect profiles of the various flows normalized with $U_m$ are presented in fig. 4. The defect profiles are not identical, as expected, but they are all regrouped which indicates that the choice of $U_m$ and $\delta$ as common outer scales is indeed adequate to compare velocity statistics between these flows. With the exception of the ZPG TBL, the velocity defect profiles of all other flows are actually very similar on the high-speed side, $y > 0.5\delta$. In agreement with previous studies [22, 23], $U_{zs}$ and $\delta$ are also found to be appropriate outer scales for the mean velocity defect of all boundary layers (not shown), but $U_{zs}$ cannot be used for the mixing layer.

The streamwise Reynolds normal stress and the Reynolds shear stress of the various flows are shown respectively in figs. 5(a) and 5(b). The Reynolds stresses are normalized with the common outer scale $U_m$ in order to be able to compare the mixing layer and the boundary layers. For all Reynolds stresses, even $\langle v^2 \rangle$ and $\langle w^2 \rangle$ not shown, the general trend observed is one of a decrease of the stresses with increasing mean velocity defect throughout the boundary layer but especially near the wall. Since $U_m$ is proportional to the mean strain rates present in the outer region, the fact that all the Reynolds stresses normalized with $U_m$ decrease with velocity defect throughout the outer region implies that large-defect TBLs are less efficient in transferring energy from the mean flow to turbulence.

The behaviour of the Reynolds stresses of the TBLs subjected to a strong APG shown in figs. 5(a) and 5(b) is typical of that seen in other strongly decelerated flows. The maximum of all Reynolds stresses is in the outer region and it shifts further away from the wall with increasing velocity defect. Furthermore, the near-wall peak of $\langle u^2 \rangle$ is not present or has almost vanished.

The comparison of the three large-defect flows indicates that when the shape factor is identical or similar, the level of the various Reynolds stresses normalized with $U_m$ is also similar. For instance, the Reynolds stress profiles at $H = 2$ from the present flow and from SK (green) differ in shape, with maxima at different locations, but are at comparable levels for all Reynolds stresses. In the case of the profiles at separation, the Reynolds stress profiles of the present

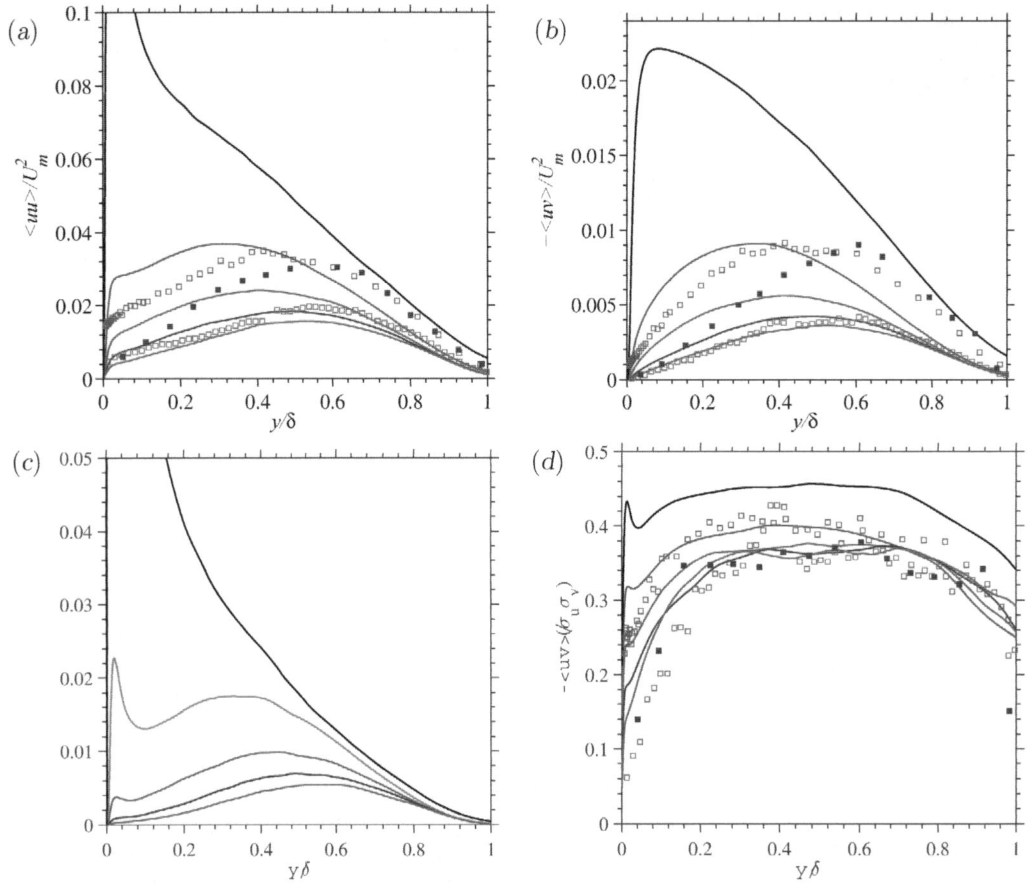


**Figure 5.** (a) Streamwise Reynolds normal stress and (b) Reynolds shear stress normalized with $U_m$. (c) Production of turbulent kinetic energy normalized with $U_m$ and $\delta$. (d) $uv$-correlation factor. Lines and symbols as in fig. 3.

flow and of the MRL flow are actually quite similar. These results suggest that the effect of the upstream history is not as important as the effect of the local mean shear, at least for these three very different large defect TBLs. Interestingly, the Reynolds stress profiles of the mixing layer are similar in shape to the profiles at separation of the boundary layers. The Reynolds stresses are however stronger in the mixing layer. Since the mean shear distribution is similar between these two types of flows (see fig. 4) the coherent structures are necessarily affected by one of the following factors or, most probably, by both of them: the presence or not of a wall and the upstream history of the flow.

The difference between ZPG and large-defect APG TBLs is even more pronounced for the production of turbulent kinetic energy as shown in fig. 5(c). As the velocity defect increases, the near-wall production peak decreases very rapidly and vanishes near separation. A production maximum appears in the outer region and its shift away from the wall follows that of the maximum of the Reynolds stresses. Figure 5(c) confirms that the present large-defect TBL is globally less efficient in extracting turbulent energy from the mean flow than the ZPG one. This has also been shown by Gungor et al. [8] for two other types of large-defect TBLs. They had also shown that the turbulent kinetic energy budget terms of the mixing layer are qualitatively similar to those of the large-velocity-defect TBLs.

Figure 5(d) shows the $uv$ correlation factor for the various flows, where $\sigma_u$ and $\sigma_v$ are respectively the local standard deviations of $u$ and $v$. The level of correlation between $u$ and $v$ is higher in the ZPG TBL than in all other flows. In the case of the APG TBLs, the correlation of $u$ and $v$ is seen to deteriorate as the mean velocity defect increases, especially near the wall. Lower correlation levels for large-defect TBLs had already been noted by [24] and [8]. The correlation profiles of the mixing layer and of the APG TBL of MRL at separation are strikingly similar. They suggest that turbulence significantly looses its coherency in the low-speed side of these flows.

### 3.3. Two-point correlations

The spatial organization of $u$ and $uv$ is now investigated using three-dimensional spatial correlations. In physical space, the two-point cross-correlation coefficient for two generic variables $a$ and $b$ is defined as

$$C_{ab}(\vec{r}, \vec{r'}) = \frac{\langle a(\vec{r}) b(\vec{r'}) \rangle}{\sigma_a(\vec{r}) \sigma_b(\vec{r'})}, \quad (1)$$

where $\vec{r}$ is the reference position and $\vec{r'}$ the moving one. The averaging is over time and spanwise direction. The correlation functions are actually computed in Fourier space in the homogeneous spanwise direction and then Fourier transformed.

Figure 6 presents three-dimensional views of $C_{uu}$ with the reference point at $y = 0.6\delta$ for three streamwise positions of the APG TBL corresponding to $H = 2$, 2.5 and 3.45 (separation). The three-dimensional representation of Sillero et al. [25] for their ZPG TBL at $Re_\theta = 4850$ is also included in the figure for comparison. The four plots are for a domain of size $6\delta \times 1.2\delta \times 2\delta$ in $x$, $y$ and $z$ respectively, centered at the reference point. The isosurfaces $C_{uu} = 0.09$ (turquoise) and $C_{uu} = -0.09$ (white) clearly show that the regions of positive and negative coherence of $u$ are shorter and more inclined with respect to the wall for the large-defect TBL in comparison to a ZPG TBL. It suggests that large-scale $u$ structures are shorter and more inclined in this large-defect TBL. The streamwise extent of the isosurface $C_{uu} = 0.09$ is about $4\delta$ for the ZPG TBL and $2\delta$ for the APG TBL. Moreover, both the positive and negative regions become smaller as the velocity defect increases. The negative-positive-negative spanwise trend has almost disappeared at separation. Streamwise-spanwise sections of $C_{uu}$ at various heights in the outer region confirm that the negative $C_{uu}$ contours are much smaller than in the ZPG TBL and they shrink in size as the velocity defect increases. In the case of the APG TBL of Rahgozar and Maciel [13], the contour $C_{uu} = -0.02$ has disappeared at separation at $y = 0.2\delta$. These results suggest that the streaky pattern of large-scale $u$ structures occurs less frequently as the defect increases.

Streamwise-wall-normal maps of $C_{vu}$ are shown in fig. 7 for a reference point at $y = 0.4\delta$. At that wall-normal height, the Reynolds shear stress is high in the APG TBL as shown in fig. 3(c). For the APG TBL, the low-level contours are noisy due to the insufficient number of statistically independent flow realizations. As expected, the negative value of the cross-correlation indicates that Q2 and Q4 motions are dominant. It is interesting to note that for all the cases shown in fig. 7 there is a different behaviour between the high negative level contours close to the reference point and the low negative level contours further away. The contours near the reference point are compact and inclined in the upstream direction. They reflect the character of smaller-scale intense Q2 and Q4 motions and the inclination angle is similar to what Adrian et al. [26] have observed for strong Q2 and Q4 motions in a ZPG TBL. This inclination angle is maintained as the velocity defect increases. The weak larger-scale contours are streamwise elongated in the case of the ZPG TBL and inclined in the downstream direction. They resemble the $C_{uu}$ contours in the same plane (not shown). Sillero et al. [25] suggested that they probably reflect the combination of long $u$ streaks with a smaller sweep or ejection motion. In the case of

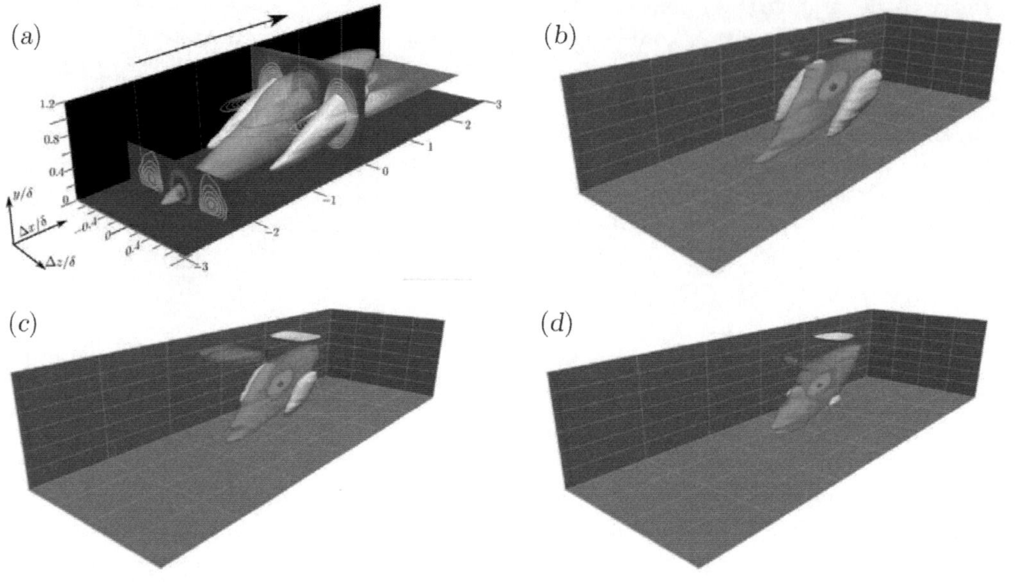

**Figure 6.** Three-dimensional views of $C_{uu}$ with reference point at $y = 0.6\delta$. (a) ZPG TBL of Ref. [15] at $Re_\theta = 4850$ (fig. 1 of [25] with permission from AIP). APG TBL at positions corresponding to (b) $H = 2$ (c) $H = 2.5$ and (d) $H = 3.45$. Isosurfaces at $C_{uu} = -0.09$ (white), $+0.09$ (turquoise), $+0.4$ (yellow) and $+0.8$ (blue). Same axes ranges in all four plots

the APG TBL, the streamwise elongation of the weak large-scale contours is reduced and the inclination with respect to the wall is increased. The large-scale $C_{uu}$ contours exhibit similar trends (not shown).

### *3.4. Q structures*

Although they provide valuable statistical information on the structure of turbulence, two-point correlations mix the signatures of various types of coherent structures of a wide spectrum of sizes. For this reason, we now extract and directly analyze the intense three-dimensional $uv$ structures (Qs) in both the ZPG TBL of Sillero et al. [15] and in the present APG TBL. The procedure adopted to identify the Qs follows the method used by Lozano-Durán et al. [27] (hereafter denoted LFJ). The Qs are defined as regions of connected points that satisfy simultaneously two conditions. The first condition is $|u(\vec{r})v(\vec{r})| > H^*\sigma_u\sigma_v$, where $H^*$ is the threshold constant discussed below, also called hyperbolic-hole size. LFJ give a detailed justification of the choice of $H^*\sigma_u\sigma_v$ as the threshold function. The second condition is that all points within a Q structure are in the same quadrant of the $u$, $v$ space. Point connectivity is defined with the six orthogonal neighbours. Following the notation of LFJ, the Q2 and Q4 structures will be referred to as $Q^-$s, and the Q1s and Q3s as $Q^+$s.

Table 3 presents the parameters of the subdomains used for the extraction of the Q structures. For the APG TBL, the extraction box covers the zone of large velocity defect of the flow prior to separation. The streamwise extent of the box is $5\delta_a$ where $\delta_a$ is the average boundary layer thickness inside the box. The streamwise position of the box for the ZPG TBL gives ranges of the three Reynolds numbers comparable to those of the APG TBL. $Re_m = U_m\delta/\nu$ and $Re_{zs} = U_{zs}\delta/\nu$ are Reynolds numbers of the outer region. Both extraction boxes cover the full width of the simulation domain. The same wall-normal height of $2\delta_a$ was chosen for both

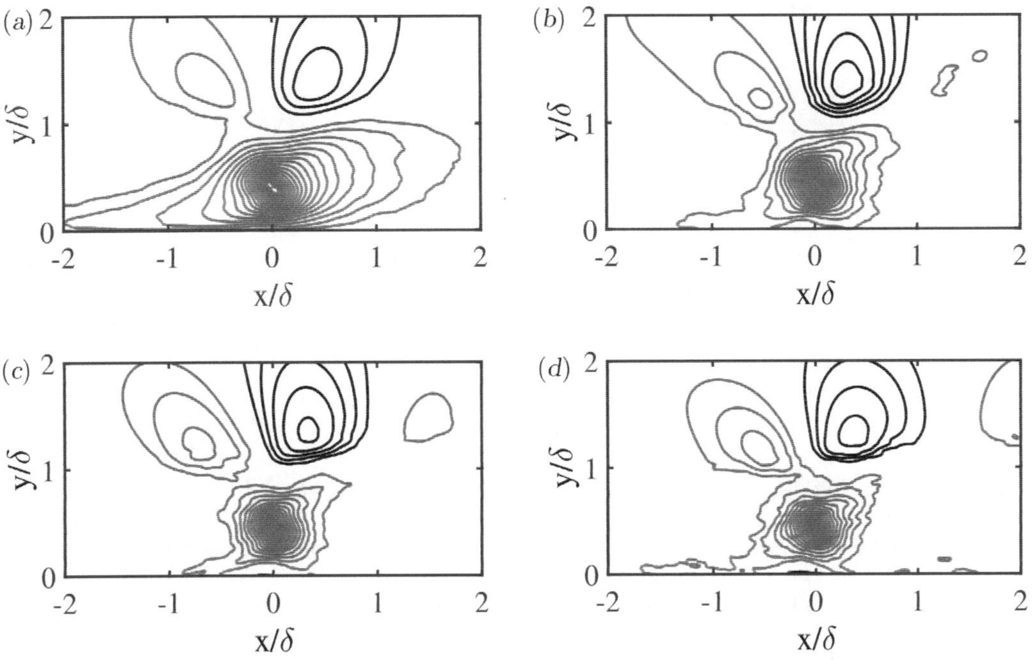


**Figure 7.** Streamwise-wall-normal maps of $C_{vu}$ with reference point at $y = 0.4\delta$. (a) ZPG TBL of Ref. [15] at $Re_\theta = 4850$. APG TBL at positions corresponding to (b) $H = 2$ (c) $H = 2.5$ and (d) $H = 3.45$. Positive contours (black) start at 0.03 by increments of 0.02. Negative contours (red) start at -0.03 by increments of -0.02

boxes in order to be able to compare the volume occupied by the Qs in the two flows. Since the extraction of the Qs is computationally intensive, the number of instantaneous fields processed is limited to 27 for the ZPG TBL (box of $2.2 \times 10^9$ grid points) and 45 for the APG TBL (box of $1.7 \times 10^8$ grid points).

As in LFJ a percolation analysis has been performed to determine a value for $H^*$ that gives an equilibrium between detecting only a few very big objects and detecting only a few small and very intense Qs. The percolation crisis (rapid change of the normalized maximum volume of the Qs while varying $H^*$) takes place in the range $1.2 \leq H^* \leq 2.5$. A hyperbolic hole size of $H^* = 1.75$ is chosen because it is in the middle of this range and it maximizes the number of $Q^-$s. It is the same value as used by LFJ in turbulent channel flows.

The linear dimensions of the Qs are defined with a rectangular box circumscribing them, the sides of this box being denoted as $\Delta x$, $\Delta y$ and $\Delta z$ and the midheight position of the box $y_c$. Structures that are as long as the streamwise length of the extraction box, $\Delta x = B_x$, are

**Table 3.** Parameters of the Q extraction zones. $B_x$, $B_y$ and $B_z$ are the box dimensions along the three axes and $\delta_a$ is the average boundary layer thickness inside the box. $N_f$ is the number of flow fields used

| Flow | $Re_\theta$ | $Re_m$ | $Re_{zs}$ | $H$ | $(B_x, B_y, B_z)/\delta_a$ | $N_f$ |
|---|---|---|---|---|---|---|
| ZPG TBL | 4544-5801 | 10195-10811 | 6251-6599 | 1.38-1.37 | 4.51, 2.00, 10.57 | 27 |
| APG TBL | 2577-3916 | 9204-27889 | 5084-13413 | 1.97-3.42 | 4.99, 2.00, 4.09 | 45 |

disregarded because the length of these structures is undetermined. Similarly, very small Qs with a volume $V < (3\Delta x_g)^3$ are rejected because their sizes are not well resolved on the numerical grid.

With the present extraction procedure, a total of 184576 Qs are identified in the APG TBL and 441217 Qs in the ZPG TBL. The difference in number corresponds approximately to the difference in volume of the extraction boxes. As LFJ found in turbulent channel flows, $Q^+$s are less frequent than $Q^-$s in both TBLs, and they occupy a much smaller fraction of the space, less than 1% of the box volume against 4 to 5% for $Q^-$s. $Q^-$ structures represent 57% and 52% of all Qs in the ZPG and APG TBL respectively. Table 4 summarizes the number and volume proportions for the $Q^-$s. Although the number fraction of ejections and sweeps is higher in the ZPG TBL than in the large-defect TBL, the volume occupied by these structures is smaller (4.2% of the box volume against 5.3%). The reason lies in the different wall-normal distribution of these sweep and ejection motions in the two flows. In the ZPG TBL, 43% of all $Q^-$s are small near-wall structures (near-wall Qs are defined here as structures whose center is below $0.05\delta$) which is consistent with the fact that the mean shear is the strongest there [28]. In the large defect APG TBL, who has a very different mean shear distribution, only 10% of all ejections and sweeps are small near-wall structures.

Figure 8 shows the joint probability density function (pdf) of the minimum and maximum wall distances for the ejections and sweeps. The structures separate into two groups: structures reaching the wall region (wall-attached) and not reaching it (wall-detached). In both flows, the wall-attached $Q^-$s form the narrow vertical band of the joint pdf with $y_{min} < 0.05\delta$ , while wall-detached structures form the wide diagonal band. Note that the name *wall-attached* can be misleading as it seems to imply that the structure remains attached to the wall during its whole life, which is usually not the case. In turbulent channel flows, Lozano-Durán et al. [28] have shown that wall-attached ejections are generally born with their base near the wall, but remain attached only for approximately 2/3 of their lives. Attached sweeps are the mirror image of attached ejections. They usually start as detached structures but become attached at about 1/3 of their lives and remain so thereafter.

It is seen from fig. 8 that the height of $Q^-$s can exceed the boundary layer thickness. Wall-attached $Q^-$s almost as tall as $2\delta$ are found in both flows. In the present APG flow the probabilities are higher everywhere, except near the origin. This implies that, as was discussed above, there are definitely less small $Q^-$s close to the wall, in proportion, in the APG TBL than in the ZPG TBL. For the detached $Q^-$s (diagonal band), the contours are wider in the vertical direction for the APG TBL, which indicates that the detached structures are generally taller in that flow. They are in fact bigger in all directions as it will be confirmed below.

Wall-attached Q2s and Q4s represent 35% of the total number of $Q^-$s and 58% of their volume (see Table 4). In the ZPG TBL, these percentages are respectively 51% and 73%. The number and size proportions of attached ejections and sweeps are therefore considerably reduced in a

**Table 4.** Number and volume proportions of the $Q^-$s (Q2s and Q4s)

| Case | ZPG TBL | APG TBL |
|---|---|---|
| All $Q^-$s | 57% of all Qs<br>4.2% of box volume | 52% of all Qs<br>5.3% of box volume |
| Near-wall $Q^-$s<br>($y_c < 0.05\delta$) | 43% of all $Q^-$s<br>2% of total volume of $Q^-$s | 10% of all $Q^-$s<br>1% of total volume of $Q^-$s |
| Wall-attached $Q^-$s<br>($y_{min} < 0.05\delta$) | 51% of all $Q^-$s<br>73% of total volume of $Q^-$s | 35% of all $Q^-$s<br>58% of total volume of $Q^-$s |

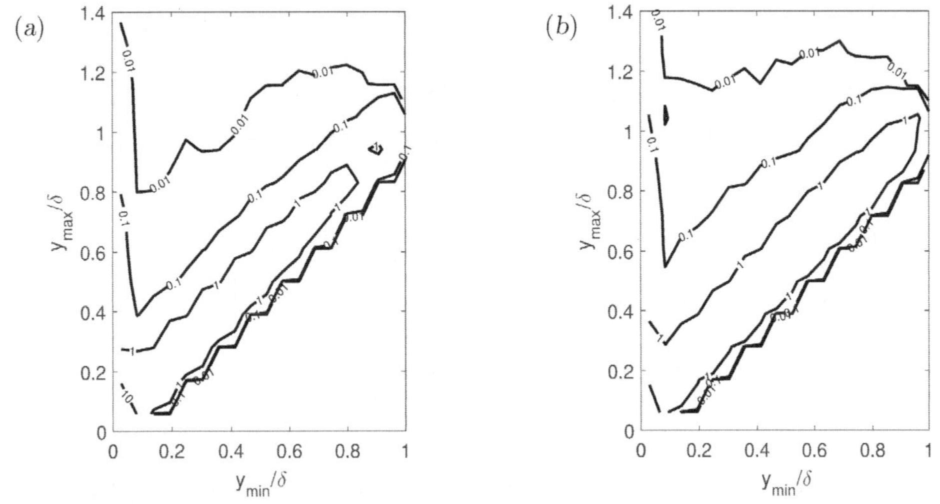


**Figure 8.** Joint pdf of $y_{min}/\delta$ and $y_{max}/\delta$ for Q$^-$s. (a) ZPG TBL, (b) APG TBL. Contour levels are 0.01, 0.1, 1 and 10

large-velocity-defect boundary layer. This is consistent with the fact that the turbulent activity is very small near the wall.

Figures 9 and 10 presents joint pdfs of the logarithms of the streamwise and wall-normal sizes, normalized with $\delta$, of the boxes circumscribing Q2s and Q4s for attached and detached structures respectively. As in LFJ for turbulent channel flows, the Q2 and the Q4 structures have similar sizes, with Q2s slightly bigger. The very small attached Q4s seen in the ZPG TBL (fig. 9) are streaky sweeps flattened against the wall [28]. They do not exist in the APG TBL.

The Q$^-$s in the APG TBL are generally bigger in all directions. But the largest structures are found in the ZPG TBL with attached Q2s that can reach the length of the extraction box $\Delta_x \approx 5\delta$ while being 3 times longer than they are tall and wide. For channel flows, LFJ found very long attached Q2s reaching $\Delta_x \approx 20h$ and $\Delta_y \approx \Delta_z \approx 2h$. The rapid streamwise variations and strong non-equilibrium state of the present APG TBL probably prohibit the existence of such long motions. In this flow, the longest attached Q2s rarely exceed a length of $3\delta$.

In the ZPG TBL, the detached Q$^-$s are globally smaller than their attached counterparts. LFJ observed that in turbulent channel flows, detached Q$^-$s and Q$^+$s are often background fluctuations of small size, of the order of a few Kolmogorov lengths, whose contributions to the Reynolds shear stress almost cancel. In the APG TBL, the situation is very different. The size distributions of the detached Q$^-$s are not significantly different from those of the attached Q$^-$s.

In terms of aspect ratio, fig. 9 shows that the attached structures tend to be streamwise elongated but much more so for the ZPG TBL with a linear law $\Delta_x \approx 3\Delta_y \approx 3\Delta_z$ identical to the law found by LFJ in channel flows. In the APG TBL, the law is $\Delta_x \approx 1.5\Delta_y \approx 1.5\Delta_z$. In both flows, the detached structures tend to be only slightly longer than they are tall and wide with $\Delta_x \approx 1.2\Delta_y \approx 1.2\Delta_z$.

## 4. Conclusion

The effects of strong adverse pressure gradients on the $u$ and $uv$ structures, as well as on the one-point velocity statistics of TBLs have been investigated using a newly simulated APG TBL. The DNS was designed to produce a flow with an important streamwise increase of the inner and outer pressure gradient parameters leading eventually to separation in the form a very thin

Journal of Physics: Conference Series **708** (2016) 012013 doi:10.1088/1742-6596/708/1/012013

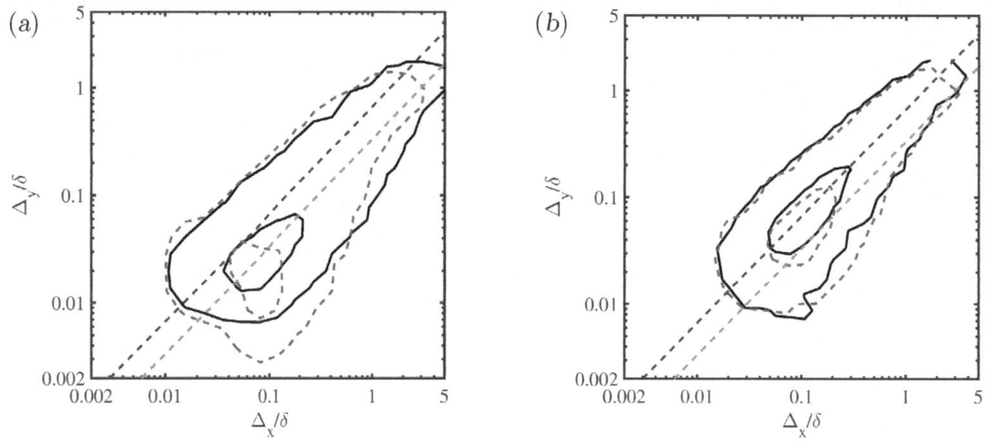


**Figure 9.** Joint pdfs $p(\Delta_x/\delta, \Delta_y/\delta)$ of the logarithms of the sizes of the boxes circumscribing wall-attached Q2s (black) and Q4s (red): (a) ZPG TBL, (b) APG TBL. The straight dashed lines are $\Delta_x = 1.5\Delta_y$ (blue) and $\Delta_x = 3.0\Delta_y$ (green)

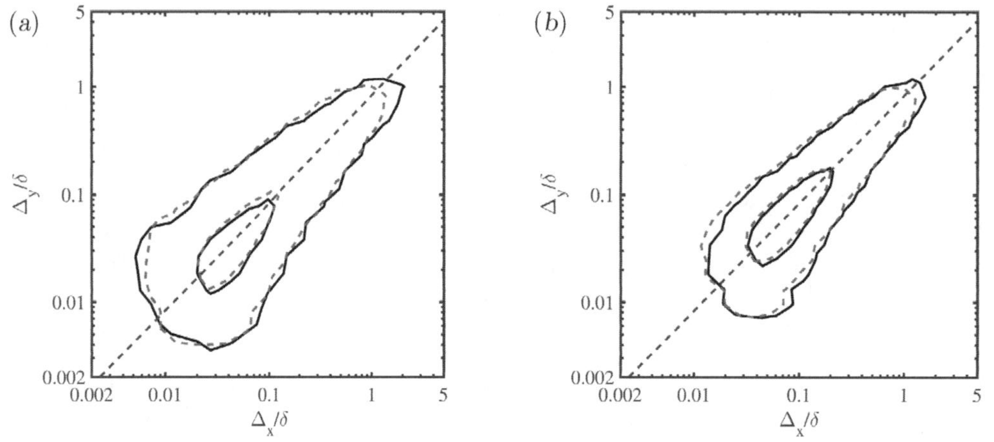


**Figure 10.** Joint pdfs $p(\Delta_x/\delta, \Delta_y/\delta)$ of the logarithms of the sizes of the boxes circumscribing wall-detached Q2s (black) and Q4s (red): (a) ZPG TBL, (b) APG TBL. The straight dashed blue lines are $\Delta_x = 1.2\Delta_y$

separation bubble. The sweeps and ejections found in the ZPG TBL of Sillero et al. [15] are also extracted and analyzed in order to establish a basis of comparison.

In terms of the one-point velocity statistics, the present study confirms behaviours of large-velocity-defect TBLs found in other studies. The present large-defect non-equilibrium APG TBL does not follow the classical law of the wall, even very close to the wall. In the case of both equilibrium and disequilibrium large-defect TBLs, the maximum turbulence activity is found to be in the middle of the boundary layer and not near the wall. The near-wall peak of streamwise Reynolds normal stress is absent or very small and that of turbulence production is rapidly decaying in the streamwise direction. These findings are consistent with the fact that near-wall streaks tend to disappear in the large-defect zone of the flow and are replaced by more disorganized $u$ motions. Near-wall sweeps and ejections are also much less numerous in the large-defect TBL in comparison to the ZPG TBL.

In the outer region of the large-defect TBL, the $u$ structures tend to be shorter, less streaky, and more inclined with respect to the wall than in the ZPG TBL. Moreover, the two-point correlations suggest that near separation, the occurrence of side-by-side low- and high-speed structures is considerably reduced. The sweeps and ejections are generally bigger with respect to the boundary layer thickness in the large-defect boundary layer, even if the biggest structures are found in the ZPG TBL. Contrary to the ZPG TBL, wall-attached and wall-detached Q structures have similar size distributions and occupy a similar space in the boundary layer. In both flows, the attached sweeps and ejections are more streamwise elongated than their detached counterparts but the difference in aspect ratio is much less for the APG TBL.

## Acknowledgments

The authors would like to thank Prof. Jiménez for organizing the First and Second Multiflow Summer Workshops. Funded in part by the Multiflow program of the European Research Council. AGG and YM were supported in part respectively by ITU-BAP and NSERC of Canada. YM thanks TUBITAK (2221 Program) for support during the collaboration stay in Turkey. The computations were made possible by generous grants of computer time from Barcelona supercomputing center and from the national center for high performance computing of Turkey. The authors would like to thank Juan Sillero and Prof. Jiménez for providing their ZPG TBL data and the programs for the two-point correlations, and Adrián Lozano-Durán and Prof. Jiménez for the processing programs for the Q structures analysis.

## References

[1] Skote M and Henningson D S 2002 *J. Fluid Mech.* **471** 107–136
[2] Melnik R E 1989 *Comput. Fluids* **17** 165–184
[3] Durbin P A and Belcher S E 1992 *J. Fluid Mech.* **238** 699–722
[4] Skåre P E and Krogstad P Å 1994 *J. Fluid Mech.* **272** 319–348
[5] Na Y and Moin P 1998 *J. Fluid Mech.* **374** 379–405
[6] Elsberry K Loeffler F, Zhou M and Wygnanski I 2000 *J. Fluid Mech.* **423** 227–261
[7] Marquillie M, Ehrenstein U and Laval J P 2011 *J. Fluid Mech.* **681** 205–240
[8] Gungor A G, Maciel Y, Simens M P and Soria J 2014 *J. Phys.: Conf. Ser.* **506** 012007
[9] Simpson R L, Strickland J and Barr P 1977 *J. Fluid Mech.* **79** 553–594
[10] Driver D M 1991 *AIAA 22nd Fluid Dyn., Plasma Dyn., Lasers Conf., Honolulu, HI* (AIAA)
[11] Chong M S, Soria J, Perry A E, Chacin J, Cantwell B J and Na Y 1998 *J. Fluid Mech.* **357** 225–247
[12] Krogstad P Å and Skåre P E 1995 *Phys. Fluids* **7** 2014–2024
[13] Rahgozar S and Maciel Y 2012 *J. Turbul.* **13** 1–24
[14] Rahgozar S and Maciel Y 2011 *Exp. Therm. Fluid Sci.* **35** 15751587
[15] Sillero J, Jiménez J and Moser R 2013 *Phys. Fluids* **25** 105102–1–16
[16] Simens M, Jiménez J, Hoyas S and Mizuno Y 2009 *J. Comput. Phys.* **228** 4218–4231
[17] Maciel Y, Rossignol K S and Lemay J 2006 *Exp. Fluids* **41** 573–590
[18] Wygnanski I and Fiedler H E 1970 *J. Fluid Mech.* **41** 327–361
[19] Stratford B S 1959 *J. Fluid Mech.* **5** 1–16
[20] Mellor G L and Gibson D M 1966 *J. Fluid Mech.* **24** 225–253
[21] Nickels T B 2004 *J. Fluid Mech.* **521** 217–239
[22] Maciel Y, Rossignol K S and Lemay J 2006 *AIAA J.* **44** 2450–2464
[23] Logdberg O, Angele K and Alfredsson P H 2008 *Phys. Fluids* **20** 075104
[24] Simpson R, Chew Y T and Shivaprasad B 1981 *J. Fluid Mech.* **113** 23–51
[25] Sillero J A, Jimnez J and Moser R D 2014 *Phys. Fluids* **26** 105109
[26] Adrian R J, Meinhart C D and Tomkins C D 2000 *J. Fluid Mech.* **422** 1–54
[27] Lozano-Durán A, Flores O and Jiménez J 2012 *J. Fluid Mech.* **694** 100–130
[28] Lozano-Durán A and Jiménez J 2014 *J. Fluid Mech.* **759** 432–471

# On determining characteristic length scales in pressure gradient turbulent boundary layers

**Ricardo Vinuesa, Ramis Örlü, Philipp Schlatter**
Linné FLOW Centre, KTH Mechanics, Stockholm, Sweden
E-mail: rvinuesa@mech.kth.se

**Abstract.** In the present work we analyze three methods used to determine the edge of pressure gradient turbulent boundary layers: two based on composite profiles, the one by Chauhan *et al.* (Fluid Dyn. Res. **41**:021401, 2009) and the one by Nickels (J. Fluid Mech. **521**:217–239, 2004), and the other one based on the condition of vanishing mean velocity gradient. Additionally, a new method is introduced based on the diagnostic plot concept by Alfredsson *et al.* (Phys. Fluids **23**:041702, 2011). The boundary layer developing over the suction side of a NACA4412 wing profile, extracted from a direct numerical simulation at $Re_c = 400,000$, is used as the test case. We find that all the methods produce robust results with mild or moderate pressure gradients, but stronger pressure gradients (with $\beta$ larger than around 7) lead to inconsistent results in all the techniques except the diagnostic plot. This method also has the advantage of providing an objective way of defining the point where the mean streamwise velocity is 99% of the edge velocity, and shows consistent results in a wide range of pressure gradient conditions, as well as flow histories. Therefore, the technique based on the diagnostic plot is a robust method to determine the boundary layer thickness (equivalent to $\delta_{99}$) and edge velocity in pressure gradient turbulent boundary layers.

## 1. Introduction & Motivation

Contrary to internal flows, where the outer length scale $\delta$ is exactly defined through the geometry, the counterpart for semi-confined flows such as boundary layers "can hardly be exactly defined" [1]. The problem is commonly circumvented by using integral quantities such as the displacement thickness $\delta^*$, *e.g.* through usage of the Rotta-Clauser outer scale[1] $\Delta = U_\infty^+ \delta^*$. In the present paper + denotes normalization with either the friction velocity $u_\tau = \sqrt{\tau_w/\rho}$ (where $\tau_w$ is the mean wall shear stress and $\rho$ is the fluid density) or the viscous length $\ell^* = \nu/u_\tau$ (with $\nu$ being the fluid kinematic viscosity). Other common work-arounds include employing analytical expressions for the wake, *i.e.*, the law of the wake, from which $\delta$ and other boundary layer parameters are extracted through fits of the data. For a detailed discussion on the employment of the law of the wake or composite velocity profile descriptions see the seminal work by Coles [3, 4] and more recent work by Nickels [5] and Nagib *et al.* [6]. The aforementioned investigations emphasize that the use of similarity laws (or ad hoc empirical relations) for the determination of boundary-layer parameters is crucial when dealing with experimental profiles, due to lack of measurements

[1] Note that Rotta himself attributes the introduction of this length scale to himself, *cf.* "the present writer proposed the introduction of the dimensionless wall distance $yu_\tau/(\delta^* U_\infty)$ instead of $y/\delta$" [2].

close to the wall and/or sparse data around the boundary-layer edge.[2] The problem is, however, also acute with numerical data even for the "simple" case of zero pressure gradient (ZPG) flat plate turbulent boundary layers (TBLs) as discussed in Schlatter and Örlü [8].

For less "simple" TBL configurations, such as those developing on curved surfaces and/or with (strong) pressure gradients, the problem becomes even more apparent, and commonly used wake descriptions are unable to accurately describe the data. One reason for the fallacy is attributed to the velocity gradient that is not necessarily zero for $y > \delta$, *i.e.*, $U(y > \delta) \neq$ constant as can be observed in Figure 1, which shows TBL profiles along the suction side of a wing [9, 10]. Due to the relatively low Reynolds numbers $Re$ and the strong pressure gradient conditions (where similarity laws usually fail) the edge of the boundary layer, its velocity $U_e$ (which is identical to the freestream velocity $U_\infty$ in the case of ZPG TBLs), the boundary layer thickness $\delta$, and the integral parameters (for which $\delta$ is needed to truncate the integrals), are all ill-defined. Another reason for the fallacy comes from the fact that composite profiles often assume that the classical or a modified logarithmic law applies, as well as a two-layer structure of the boundary layer exists. It is not clear whether these laws and layer structures are valid in the case of large-defect TBLs, in particular non-equilibrium ones, although as will be discussed in §4 available composite profiles are able to represent the data well except close to separation.

More sophisticated indicators can of course be employed such as the vorticity (enstrophy) or intermittency (see *e.g.* the discussion in Ref. [11]), but these are often not at hand (in case of most experiments) or are simply not documented. Other approaches, such as the irrotational velocity defect method, are problematic with simulations due to the fact there is not a region of irrotational flow, and with experiments where the irrotational region may be affected by blockage effects induced by the wind tunnel walls. The purpose of the present paper is to briefly review some of the commonly used methods for TBLs with pressure gradients and surface curvature and introduce a method that is based on the diagnostic plot concept by Alfredsson and co-workers [12, 13, 14] which — as will be shown here — is believed to provide a robust method for the determination of the edge velocity and boundary layer thickness, in particular, when it comes to low $Re$ and/or strong pressure gradients. Moreover, this method only requires knowledge of the mean and fluctuating streamwise velocity profiles, which also in the case of experiments are accurately measured close to the freestream.

The present article is organized as follows: in §2 a description of the analyzed datasets is provided; in §3 the techniques considered for the analysis are described in detail; in §4 we present the results and discussion; and in §5 we summarize the conclusions of the present work.

## 2. Employed data sets

The main test case considered in the present study was the flow around a NACA4412 wing section at a moderate Reynolds number of 400,000 (based on chord length and freestream velocity), which is shown in Figure 1. This database was obtained through direct numerical simulation (DNS), using the spectral element code Nek5000 [15], and a full description of the setup is given by Vinuesa *et al.* [9]. This flow is a good test case to assess the merits and deficiencies of the various methods, due to the fact that the boundary layers developing over the wing are subjected to the effect of wall curvature, as well as pressure gradients. More precisely, the boundary layer developing over the suction side of the wing experiences conditions from very weak adverse pressure gradients (APG), up to very strong APGs and incipient separation close to the trailing edge. The TBL on the pressure side experiences a mild favorable pressure gradient (FPG) throughout the whole wing. The maximum Reynolds number in this database was $Re_\theta \simeq 2,800$,

[2] This is mainly caused by the habit to measure in the wall-normal direction with a logarithmic spacing, see *e.g.* the relevant comment by George [7] with regard to some well-known data sets: "... experiments are lacking sufficient detail in the outer 50% of the boundary layer to determine the crucial outer length scale parameters ($\delta_{99}$ or even $\delta_{95}$) to within even a few percent."

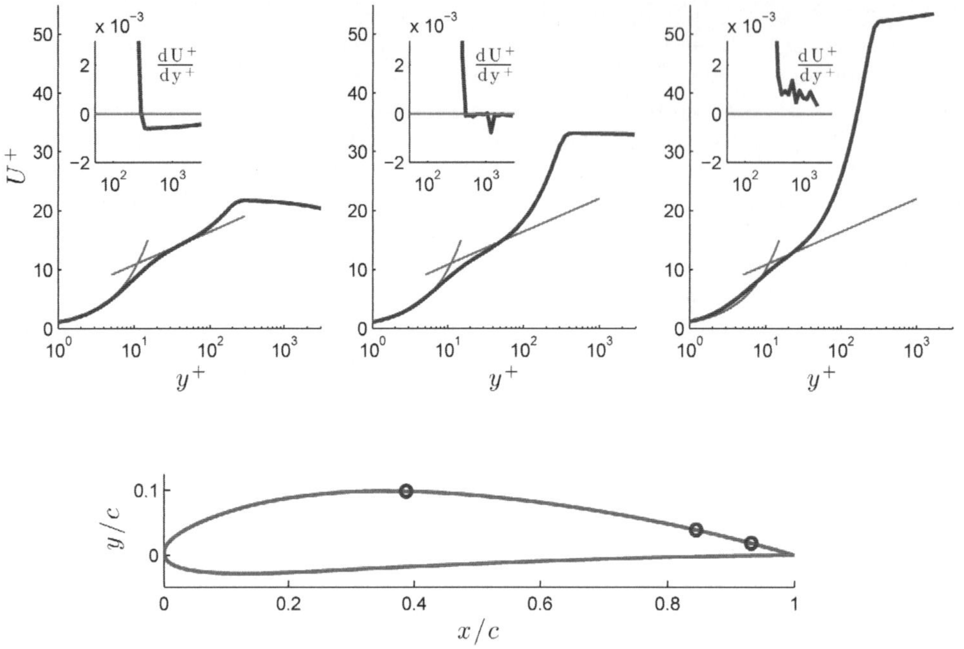


**Figure 1.** Turbulent boundary-layer profiles in inner-units at the three highlighted positions along the suction side of a wing. Data extracted from a DNS of the flow around a NACA4412 wing profile, at a Reynolds number based on freestream velocity and chord length of 400,000 [9, 10]. The linear and logarithmic profiles (with $\kappa = 0.41$ and $B = 5.2$) are shown to ease comparison and emphasize their fallacy for strong pressure gradient conditions (and low $Re$). The insets show the respective velocity gradients around the boundary-layer edge.

where $\theta$ is the momentum thickness. The analysis of the boundary layers developing over the wing was supplemented with the assessment of wind tunnel measurements at higher Reynolds numbers, namely the measurements at the National Diagnostic Facility (NDF) from the Illinois Institute of Technology (IIT) by Nagib *et al.* [16]. In this database, simple APG and FPG configurations were considered in the Reynolds number range $10,000 < Re_\theta < 40,000$.

## 3. Techniques

### *3.1. Common techniques*

Since the introduction of the law of the wake [3], various alterations and improvements for the original description of the velocity-deficit have been proposed. A "well-known and useful property of the defect law is that it avoids the awkward problem of defining the thickness $\delta$ for a boundary layer" [3]. Composite velocity profiles, *i.e.*, linear combinations of law of the wall and law of the wake descriptions, are widely used and are useful for similarity laws. Two recent ones, by Nickels [5] and by Chauhan *et al.* [17], will be employed here. Note that the first one was developed in particular for pressure gradient TBLs, and the second one has been successfully applied to several canonical wall-bounded turbulent flows. In particular, they will also serve as a validation of the diagnostic plot concept for the flow cases in which the composite profiles are known to work well.

The main advantage of composite profiles is the fact that they can help to alleviate certain problems in experimental data, namely when near-wall measurements are not available or are inaccurate, or when measurements are sparse across the boundary layer. These issues may lead to problems when evaluating boundary-layer parameters [18], and recent studies by Rodríguez-López *et al.* [19] and Vinuesa and Nagib [20] have shown the potential of composite profiles in developing correcting schemes for high-*Re* boundary layer measurements. However, the use of composite profiles at very low *Re* or under strong APGs requires pre-processing steps, which complicates their straightforward application in such cases. Nevertheless, and as will be discussed in §4, composite profiles provide consistent results when used to determine the boundary layer edge in boundary layers subjected to mild pressure gradients and at moderate Reynolds numbers.

Another technique investigated in the present study is the condition of a vanishing mean velocity gradient, $\mathrm{d}U/\mathrm{d}y \simeq 0$, which has been shown to be effective in defining the boundary layer edge in ZPG and mild pressure gradient boundary layers (even in consecutive sequences of APG and FPG [21]). It is important to note that this technique cannot be used in flows with strong streamline curvature due to the fact that the streamwise velocity necessarily varies in the wall-normal direction within the irrotational flow region. Moreover, in §4 we show that this method also exhibits problems with strong pressure gradients due to the fact that the streamwise velocity is not constant and equal to $U_\infty$ beyond the boundary layer edge as shown in Figure 1, and therefore $\mathrm{d}U/\mathrm{d}y$ does not reach 0. This leads to the introduction of arbitrary thresholds to define $\delta$, which is also not desirable when analyzing wide ranges of pressure gradient conditions.

### *3.2. The 'diagnostic plot' concept*

An alternative method to determine the boundary layer edge is in the following introduced and based on the diagnostic-plot concept [13], in which the streamwise velocity root mean square (rms) $u'$ value is plotted against the mean velocity $U$ normalised by the boundary-layer edge velocity $U_e$ rather than the wall distance. Although the diagnostic plot was originally proposed as a means for testing the adequacy of experimental data [12], it soon showed promising results to scale ZPG TBL data (but also pipe and channel flows) covering a wide *Re* range throughout the logarithmic and wake layers [14]. This is made apparent in Figure 2 (left), where ZPG TBL data from DNS [8] and experiments [22] covering a range of $670 \leq Re_\theta \leq 4{,}300$ and $2{,}500 \leq Re_\theta \leq 19{,}000$, respectively, are brought to scale *Re*-independent and linearly dependent on $U/U_e$. As apparent, the *Re*-independent region extends with increasing Reynolds number. The success of the diagnostic scaling could also be extended to turbulent boundary layers on rough walls when taking the roughness function into the scaling [23], and with pressure gradients[3] when accounting for the shape factor [26], as well as to free-shear flows such as mixing layers, jets and wakes [24].

A useful feature of the diagnostic plot is that it is constructed such that the turbulence intensity diminishes in the free stream, *i.e.* $u'/U \to 0$ for $y/\delta \to \infty$. For the purpose of the present paper, the velocity ratio of $U/U_e = 0.99$ is of particular interest, since it relates to the wall-normal location $\delta_{99}$ commonly employed as the boundary-layer thickness $\delta$. The straightforward idea is therefore to use the diagnostic plot concept, which is free of differentiation or fitting operations and employs merely the turbulence intensity, to determine the boundary-layer thickness, by simply finding the local turbulence intensity that corresponds to $U/U_e = 0.99$. This value is around 2% when scaled as in Figure 2 (right).

## 4. Results and Discussion

The scaling of the boundary layer developing over the suction side of the wing is analyzed in Figure 3, which reveals interesting similarities between low-*Re* data subjected to a strong APG and high-*Re* ZPG data. In this figure we also show that an appropriate scaling for both the mean velocity defect $U_e - U$ and the streamwise variance $u'^2$ is the one proposed

[3] Castro [24] had made similar observations when considering equilibrium APG data near separation [25].

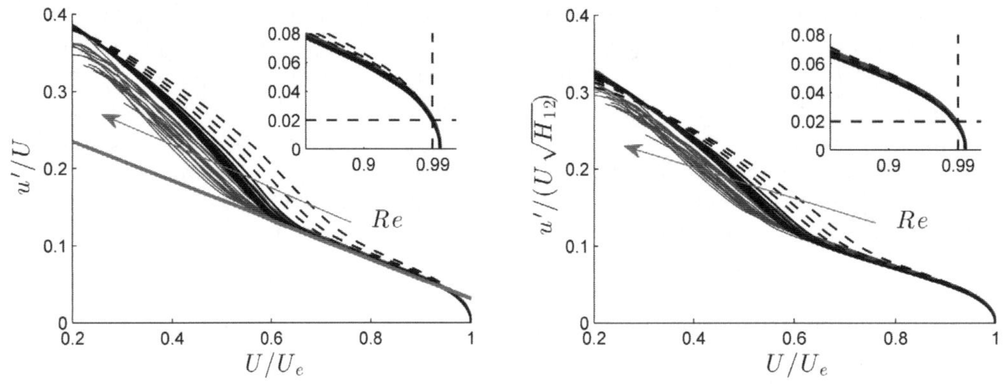


**Figure 2.** (Left) ZPG TBL data from DNS [8] and experiments [22] covering a range of 670 $\leq Re_\theta \leq$ 4,300 (black) and 2,500 $\leq Re_\theta \leq$ 18,700 (blue). Arrow points in direction of increasing $Re$. Red line depicts a straight line of form $u'/U = a - b(U/U_e)$, with $a = 0.286$ and $b = 0.255$ [14]. Lowest three $Re$ profiles from DNS (*i.e.* $Re_\theta < 1500$) are shown as dashed lines. Zoom into the wake region is shown in the inset. (Right) Same as (left), but with the shape factor $H_{12}$ taken into account in order to compare with the diagnostic plots in case of pressure gradient effects [26].

by Zagarola and Smits [27], denoted here as 'ZS', in which $U_{ZS} = U_e\delta^*/\delta$ is considered as the velocity scale, and the boundary-layer thickness $\delta$ is taken as the length scale. This scaling was also shown to be adequate for APG boundary layers by Maciel *et al.* [28]. The velocity profiles shown in Figure 3 are expressed in the direction tangential to the wing surface at each location, and the $y$ coordinate is measured in the direction normal to the wall. Profiles are shown between $x/c = 0.15$ and 0.98, which according to Vinuesa *et al.* [9] corresponds to a Reynolds number range $500 < Re_\theta < 2,800$. The pressure gradient can be quantified in terms of the so-called Rotta–Clauser pressure-gradient parameter $\beta = -\Delta/u_\tau \mathrm{d}U_e/\mathrm{d}x_t$ (where $\mathrm{d}U_e/\mathrm{d}x_t$ is the derivative of the tangential edge velocity with respect to the tangential direction $x_t$), which is in the range $0 < \beta < 85$ in this particular dataset. Figure 3 (left) shows that the outer region of the mean velocity defect profiles tend to be regrouped as higher Reynolds numbers and larger APGs are reached, and in Figure 3 (center) it is possible to observe how the scaled outer region of the streamwise variance shows a decreasing trend with $Re$ and $\beta$. Interestingly, Figure 3 (right) also shows the collapse of the diagnostic plot in the outer region at higher $Re$ and larger APG, when appropriately scaled with the shape factor $H_{12} = \delta^*/\theta$, reminiscent of the behavior at high Reynolds numbers in ZPG boundary layers. The purpose of the present study is to develop a robust method to determine the boundary layer edge, or more precisely, a practical representation of the boundary layer thickness such as $\delta_{99}$. Note that the edge velocity $U_e$ (and for that matter $H_{12}$) is not know a priori since that requires knowing $\delta_{99}$, but if one uses as a first approximation $U_e \simeq \max(U)$, then the values of $u'/\left(U\sqrt{H_{12}}\right)$ at $U/U_e = 0.99$ are around 0.02, see inset in Figure 2 (right). This motivated the use of an iterative method, where the diagnostic plot is initially computed for all the profiles using $U_e \simeq \max(U)$, then $\delta_{99}$ is defined as the location where $u'/\left(U\sqrt{H_{12}}\right) = 0.02$. Note that this location corresponds to $U/U_e = 0.99$, therefore a new $U_e$ can be defined for the next iterative step. It should be noted that the value of 0.02 was deduced from ZPG TBL data in order not to be biased by the ambiguities related to $\delta_{99}$, $U_e$ and $H_{12}$. This leads to the fact that, as apparent from the intersection highlighted in the inset from Figure 3, the profiles for the pressure gradient cases

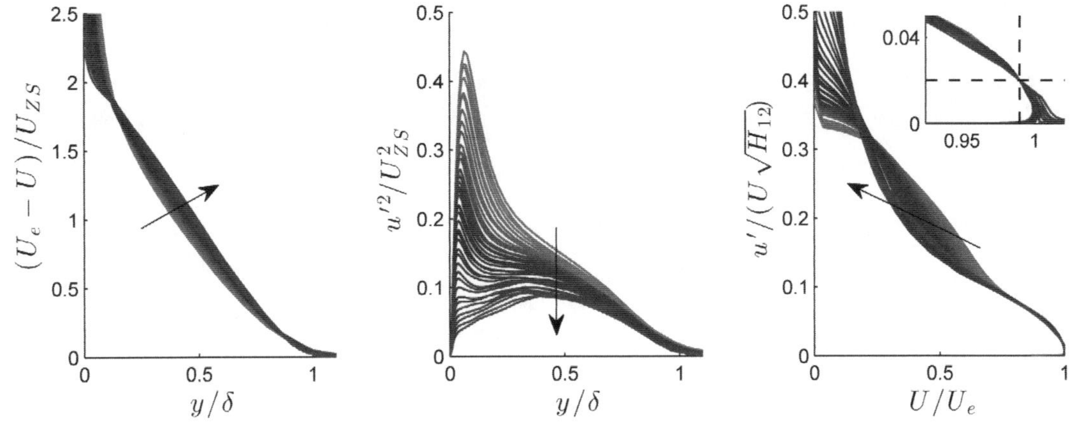


**Figure 3.** (Left) Mean velocity defect and (center) streamwise variance scaled with $U_{ZS}$ and $\delta$; (right) diagnostic plot scaled with the shape factor $H_{12}$. Profiles extracted from the suction side of the wing, in the range $0.15 < x/c < 0.98$, with red indicating data closer to the leading edge and blue profiles towards the trailing edge of the wing. The arrow also shows increasing downstream location, which is associated with higher Reynolds number but also larger $\beta$. Note the similarities between the high-$Re$ TBL data shown in Figure 2 and the high-$\beta$ cases presented here.

are forced through the point $u'/\left(U\sqrt{H_{12}}\right) = 0.02$ and $U/U_e = 0.99$. This scheme converged to the final $H_{12}$ value after one iteration in most of the profiles, and after few iterative steps in the profiles near separation. Therefore, it is a robust criterion to determine the boundary layer thickness $\delta_{99}$. It is also important to note that $\delta_{99}$ is an arbitrary representation of the boundary layer thickness and it is problematic when $Re \to \infty$, but it is widely used in the literature for practical purposes, and that is why it is also adopted in the present study.

The results shown for the wing at $Re_\theta$ up to 2,800 are extended to higher Reynolds numbers up to $Re_\theta = 40,000$ in Figure 4, where the experimental data by Nagib *et al.* [16] is shown. These measurements include APG, FPG and strong FPG measurements approximately in the range $-0.5 < \beta < 0.5$, and as it can be seen in Figure 4 (left) and (center) the Zagarola–Smits scaling is also adequate for these higher $Re$ cases since the outer region of the mean velocity defect and the streamwise variance profiles tend to be regrouped with increasing Reynolds number. In addition to this, Figure 4 (right) also shows very good collapse of the diagnostic plot scaled with $H_{12}$, Note that although the slope may change slightly depending on the pressure gradient (as also observed by Castro [24]) and the history effects of the flow, the value 0.02 is robust and accurate enough for the purposes of the present study. This is a very reassuring test for the use of the diagnostic plot concept to determine the boundary-layer edge in pressure-gradient boundary layers, due to the wide range of pressure gradient and Reynolds-number conditions, but also the very different history effects exhibited by the datasets under consideration.

The various methods are then used to determine the boundary-layer thickness of the profiles developing over the suction side of the wing in the range $0.15 < x/c < 0.98$, and the results are shown in Figure 5. For the two composite profile methods, *i.e.*, the one by Nickels [5] and the one by Chauhan *et al.* [17], previous processing of the data with the diagnostic plot (or any other pre-treatment) is necessary in order to provide initial estimations for the boundary layer thickness. Note that in both cases we performed curve fits without prescribing any variable, and the boundary layer thickness $\delta$ was one of the fitting parameters. It is also important to

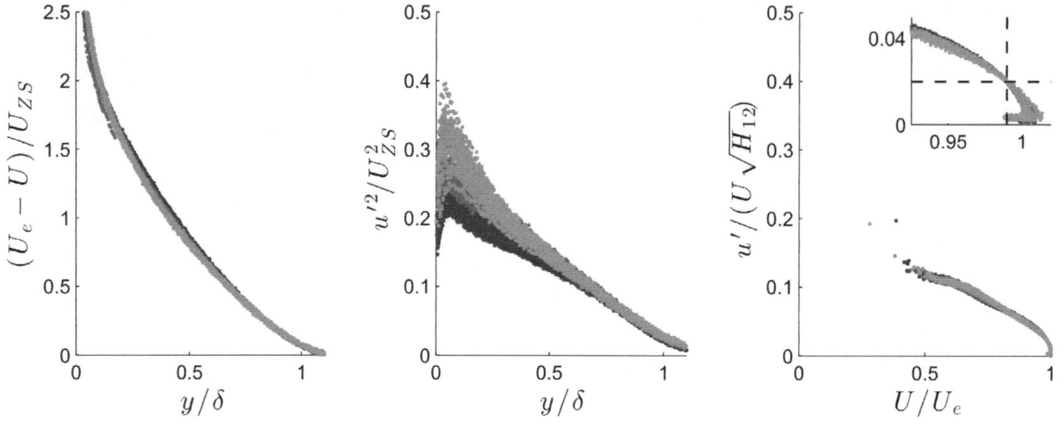


**Figure 4.** (Left) Mean velocity defect and (center) streamwise variance scaled with $U_{ZS}$ and $\delta$; (right) diagnostic plot scaled with the shape factor $H_{12}$. Experimental results by Nagib *et al.* [16] in several pressure gradient configurations, where blue corresponds to APG, red to FPG and green to strong FPG cases.

note that, as can be observed in Figure 5 (left), in the composite profile by Chauhan *et al.* the fitting parameter $\delta$ corresponds to $\delta_{100}$, whereas in the case of Nickels it is $\delta_{95}$. With adequate initial estimations of the boundary-layer thickness, both composite profile approaches provide consistent results up to around $x/c \simeq 0.85$, after which the APG is strong (with $\beta$ values larger than 7), and especially the composite law of Chauhan *et al.* [17] leads to scattered values of $\delta$. Let us recall that the composite law of Nickels [5] was developed for pressure gradient boundary layers, whereas the one by Chauhan *et al.* [17] inherits a number of parameters in the wake function specifically developed for ZPG boundary layers. The method based on the mean velocity gradient $\mathrm{d}U/\mathrm{d}y$, with an arbitrary threshold of $10^{-3}$, leads to results similar to the diagnostic plot up to $x/c \simeq 0.8$, which means that the boundary layer thickness it provides is approximately $\delta_{99}$ in this range. Beyond this point the gradient does not decrease to $10^{-3}$, and therefore the position of the minimum of $\mathrm{d}U/\mathrm{d}y$ is taken as the boundary-layer edge. This lead to values of $\delta$ different from the ones calculated from the diagnostic plot after $x/c = 0.8$, which also supports the need to avoid arbitrary thresholds in order to obtain robust results.

Finally, the results from the method based on the diagnostic plot are also shown in Figure 5 (left). As shown in Figure 3 (right), the point where $u'/\left(U\sqrt{H_{12}}\right) \simeq 0.02$ is found for each profile, which corresponds to the location where $U/U_e \simeq 0.99$, and therefore the boundary layer thickness obtained with this approach corresponds to $\delta_{99}$. Figure 5 (left) shows that the results are robust throughout the whole suction side of the wing, and since this method is based on an objective way of determining the point where $U/U_e \simeq 0.99$, we believe that it is the best method to determine the boundary layer thickness in pressure gradient TBLs. To further show the consistency of the results, in Figure 5 (right) we show the values of $\delta_{99}$ obtained from the two composite approaches, and compare them with the diagnostic plot and the velocity gradient. The four methods exhibit excellent agreement up to $x/c \simeq 0.8$, which is the region where the APG is mild or moderate, and beyond this point the diagnostic plot and the composite from Nickels [5] (provided an initial $\delta$ is used, which is here obtained through the diagnostic plot) are still in very good agreement until almost the trailing edge.

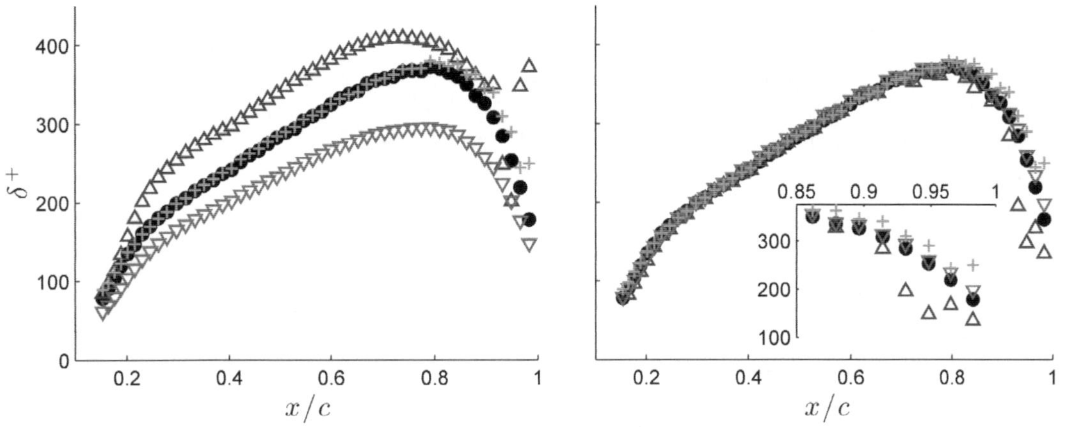


**Figure 5.** (Left) Inner-scaled boundary layer thickness $\delta^+$ extracted from the four methods, throughout the suction side of the wing. Blue triangles correspond to the composite law of Chauhan *et al.* [17], red triangles to the composite law of Nickels [5], green crosses to the velocity gradient method and black dots to the diagnostic plot method. (Right) Same results, but with the boundary-layer thickness $\delta_{99}$ (instead of the inherent $\delta$ corresponding to $\delta_{100}$ and $\delta_{95}$, respectively) extracted from the two composite profiles. Inset enlarges the region close to the trailing edge.

## 5. Conclusions

The present study is focused on the determination of the edge of pressure-gradient turbulent boundary layers, *i.e.* $U_e$ and $\delta$, which can be quite problematic due to the fact that in such boundary layers the mean velocity is not necessary constant beyond the layer edge. The merits and shortcomings of the following four methods are analyzed: (i) curve fit to the composite profile by Chauhan *et al.* [17], developed for ZPG TBLs; (ii) fit to the composite profile by Nickels [5], which was developed for pressure gradient TBLs; (iii) find the point where the mean velocity gradient $\mathrm{d}U/\mathrm{d}y$ is approximately zero, with a threshold of $10^{-3}$ in inner units (in case that the gradient does not go below the threshold, then the position of its minimum is considered as the edge); (iv) use of the diagnostic plot to find $U/U_e \simeq 0.99$, which is given by the value $u/\left(U\sqrt{H_{12}}\right) \simeq 0.02$. Note that methods (i) and (ii) require an initial estimation for $\delta$, which is taken as the value obtained from method (iv). Without such a pre-treatment the composite profiles would not result in such smooth curves as presented in Figure 5 towards the stronger pressure gradient cases.

All the methods are tested on turbulent boundary-layer profiles developing over the suction side of a NACA4412 wing profile, extracted from a recent DNS at Reynolds number 400,000 [9, 10]. While the four methods provide robust results when the pressure gradient is mild or moderate, stronger pressure gradients ($\beta > 7$) lead to inconsistent results in all the techniques except the diagnostic plot one. It is important to note that although the composite profile approaches produce acceptable results, they require initial estimations of $\delta$ which have to be obtained by other means (in this case, they were obtained with the diagnostic plot method). Moreover, in the strong pressure gradient cases the composite profiles do not fit the velocity profiles, although the values of $\delta$ they provide are acceptable. This is due to the particular formulation of the wake function, which is not general enough to represent the wake encountered in such pressure gradient TBLs. The diagnostic plot method has the additional advantage of determining the point where $U$ is 99% of the edge velocity in an objective way, and shows consistent results over a wide range of pressure gradient conditions, as well as flow histories. The robustness of the diagnostic plot was also checked with the pressure gradient TBL experiments by Nagib *et al.* [16]. In addition to this, it is based on data that is usually available in most experimental studies, namely mean and streamwise turbulence intensity profiles. Therefore, the method based on the diagnostic plot described in the present study is a robust technique to determine the boundary layer thickness in pressure gradient TBLs.

**Acknowledgments**
This work was supported in part by the Multiflow program of the European Research Council, the Knut and Alice Wallenberg Foundation and the Swedish Research Council (VR). The authors would like to thank Prof. Yvan Maciel for his insightful comments on the manuscript. Computer time was provided by SNIC (Swedish National Infrastructure for Computing).

## References

[1] Rotta J C 1953 *NACA Tech. Rep. 1344*
[2] Rotta J C 1962 *Prog. Aerosp. Sci.* **2** 1–219
[3] Coles D E 1956 *J. Fluid Mech.* **1** 191–226
[4] Coles D E 1968 *AFOSR-IFP-Stanford Conf. on Computation of turbulent boundary layers, D. E. Coles and E. A. Hirst (Eds.), pp. 1–45*
[5] Nickels T B 2004 *J. Fluid Mech.* **521** 217–239
[6] Nagib H M and Chauhan K A 2008 *Phys. Fluids* **20** 101518
[7] George W K 2006 *AIAA J.* **44** 2435–2449
[8] Schlatter P and Örlü R 2010 *J. Fluid Mech.* **659** 116–126
[9] Vinuesa R, Hosseini S M, Hanifi A, Henningson D S and Schlatter P 2015 *Proc. 9th Intl Symp. on Turbulence and Shear Flow Phenomena, Melbourne, Australia* 1–6
[10] Hosseini S M, Vinuesa R, Schlatter P, Hanifi A and Henningson D S 2015 *Proc. 23rd Annual Conference of the Computational Fluid Dynamics Society of Canada, Waterloo, Canada* 1–6
[11] Spalart P R and Watmuff J H 1993 *J. Fluid Mech.* **249** 337–371
[12] Alfredsson P H and Örlü R 2010 *Eur. J. Mech. B-Fluid* **29** 403–406
[13] Alfredsson P H, Segalini A and Örlü R 2011 *Phys. Fluids* **23** 041702
[14] Alfredsson P H, Örlü R and Segalini A 2012 *Eur. J. Mech. B-Fluid* **36** 167–175
[15] Fischer P F, Lottes J W and Kerkemeier S G 2008
[16] Nagib H M, Christophorou C and Monkewitz P A 2004 *IUTAM Symposium on one hundred years of boundary layer research, G. E. A. Meier and K. R. Sreenivasan (Eds.), Göttingen, Germany* 383–394
[17] Chauhan K A, Monkewitz P A and Nagib H M 2009 *Fluid Dyn. Res.* **41** 021404
[18] Vinuesa R, Schlatter P and Nagib H M 2014 *Exp. Fluids* **55** 1751
[19] Rodríguez-López E, Bruce P J K and Buxton O R H 2015 *Exp. Fluids* **56** 68–16
[20] Vinuesa R and Nagib H M 2015 *Eur. J. Mech. B-Fluid*
[21] Vinuesa R, Rozier P H, Schlatter P and Nagib H M 2014 *AIAA J.* 1–17
[22] Örlü R and Schlatter P 2013 *Exp. Fluids* **54** 1547
[23] Castro I A, Segalini A and Alfredsson P H 2013 *J. Fluid Mech.* **727** 119–131
[24] Castro I A 2015 *J. Fluid Mech.* **770** 289–304
[25] Skåre P E and Krogstad P Å 1994 *J. Fluid Mech.* **272** 319–348
[26] Drózdz A, Elsner W and Drobniak S 2015 *Eur. J. Mech. B-Fluid* **49** 137–145
[27] Zagarola M V and Smits A J 1998 *J. Fluid Mech.* **373** 33–79
[28] Maciel Y, Rossignol K S and Lemay J 2006 *AIAA J.* **44** 2450–2464

# Inception and evolution of coherent structures in under-expanded supersonic jets

**P. C. Stegeman** [1], **J. M. Pérez**[2], **J. Soria**[1,3], **V. Theofilis**[2]

[1]Department of Mechanical and Aerospace Engineering, Monash University, Melbourne, Wellington Road, Clayton, Victoria 3800, Australia
[2]School of Aeronautics, Universidad Politécnica de Madrid, Plaza Cardenal Cisneros 3, E-28040 Madrid, Spain [3]Department of Aeronautical Engineering, King Abdulaziz University, Jeddah, Kingdom of Saudi Arabia

E-mail: Paul.Stegeman@monash.edu

**Abstract.** The purpose of this paper is to examine the generation and nature of the large coherent structures observed experimentally in an under-expanded supersonic impinging jet. More specifically, the questions to answer are: What mechanisms govern the receptivity process at the nozzle lip?, how does the underlying flow field affect the evolution of the large-scale coherent structure generated from the initial instability? and what are the interactions between the large-scale (forced) coherent structures and the developing turbulence in the jet shear layer? In order to answer some of these questions both alternatives, that these structures come from global modal flow instabilities or from convective instabilities, the latter, are considered in this work.

The stability analysis considered in the former case is performed in this work near the nozzle around the temporal average of the flow obtained by using an in-house LES (Large Eddy Simulation) code. The flow in this region is considered laminar, steady and without non-linear effects. The well known feedback loop in the impinging jet, according to which acoustic waves propagate upstream and excite the jet shear-layer (see Figure 2), advises against some of the hypothesis considered previously in the global stability analysis (ie. non-linear approximation). However the acoustic waves are orders of magnitude smaller than the hydrodynamic waves and should be smoothed out in the temporal average used in the calculation of the mean flow.

The results show that both, axisymmetrical ($m = 0$) and azimuthal modes ($m \geq 1$) are stable to global modal analysis and only convective instability could justify the instabilities observed in experiments in the shear layer. A study on the receptivity problem confirms that external disturbances may enter and excite the shear layer, being responsible of the instabilities observed in both experiments and direct numerical simulations.

## 1. Introduction

The flow structure of an under-expanded supersonic impinging (USI) jet is presented in Figure 1. Here the flow exits from the nozzle at Mach number $Ma = 1$ towards an impingment wall placed normal to the jet axis at a distance of $z/d = 5$ vertically above the nozzle, where $d$ is the nozzle diameter.

USI jets characteristically produce large-scale structures at discrete frequencies due to the development of an acoustic feedback mechanism. In this mechanism acoustic waves force an instability at the nozzle lip which grows within the shear layer as it travels downstream. These structures interact with the stand-off shock which produces a disturbance in the wall jet. An

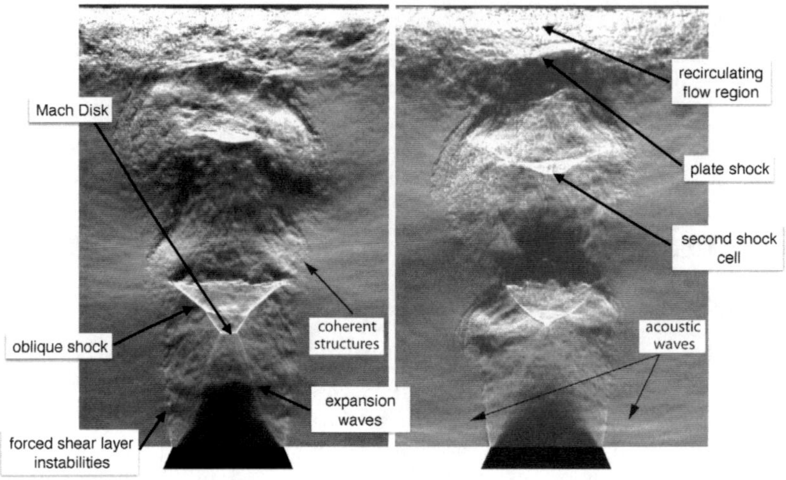


Figure 1: Instantaneous flow obtained from experiment with the physical description of regions and structures presented in a typical under-expanded supersonic impinging jet (LTRAC) [1].

acoustic wave is then generated in the wall jet which travels upstream and forces a new jet shear layer instability. This process, seen in Figure 2, is often referred to an acoustic feedback loop and it is a multi-physical complex problem which is not presently understood sufficiently. As impinging jets provide an effective way to transfer mass and energy in industrial applications, the prediction and control of the instabilities that arise in this configuration is of both fundamental and practical interest.

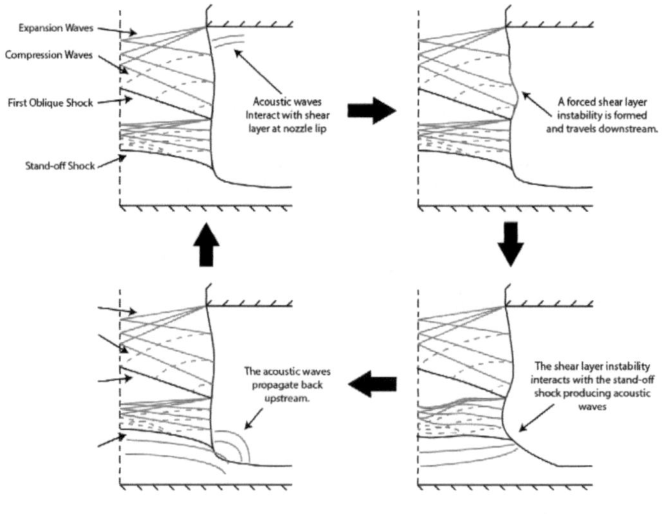


Figure 2: Feedback lop mechanism: Shear layer instability grows and interacts with the standoff shock to produce an acoustic wave. Then acoustic wave reflects and travels back upstream and excite the shear layer.

In this problem the jet emerges from a nozzle with a velocity and a temperature profile, that depends on the nozzle configuration, which in turn determine the characteristics of the flow upstream. This flow presents a characteristic pattern of shock waves, expansion waves, coherent structures, etc (see Figure 1). Finally the flow impinges against the wall which is perpendicular to the driving direction. At high Reynolds numbers, the shear layer generates flow instabilities that look like classical Kelvin–Helmholtz waves. Theses instabilities have been observed in experiments and numerical simulations but their nature is not sufficiently well understood. For example, Powell [5] identified four frequencies associated with different instability modes in a jet, naming them A, B, C and D. Later, Davies & Oldfield [6] identified the nature of these modes: A1 and A2 are axisymmetric, B is sinuous or lateral, C is helical and D is another lateral mode. Dimensionless parameters important to these flows are: the impingement wall stand-off distance $z/d$, the nozzle exit Mach number $Ma$ and the jet Reynolds number Re defined in section 3.

Investigation of the hydrodynamic and acoustic instability mechanisms in these types of flows are essential for the understanding of the transition process from a stable steady laminar flow to transitional and turbulent flow. A modal linear stability analysis, focusing on the spectrum of the flow, and a non-modal analysis, for short-time perturbations, can provide a complete insight of the physical mechanisms responsible of the transition to turbulent flows.

## 2. Theory

### 2.1. Governing equations

The dimensionless Navier-Stokes equations representing the conservation of mass, momentum and energy that govern the dynamics of such jet flows are defined respectively as

$$\frac{\partial \rho}{\partial t} + \nabla \cdot (\rho \mathbf{u}) = 0, \tag{1}$$

$$\frac{\partial \rho \mathbf{u}}{\partial t} + \nabla \cdot (\rho \mathbf{u} \otimes \mathbf{u} + p\mathbb{I} - \mu \mathbb{T}) = 0, \tag{2}$$

and

$$\frac{\partial \rho e}{\partial t} + \nabla \cdot (\mathbf{u}(\rho e + p) - k\nabla T - \mu \mathbb{T} \cdot \mathbf{u}) = 0. \tag{3}$$

where the deviatoric stress tensor ($\mathbb{T}$) is defined as

$$\mathbb{T} = \left[\nabla \mathbf{u} + (\nabla \mathbf{u})^T + \left(\frac{\mu_b}{\mu} - \frac{2}{3}\right)(\nabla \cdot \mathbf{u})\,\mathbb{I}\right], \tag{4}$$

$\rho$ is the density, $\mathbf{u}$ is the velocity vector, $e$ is the total energy per unit mass, $p$ is the pressure, $T$ is the temperature, $\mu$ is the viscosity, $\mu_b$ is the bulk viscosity, $k$ is the thermal conductivity and $\mathbb{I}$ is the identity tensor. In these equations length, density, velocity, time, temperature, energy and viscosity have been non dimensionalised by $L_0$, $\rho_0$, $u_0$, $t_0 = L_0/u_0$, $T_0$, $\rho_0 u_0^2$ and $\mu_0$ respectively. This results in the dimensionless Reynolds number

$$Re_0 = \frac{\rho_0 u_0 L_0}{\mu_0} \tag{5}$$

and Mach number

$$M_0 = \frac{u_0}{\sqrt{RT_0}}. \tag{6}$$

where $R = 287Jkg^{-1}K^{-1}$ for air. For an ideal gas the total energy per unit mass and equation of state are defined as

$$e = \frac{1}{2}\|\mathbf{u}\|^2 + \frac{T}{M_0^2(\gamma - 1)} \tag{7}$$

and

$$p = \frac{\rho T}{M_0^2} \tag{8}$$

Journal of Physics: Conference Series **708** (2016) 012015 doi:10.1088/1742-6596/708/1/012015

respectively. The viscosity and thermal conductivity are determined using the Power law [11]

$$\mu = \frac{1}{Re_0} T^{0.76} \tag{9}$$

and

$$k = \mu \frac{\gamma}{\gamma - 1} \frac{1}{M_0^2 Pr} \tag{10}$$

respectively, where for air $\gamma = 1.4$ and $Pr = 0.7$.

Locally one-dimensional inviscid compressible boundary conditions defined in [10] are used for the adiabatic walls and outflow regions. The jet inlet velocity profile was modeled using the hyperbolic-tangent function found in [12] while the temperature profile was determined using the Crocco-Busemann relationship [13]. No inlet turbulence has been applied.

### *2.2. Linear instability analysis*

The three-dimensional compressible Navier-Stokes equations can be written in a compact form as,

$$\frac{\partial \mathbf{q}}{\partial t} = \mathbf{F}(\mathbf{q}) \,, \tag{11}$$

where $\mathbf{q} = (\rho, u, v, w, T)$ and $\mathbf{F}$ is the right hand side of the system of equations (1-3) when these are written in primitive variables.

Linear stability theory studies the temporal evolution of small amplitude disturbances $\mathbf{q}'$ superimposed upon a steady base flow $\overline{\mathbf{q}}$,

$$\mathbf{q}(x, y, z, t) = \overline{\mathbf{q}}(x, y, z) + \epsilon \mathbf{q}'(x, y, z, t) \,, \tag{12}$$

where $\epsilon \ll 1$ is an small parameter. A linearized problem can be written by introducing this decomposition in the Navier-Stokes equations (11) and retaining the terms up to $O(\epsilon)$. Then two approaches could be considered: (a) Modal linear instability theory where modal perturbations $q'(x, r, \theta, t) = \hat{q}(x, r) e^{i(m\theta - \omega t)}$ are assumed. In this case a generalized eigenvalue problem must be solved,

$$\mathcal{A}(Re, Ma, \overline{q}, m)\hat{q} = \omega \mathcal{B}(Re, Ma, \overline{q}, m)\hat{q} \,, \tag{13}$$

where $Re$ and $Ma$ are the Reynolds and Mach number based on the nozzle conditions, respectively, while $m$ represent additional parameters that arise from the modal analysis approach considered. The other possibility is, (b) non-modal linear instability analysis, where a initial value problem must be solved,

$$\mathcal{B}(Re, Ma, m, \overline{q}) \frac{dq'}{dt} = \mathcal{A}(Re, Ma, m, \overline{q}) q' \,,$$

by computing the singular value decomposition of the evolution operator $\Phi = e^{(\mathcal{B}^{-1}\mathcal{A})t}$.

Different levels of modelling can be considered in the modal analysis (see Table 1). So for example, the basic flow deepens only on one variable in the ODE model, two variables in the BiGlobal model and three variables in the TriGlobal case, while the rest of homogeneous variables, including time, are considered in a exponential function (ansatz). In our case, we used the temporal BiGlobal stability model which $\beta$ is the wavenumber in the azimuthal direction and $\omega = \omega_r + i \cdot \omega_i$ is a complex number which real part is the frequency of the perturbation and the imaginary part the amplification or damping rate of this. Therefore, a large-scale eigenvalue problem is solved by an iterative approach [4].

| | Assumptions | Base | Amplitude | Phase Function $\Theta$ ( $Exp(i\Theta)$ ) |
|---|---|---|---|---|
| TriGlobal | - | $\overline{\mathbf{q}}(x,r,\theta)$ | $\hat{q}(x,r,\theta)$ | $-\omega t$ |
| PSE-3D | $\partial_\theta\overline{\mathbf{q}} \ll \partial_x\overline{\mathbf{q}}, \partial_r\overline{\mathbf{q}}$ | $\overline{\mathbf{q}}(x,r,\theta^*)$ | $\hat{q}(x,r,\theta^*)$ | $\int m(\theta')d\theta' - \omega t$ |
| BiGlobal | $\partial_\theta\overline{\mathbf{q}} = 0$ | $\overline{\mathbf{q}}(x,r)$ | $\hat{q}(x,r)$ | $m\theta - \omega t$ |
| PSE | $\partial_x\overline{\mathbf{q}} \ll \partial_r\overline{\mathbf{q}}$<br>$\partial_\theta\overline{\mathbf{q}} = 0$ | $\overline{\mathbf{q}}(x^*,r)$ | $\hat{q}(x^*,r)$ | $\int \alpha(x')dx' + m\theta - \omega t$ |
| Local | $\partial_x\overline{\mathbf{q}} = \partial_\theta\overline{\mathbf{q}} = 0$ | $\overline{\mathbf{q}}(r)$ | $\hat{q}(r)$ | $\alpha x + m\theta - \omega t$ |

Table 1: Levels of modeling can be considered in the modal analysis for problems defined in cylindrical coordinates. The asterisk denotes a slowly-varying spatial direction.

## 3. Numerical Implementation

### *3.1. Mean flow computation*

Datasets used in the following analyses were computed using an in-house compressible LES solver implementing a non-uniform three-dimensional cylindrical finite difference, time-accurate scheme with hybrid spatial discretisation. For the hybrid spatial discretisation the spatial fluxes are calculated in smooth regions by utilising a 6th order dispersion relation preserving (DRP) scheme and a 5th order weighted essentially non-oscillatory scheme with local Lax-Friedrichs flux splitting in discontinuous regions. Temporal integration is performed using a fourth order five step Runge-Kutta scheme. The sub-grid scale terms were computed using Germano's dynamic model with the adjustments made by Lilly [9]. The domain consists of approximately 30 million nodes with the spatial extent of $15d$ the radial direction.

The mean flow components, temperature, stream-wise and radial velocity are illustrated in Figure 3. These fluid variables are scaled by using the flow properties at the nozzle exit. The values of the dimensionless parameters used in the simulations are: $Re = u_j d/\nu_j = 50000$, $Ma = u_j/\sqrt{\gamma R T_j} = 1$, the ratio between the stagnation pressure measured in the jet plenum and the ambient pressure $NPR = 3.4$ and $z/d = 5$, where $u_j$ and $T_j$ are the velocity and temperature in the nozzle at $r = 0$. The velocity gradients in the jet, near to the nozzle lip, create a lateral transfer momentum, the jet loses energy and the velocity profile expands laterally and decreased in magnitude. On the other hand, the temperature of the supersonic jet decreases as it exits the nozzle due to the expansion and increases at the shocks. The temperature is higher near the impinging wall and over the shear layer.

Figure 4 shows the complex growth/decay dynamics of the acoustically forced instabilities. The left Figure is the temporal spectrum of the turbulent kinetic energy (integrated in the azimuthal direction) as a function of position along the shear layer. The central Figure shows the same information particularized for a given position in the shear layer, $x = 0.05d$ (marked with a discontinuous white line in the left figure), where two peaks corresponding a two forced modes are obtained: $St_\infty = 0.12$ (red line) and $St_\infty = 0.54$ (green line). Finally, the right Figure shows how the peak broadband turbulence frequency corresponding to the previous modes moves from high (St 3-5) to low (St 0.25) along the shear layer.

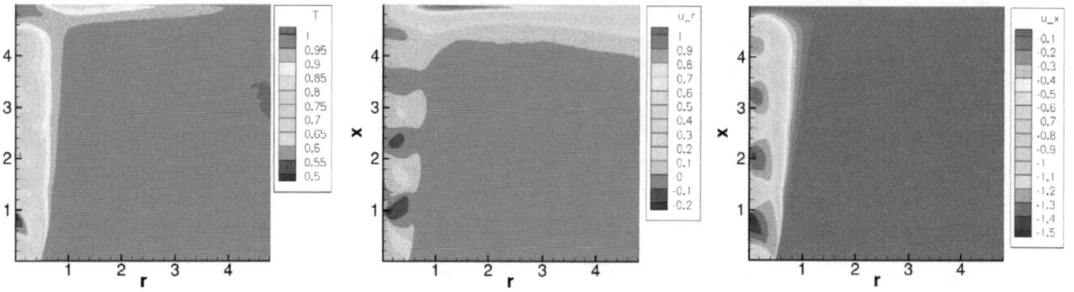


Figure 3: Mean flow components obtained by the DNS code at $Re = 50000$, $Ma = 1$, $NPR = 3.4$ and resolution equal to $nx \times nr = 360 \times 608$. (*left:*) Temperature, (*center:*) radial velocity and (*right:*) streamwise velocity.

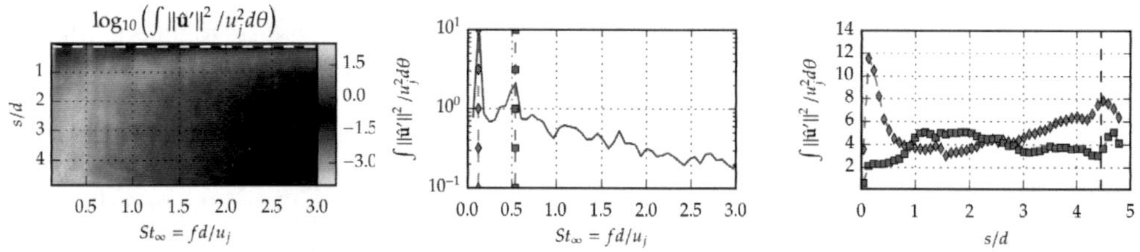


Figure 4: Velocity fluctuation spectra for $z/d = 5$, $Re = 50000$, $Ma = 1$ and $NPR = 3.4$ where two forced modes with Strouhal number, $St_\infty$ equal to 0.12 and 0.54 are obtained, respectively.

### *3.2. Global instability analysis*

Global stability analysis near the nozzle on supersonic impinging jet is performed in this section. This analysis was used by Theofilis and Colonius [3] in the case of compressible flows and it is described in detail by Theofilis [4]. The two-dimensional generalized eigenvalue problem for temporal three-dimensional perturbations defined in Equation (13) is solved near to the nozzle by using an in-house code [7] based on a FD-q scheme, where $x \in [0, 1.5]$ and $r \in [0, 3]$. The variables were dimensionless with the displacement thickness of the shear layer at $x = 0.5$. The Reynolds number based on this length and the nozzle conditions is approximately equal to 6800.

Homogeneous Dirichlet at the wall and nozzle and axial boundary conditions at $r = 0$ are considered. The later is described in detail by Malik et al. [2], and depends on the value for the azimuthal wave number,

$$\begin{aligned}
m = 0 \quad &: \quad \partial_r u' = v' = \partial_r T' = 0\,,\\
|m| = 1 \quad &: \quad u' = T' = \partial_r v' = 0 \text{ and } v' + imw' = 0\,,\\
|m| > 2 \quad &: \quad u' = v' = w' = T' = 0\,.
\end{aligned}$$

where prime denotes perturbation variables. The density is not imposed at the boundaries in the analysis, instead this is solved by using the continuity equation in the inner points. Homogeneous Dirichlet boundary condition for the perturbations at $r_{max}$ and homogeneous Neumann at $x_{max}$ are considered in the other boundaries.

Figures 5-7 shown the convergence of the spectrum for different values of the azimuthal wavenumber $m$ where $\omega_i$ is the growth/damping rate of the temporal disturbances and $w_r$ is its frequency (see third row of Table 1 for more details). In all the cases, left Figure shows the convergence of the spectrum as function of the resolution for a given order of the method (4 and 8) while right Figure shows the convergence of the spectrum as function of the order of the method for a given resolution. In both cases the leading mode is reasonably well converged.

The amplification function of this mode is represented in the following figures. Figure 8 shows the real and imaginary part of the amplitude functions of the leading mode for $m = 0$. In this case the azimuthal component is negligible compared to other components and then is not represented. The same information for the leading mode for $m = 1$ and 2 is shows in Figures 9 and 10, respectively. As it can been see in these two cases the leading mode has qualitatively the same structure. On the other hand, in all the cases the system is stable to linear global modes.

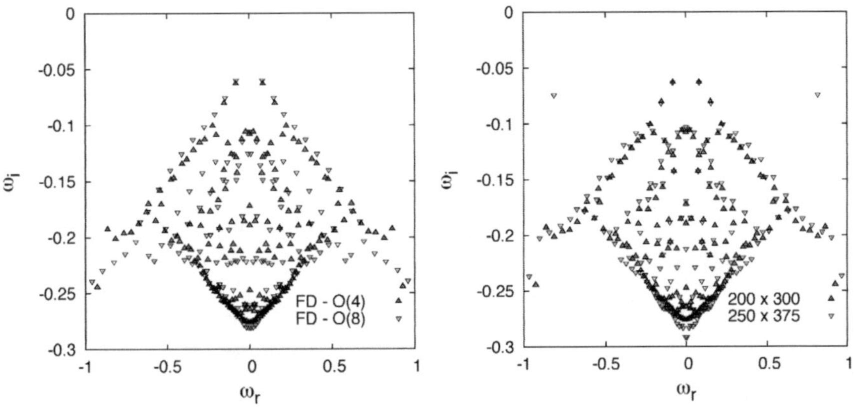


Figure 5: Spectrum for $m = 0$. (*Left:*) Convergence of the solution in function of the order of the discretization, 4 and 8, (*Right:*) Convergence of the solution in function of the resolution: $n_x \times n_r = 200 \times 300$ and $250 \times 375$.

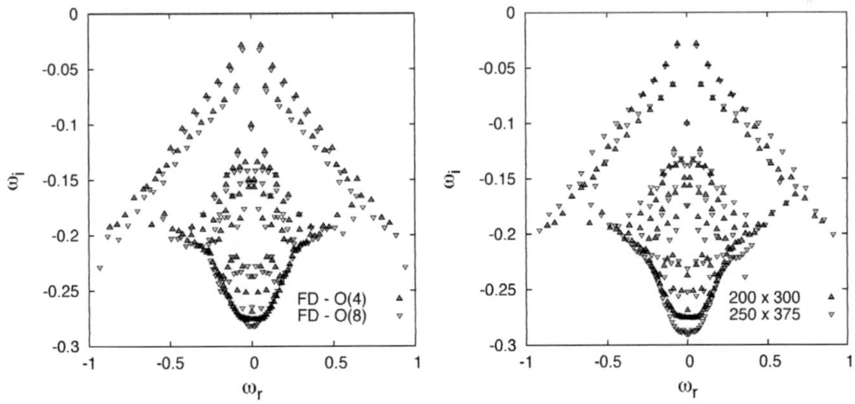


Figure 6: Spectrum for $m = 1$. (*Left:*) Convergence of the solution in function of the order of the discretization, 4 and 8, (*Right:*) Convergence of the solution in function of the resolution: $n_x \times n_r = 200 \times 300$ and $250 \times 375$.

## 4. Receptivity

In this section a receptivity analysis of a solution of the linearised compressible Navier-Stokes (LCNS) equations applied to the jet is presented. The receptivity analysis tries to quantify the system response to external forcing of the shear layer at the nozzle lip by acoustic waves. The

Journal of Physics: Conference Series **708** (2016) 012015 doi:10.1088/1742-6596/708/1/012015

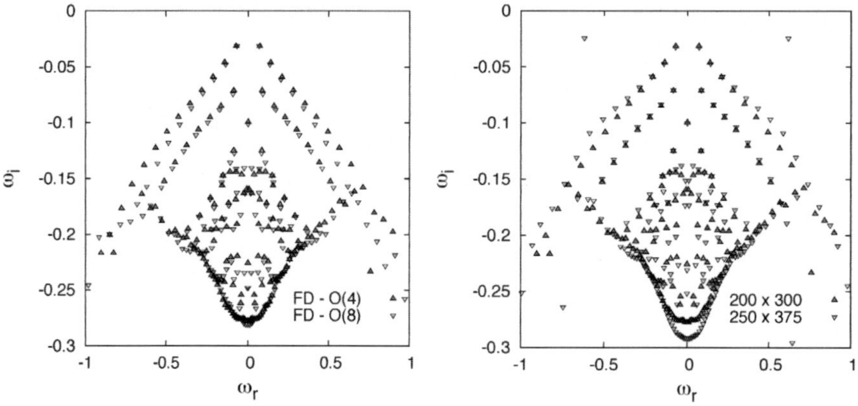


Figure 7: Spectrum for $m = 2$. (*Left:*) Convergence of the solution in function of the order of the discretization, 4 and 8, (*Right:*) Convergence of the solution in function of the resolution: $n_x \times n_r = 200 \times 300$ and $250 \times 375$.

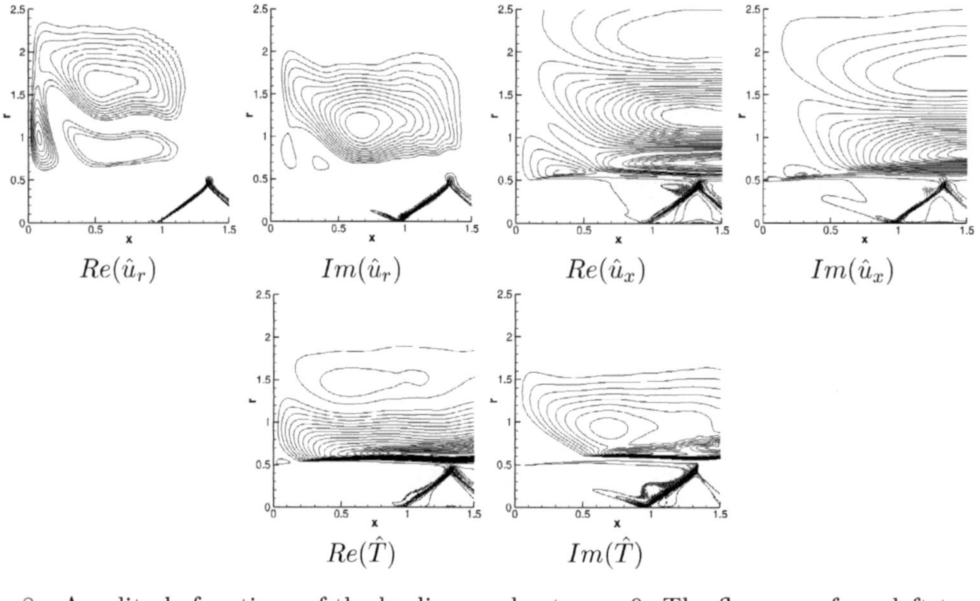


Figure 8: Amplitude functions of the leading mode at $m = 0$. The flow goes from left to right. $r = 0$ correspond with the jet center-line.

LCNS equations are

$$\frac{\partial \rho'}{\partial t} + \nabla \cdot (\rho \mathbf{u})' \tag{14}$$

$$\frac{\partial (\rho \mathbf{u})'}{\partial t} + \nabla \cdot \left(\overline{\rho \mathbf{u}} \otimes \mathbf{u}' + \rho \mathbf{u} + \rho \mathbf{u}' \otimes \overline{\mathbf{u}}'\right) + \nabla p' = \overline{\mu} \nabla \cdot \mathbb{T}' \tag{15}$$

$$\frac{\partial (\rho e)'}{\partial t} + \nabla \cdot \left(\overline{(\rho e + p)} \mathbf{u}' (\rho e + p)' \overline{\mathbf{u}}\right) = \overline{k} \nabla^2 T' + \overline{\mu} V_{KE} \tag{16}$$

where

$$V_{KE} = \left(\left(\nabla \cdot \mathbb{T}'\right) \cdot \overline{\mathbf{u}} + \left(\nabla \cdot \mathbb{T}'\right) \cdot \mathbf{u}' + \mathbb{T}' : \nabla \overline{\mathbf{u}} + \overline{\mathbb{T}} : \nabla \mathbf{u}'\right), \tag{17}$$

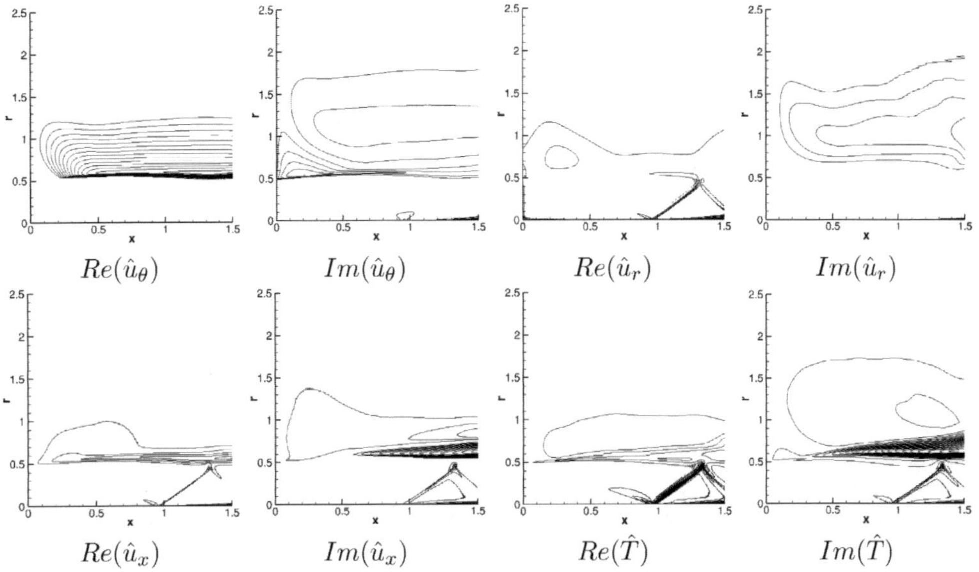


Figure 9: Amplitude functions of the leading mode at $m = 1$. The flow goes from left to right. $r = 0$ correspond with the jet center line.

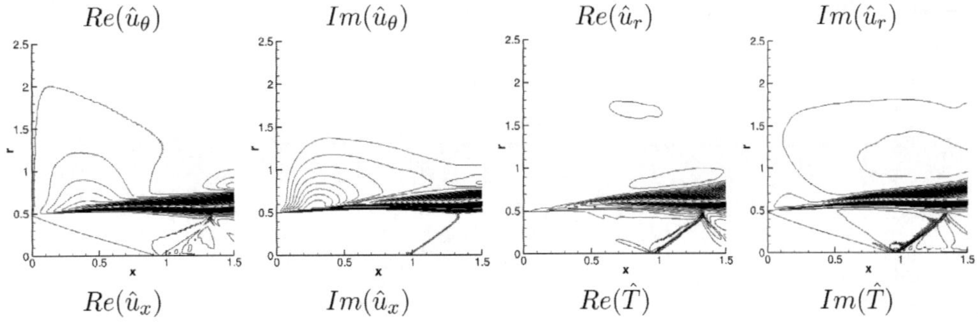


Figure 10: Amplitude functions of the leading mode at $m = 2$. The flow goes from left to right. $r = 0$ correspond with the jet center line.

$$\mathbb{T} = \nabla \mathbf{u} + (\nabla \mathbf{u})^T + \left(\frac{\mu_B}{\mu} - \frac{2}{3}\right)(\nabla \cdot \mathbf{u})\,\mathbb{I}, \tag{18}$$

$$(\rho e)' = \frac{1}{2}\rho \|\mathbf{u}'\|^2 + \frac{p'}{\gamma - 1}, \tag{19}$$

and the linearised version of the equation of state

$$p' = \frac{\overline{\rho} T' + \rho' \overline{T}}{M_0^2}. \tag{20}$$

Finally, viscosity and thermal conductivity are given by,

$$\overline{\mu} = \frac{\overline{T}^{0.76}}{Re} \quad \text{and} \quad \overline{k} = \overline{\mu}\frac{\gamma}{\gamma - 1}\frac{1}{M_0^2 Pr} \tag{21}$$

where the non dimensional quantities are

$$Re = \frac{\rho_\infty d u_j}{\mu_\infty} \quad \text{and} \quad M_0 = \frac{a_\infty}{\sqrt{RT_\infty}}. \tag{22}$$

The mean fields input into the linear simulation are calculated by taking the mean of the jet time series data generated by the LES solver detailed in section 3.

To establish the effect of the receptivity the transfer function between the input fluctuating acoustic velocity and output fluctuating hydrodynamic axial and radial velocities are determined. An initial acoustic wave is generated impulsively at the beginning of the linear simulation as a Gaussian pressure disturbance in the free-stream and is defined as

$$p' = A p_\infty e^{-\frac{(x - R_a \cos(\theta_a))^2 + (r - R_a \sin(\theta_a))^2}{2\sigma^2}} \quad \text{and} \quad R_a = \overline{R}_a + R'_a \cos(2\pi k_\theta \theta) \tag{23}$$

where $A$ is the amplitude ratio between the initial pulse peak and free-stream pressure, $x$ is the axial position from the nozzle outlet, $r$ is the radial position from the jet axis, $\theta$ is the azimuthal position, $\sigma$ controls the initial width of the pulse, $\theta_a$ is the angle between the initial pulse and the x-axis taken from the nozzle lip, $R_a$ is the distance between the nozzle lip and the pulse centre, $\overline{R}_a$ the azimuthal mean position of $R_a$, $R'_a$ is the amplitude of the variation of $R_a$ in the azimuthal direction and $k_\theta$ is the azimuthal mode of the initial pulse. Each of these variables as they relate to the geometry of the jet may be found in Figure 11(a). The parameter values chosen for the analysis are $\overline{R}_a = 2$, $R'_a = 0.06$, $\sigma = 0.05$, $\theta_a = \{15°, 25°, 35°, 45°\}$, and $k_\theta = \{0, 2\}$. $\sigma$ was chosen to be as small as possible in order to increase the initial spread of energy in the wavenumber domain while maintaining numerical accuracy.

Locations of the sample points for the incoming acoustic wave and outgoing hydrodynamic wave are shown in Figure 11(b). The distance between the incoming acoustic wave and outgoing hydrodynamic wave are $s_{in}$ and $s_{out}$ respectively. The transfer function is then defined in the frequency domain as

$$F(St, A) = \frac{\hat{A}(St)\Big|_{s=s_{out}}}{\hat{u'_a}(St)\Big|_{s=s_{in}}} \tag{24}$$

where $u'_a$ is the fluctuating velocity at the input sample point in the direction of $\mathbf{n}_a$ (seen in Figure 11(b)), $A$ represents either the fluctuating axial velocity $u'$ or fluctuating radial velocity $v'$ at the output sample point, and the hat ascent denotes the Fourier transform of the variable as $\hat{\phi}(St) = \int_{-\infty}^{\infty} \phi(t)\, e^{-2\pi i t St} dt$.

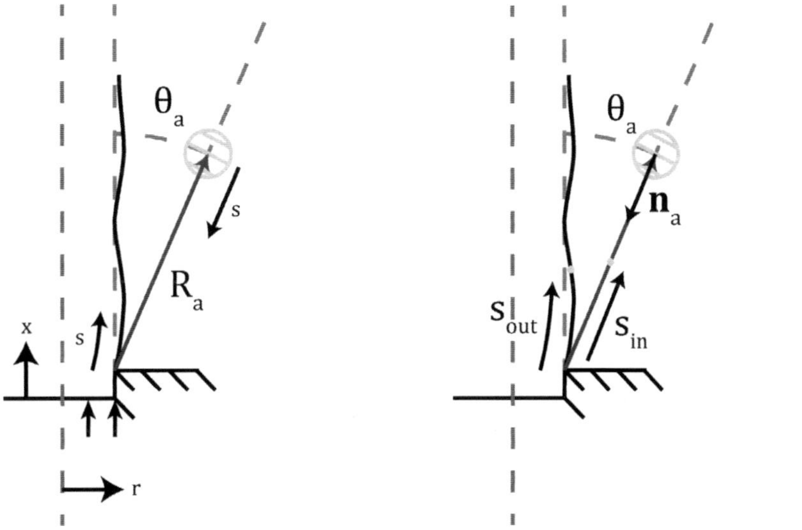


(a) Initial pulse position. (b) Input and output sample positions.

Figure 11: Layout of the linear simulation and analysis.

To obtain a representative from of the transfer function for the receptivity process then $s_{in} \to 0$ and $s_{out} \to 0$. However the minimum values of $s_{in}$ and $s_{out}$ are limited by the ability to distinguish the input and output signals at the sample points which in turn is limited by the spatial width of the initial acoustic wave and grid resolution in the simulation. For this analysis $s_{in} = 0.8d$ and $s_{out} = 0.3d$. The evolution of a single case from the initial acoustic wave, its interaction with the nozzle lip and the generation of the hydrodynamic wave in the shear layer may be seen in the contour plot sequence of Figure 12.

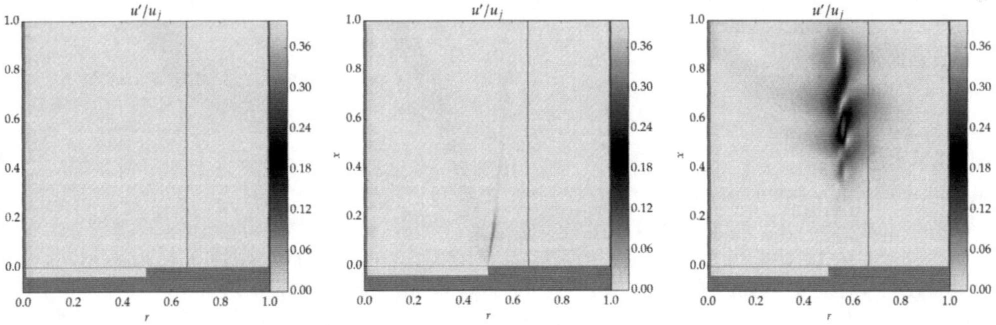

Figure 12: Sequence of figures that illustrate the receptivity problem. (*Left:*) The incoming acoustic wave. (*Center:*) Its interaction with the nozzle lip. (*Right:*) The resulting hydrodynamic wave as its amplified along the shear layer.

Effects of the acoustic reflections from the nozzle wall and radiation of acoustic waves generated by the hydrodynamic instability travelling downstream in the shear layer must be removed from the input acoustic time series. Their effect may be seen in Figure 13 in which the initial acoustic wave reaches the nozzle lip at $tu_j/d = 2$. Removal of these effects is performed by setting $u'_a = 0$ for $t > 2$. The initial acoustic pulse in the output signal, due to the acoustic wave interacting with the shear layer, is ignored as it's amplitude is approximately 0.07% of the

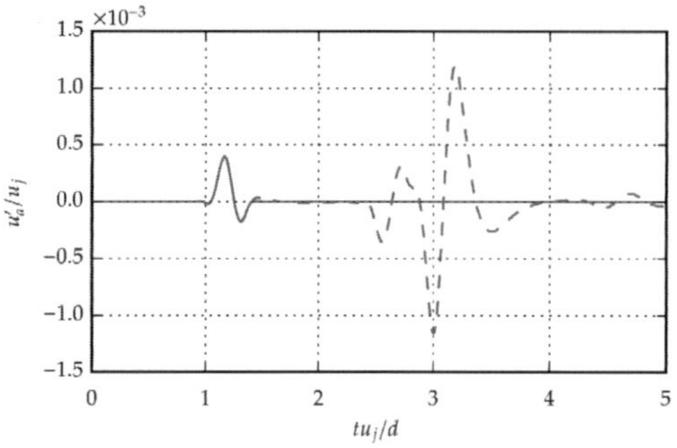


Figure 13: Temporal plot of the fluctuating acoustic velocity at the input sample point. The solid line indicates the signal from the initial acoustic pulse while the dashed line indicates the initial reflected wave and radiated acoustic wave generated by the hydrodynamic instability travelling downstream in the shear layer.

hydrodynamic instability's.

The resulting transfer function for the fluctuating axial hydrodynamic velocity is presented in Figure 14 and the fluctuating radial hydrodynamic velocity in Figure 15, both as a function of $\theta_a$ and $k_\theta$. Colours represent varying initial pusle angle ($\theta_a$), the solid lines represent the axisymmetric mode ($k_\theta = 0$), and the dashed lines represent the 2nd azimuthal mode ($k_\theta = 2$). For all $\theta_a$ and $k_\theta$ there are two spectral peaks in the magnitude response found for the axial velocity in the low frequency range, $St \approx 1.75$ and mid-frequency range $St \approx 4$ while there is only a single dominant spectral peak at $3 < St < 4$ for the radial velocity with a small peak occuring for $\theta_a = 15°$ and $\theta_a = 25°$ at $St \approx 6.75$. The phase response for all cases are linear with respect to temporal frequency. Increasing the angle of the incoming acoustic wave to the nozzle lip increases the magnitude of the resulting hydrodynamic wave velocities. The 2nd azimuthal mode also exhibits a slight increase in the amplitude of the resulting hydrodynamic velocity with small shifts in the peak spectra positions. Noise exists in the magnitude and phase response for frequencies greater than $St > 7.5$ and is believed to be due to numerical error as the high frequencies relate to smaller scale structures.

## 5. Conclusions

In conclusion, it has been shown that temporal global disturbances defined in the near region to the nozzle are globally stable near the nozzle and only convective instabilities are present for $m \leq 2$. This suggests that a transient growth analysis based on the BiGlobal approach should be undertaken. Future work on the modal analysis requires the study of the interactions of the shear layer with the turbulence and acoustic forcing present in the jet. This could be performed by using the resolvent model developed by Sharma and McKeon [8]. It is expected the resolvent analysis will produce a more detailed representation of the energy transfer processes within the jet.

An initial analysis of the receptivity process has also been performed. It has shown that the result of the interaction of the shear layer lip with the incoming acoustic wave results in two dominant temporal modes in the axial fluctuating velocity and one dominant mode in the radial fluctuating velocity of the generated hydrodynamic instability. Further analysis must be undertaken to establish how these instabilities are modified by the shear layer as they travel downstream and interact with the stand-off shock to produce a new acoustic wave that acts as the forcing phenomenon for the following cycle in the feedback loop.

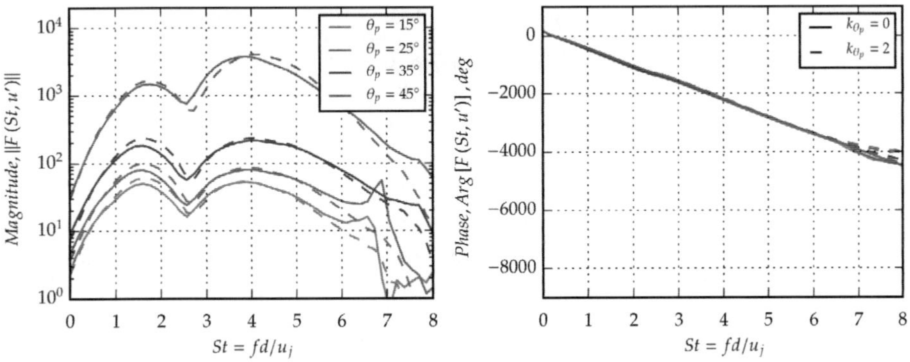


Figure 14: Magnitude and phase of the transfer function from the fluctuating acoustic input velocity $u'_a$ to the fluctuating hydrodynamic output axial velocity $u'$ as a function of initial pulse angle and azimuthal mode. Colours represent varying initial pulse angle ($\theta_p$), the solid lines represent the axisymmetric mode ($k_\theta = 0$), and the dashed lines represent the 2nd azimuthal mode ($k_\theta = 2$).

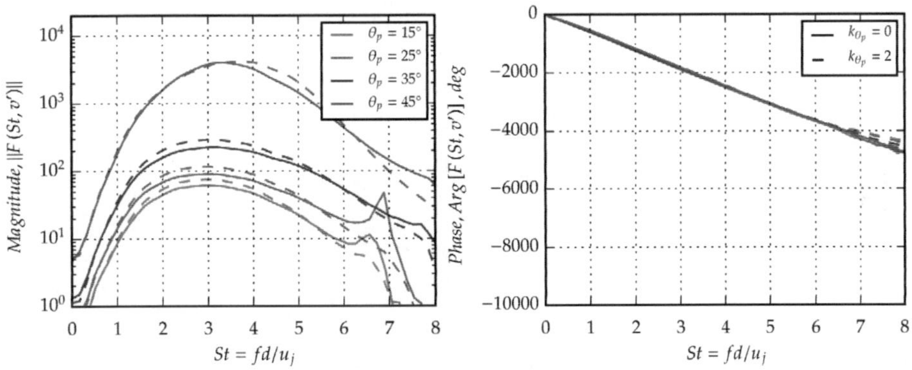


Figure 15: Magnitude and phase of the transfer function from the fluctuating acoustic input velocity $u'_a$ to the fluctuating hydrodynamic output radial velocity $v'$ as a function of initial pulse angle and azimuthal mode. Colours represent varying initial pulse angle ($\theta_p$), the solid lines represent the axisymmetric mode ($k_\theta = 0$), and the dashed lines represent the 2nd azimuthal mode ($k_\theta = 2$).

## Acknowledgements

This research was funded in part by the Multiflow program of the European Research Council and was undertaken with the assistance of resources from the National Computational Infrastructure (NCI), which is supported by the Australian Government. The support of the Australian Research Council (ARC) is also gratefully acknowledged.

## References

[1] D.M. Mitchell, D.R. Honnery and J. Soria 2012 The visualization of the acoustic feedback loop in impinging underexpanded supersonic jet flows using ultra-high frame rate schlieren *J. Visual.* **15** 333-341

[2] M.R. Malik and C.-L. Chang 2000 Nonparallel and nonlinear stability of supersonic jet flow *Comput. Fluids* **29** 327-365
[3] V. Theofilis and T. Colonius 2004 Three-Dimensional Instabilities of Compressible Flow over Open Cavities: Direct Solution of the BiGlobal Eigenvalue Problem *AIAA Paper* **2004-2544**
[4] V. Theofilis 2011 Global linear instability *Annu. Rev. Fluid Mech.* **43** 319-352
[5] A. Powell 1953 On the mechanism of choked jet noise *Proc. Phys. Soc. Lond.* **66** 10391056
[6] M. Davies and D. Oldfield 1962 Tones from a choked axisymmetric jet *Acustica* **12** 257-277
[7] P. Paredes, M. Hermanns, S. Le Clainche, and V. Theofilis 2013 Order $10^4$ speedup in global linear instability analysis using matrix formation *Comput. Method. Appl. M.* **253** 287-304
[8] A. S. Sharma and B. J. McKeon 2013 On coherent structure in wall turbulence *J. Fluid Mech.* **728** 196-238
[9] D. Lilly 1992 A proposed modification of the Germano subgrid-scale closure method *Phys. Fluids* **4** 633-635
[10] T. Poinsot and S. K. Lele 1992 Boundary conditions for direct simulations of compressible viscous flows *J. Compute. Phys.* **101** 104-129
[11] H. W. Liepmann and A. Roshko 1957 Elements of Gas Dynamics *Dover Publications*
[12] D. Bodony and S. Lele 2005 On using large-eddy simulation for the prediction of noise from cold and heated turbulent jets *Phys. Fluids* **17** 85-103
[13] H. Schlichting and K. Gersten 2000 Boundary Layer Theory *Springer*